The Political Economy of Iraq

To the men and women of five nations who served in the
Economic Section, Strategic Effects, MNF-I, Baghdad and in
C-9, MNC-I, Camp Liberty, Iraq

The Political Economy of Iraq

Restoring Balance in a Post-Conflict Society

SECOND EDITION

Frank R. Gunter

*Professor of Economics, Lehigh University, USA,
Advisory Council Member, Iraq Britain Business Council and
Colonel, U.S. Marines (Retired)*

Cheltenham, UK • Northampton, MA, USA

Published by
Edward Elgar Publishing Limited
The Lypiatts
15 Lansdown Road
Cheltenham
Glos GL50 2JA
UK

Edward Elgar Publishing, Inc.
William Pratt House
9 Dewey Court
Northampton
Massachusetts 01060
USA

Paperback edition 2023

A catalogue record for this book
is available from the British Library

Library of Congress Control Number: 2021945073

This book is available electronically in the **Elgar**online
Economics subject collection
http://dx.doi.org/10.4337/9781789906073

ISBN 978 1 78990 606 6 (cased)
ISBN 978 1 78990 607 3 (eBook)
ISBN 978 1 0353 2366 1 (paperback)

Printed and bound by CPI Group (UK) Ltd, Croydon, CR0 4YY

Contents

Figures

Tables

Preface

This is an attempt to write the book that I wish I had read before my first nighttime "corkscrew landing" into Baghdad International Airport in 2005. By that time, I had read or at least scanned several hundred books, journal articles, and research studies on the post-1955 political economy of Iraq. What I was not able to find was a single recent work that attempted to provide an integrated study of the entire political economy. I decided to write this book, not for my colleagues at the Lehigh University, but for the Iraqis and people of many nations who are working to build a better future for Iraq. As a result, I attempted to write a work that – in an integrated fashion – described, analyzed, and made policy recommendations for almost the entire political economy of Iraq. The first edition in English and Arabic found a significant readership and provided the motivation to update the work to incorporate the substantial economic, political, and social changes since the first edition was published in 2013.

My thinking on the most important challenges facing Iraq has evolved significantly over the last decade and a half. I initially thought that increasing oil exports and restoring agriculture production were the most important economic challenges facing the country. However, I am now convinced that, as important as the petroleum and agricultural sectors are, the greatest barriers to accelerating Iraqi economic development are corruption, political instability, and regulatory hostility towards private business. As a result, I consider Chapter 4 (Corruption), Chapter 5 (Preventing al Qaeda 3.0), and Chapter 11 (Entrepreneurship in Post-Conflict Iraq) to provide the most value added.

The arrangement of the book is as follows. Chapters 2 and 3 provide overviews of Iraq's real growth, unemployment, and inflation followed by discussions of health, poverty, education, and gender issues. Chapters 4 through 7 are devoted to the three dominant characteristics of the political economy of Iraq: corruption, political instability, and petroleum. Chapters 8 through 13 each focus on a sector of the Iraq economy: agriculture, financial intermediation, large industrial enterprises, entrepreneurship, infrastructure and essential services, and international trade and capital flows. Chapter 14 discusses fiscal, monetary, and exchange rate policy. And the last chapter tries to identify the major trends that will define Iraq in 2035.

Disparate individuals and groups have shaped my thinking about modern Iraq. Most importantly, I am indebted to the military, diplomatic, consultant men and women with whom I shared long workdays in Iraq for 25 months in 2005–06 and 2008–09. This book has also benefited greatly from conversations with Iraqi, British, and American businessmen, academics, government officials, consultants, entrepreneurs, and students. If I tried to list all the names of all of those individuals who provided useful insights, I would greatly exceed the publisher's word limit and probably accidently leave someone out. However, I am especially indebted to Mohaned al Hamdi who translated the first edition into Arabic, as well as Baroness Emma Nicholson of the Iraq Britain Business Council, and Alan Luxenberg of the Foreign Policy Research Institute, for their support and encouragement of my research. I am also grateful to Generals Lloyd Austin and Rick Lynch, both of the United States (US) Army, for permitting me to join their Staffs in Iraq.

This work also benefited greatly from the author's participation in seminars, workshops, and panel discussions organized by the Iraq Britain Business Council, the Foreign Policy Research Institute, the Atlantic Council and Konrad Adenauer Stifung, National Bureau of Economic Research (NBER) Economics of National Security Workshop, the US Embassy in Baghdad, the US State Department, the American University of Sharjah in the United Arab Emirates (UAE), the American University of Iraq in Sulaymaniyah, George Mason University, the Kansas State University, as well as the Eastern, Southern, and Western Economic Associations.

I also profited from conversations with many individuals, even though these conversations often ended in disagreements. Among those who provided valuable insights were Mohammed al Uzri, Renad Mansour, Hani Akkawi, Hussein al Uzri, Shwan Aziz Ahmed, Christophe Michels, Richard Cotton, Ahmed Tabaqchali, Yass al Kafaji, Christine van den Toon, Anas Morshed, Tally Helfont, Bilal A. Wahab, Amer Hirmis, Henri Barkey, Aaron Faust, Karen Puschus, Wendy Polhemus, Saad Hasan, Hoshyar Muhammed Ali, Drew Johnson, Kat Woolford, Paul Savello, June Reed, Mohsen Khairaldin Garcia, Brian Moore, Andrew Gough, Kevin Darnell, Kenneth Pollack, John Nash, Glenn Goddard, Beverley Simpson, Larry Milam, Terry Kelly, Sameeksha Desai, Frank Mulcahy, Ian Furgerson, Tony Meyer, Chris Canniff, Tim Kane, Geoffrey L. Keogh, Lauren D. Mayne, Ali al Makhzomy, Mujahed Z. Waisi, Scott Chando, John Holmes, Jeffrey Butcher, Pat Carroll, Frederick Alegre, Karl Schwartz, Miriam Lutz, Tony Daza, Martin Sierra, Seung-gu Weon, Sam Korab, Janet Rudasil-Allen, Samuel Helfont, Sandy MacMurtrie, Ali Bachani, Tim Fawcett, Tim Curran, Susan Maybaumwisniewski, Jeff Peterson, Robert Looney, Joe Banavige, Keith Crane, Mercedes Fitchett, Andrew Wallen, and many others.

I have not listed the names of some Iraqis who were extremely helpful. Several asked for confidentiality and I am concerned that others may be endangered because of their frankness about specific incidences of corruption and mismanagement.

Amanda Wakefield, Vic Froggett, Cathrin Vaughan, Katia Williford, Kate Pearce, Caroline Kracunas, and especially Alan Sturmer of Edward Elgar Publishing have been extremely helpful, professional, and patient. In addition, I am especially indebted to four colleagues who aggressively pushed me to finish this book: Eli Schwartz (now deceased), Nicolas Balabkins, Wight Martindale, and Anthony O'Brien. I am also grateful to Shin-Yi Chou, the Economics Department Chairperson at Lehigh University, who cheerfully tolerated my mid-semester trips to the Middle East. Finally, as always, I am grateful to my wife for continuing to smile whenever I bring up "land ownership in Anbar Province" as a topic for dinner conversation.

A note on spelling. There is no standard transliteration of Arabic into English. Even Government of Iraq publications will use different spellings of the same Arabic word, sometimes in the same document. For example, the Province and City in Northern Iraq can be rendered as Arbil, Arbīl, Erbil, Irbil, and so on. For the names of geographic locations – provinces, cities, rivers, and so on – I have tried to consistently use the spelling of the National Geographic Society. This usage is primarily a matter of convenience since I make constant use of their maps.

1. Iraq's lost decades

> Iraq has an altogether exceptional opportunity of achieving a development which within a few years would substantially increase her economic resources and raise her general standard of living. For she now has advantages which are rarely found in combination.
>
> (Lord Salter 1955, p. 1)

> Iraq in the years prior to its invasion of Kuwait was at the top of the per capita GDP [gross domestic product] ladder of developing countries. By 1993, real monthly earnings were lower than the monthly earnings of unskilled agricultural workers in India – [then] one of the poorest countries in the world.
> What happened to an economy noted for the wealth of its oil reserves, agricultural potential, water resources, relatively high rates of literacy and skills, vast access to foreign technology and expertise, an enviable balance-of-payments surplus and foreign reserves, and a long history of determined effort to develop and diversify the economy?
>
> (Abbas Alnasrawi 1994, p. xv)

In Iraq, history has been a cycle of brief periods of progress alternating with longer periods of economic, political, and social regression. In 1955, Lord Salter argued that an objective analysis of Iraq's political economy showed that the country would have a bright future of peace and prosperity. His vision was wrong, or very premature, as conflict, corruption, and mismanagement devastated the country. A military coup in 1958 overthrew King Faisal II and imposed a socialist republic. A leader of the Arab Ba'athist Socialist Party, Saddam Hussein, became president in 1979, purged his enemies and launched a 24-year absolutist reign that combined severe oppression at home, brutal wars with two of Iraq's regional neighbors, and the building of an extremely corrupt and bureaucratic economy. After Saddam's overthrow by a US-led coalition in 2003, Iraq was politically, economically, and physically in ruins. Since then, there have been two violent insurgencies, al Qaeda (2005–07) and ISIS (2014–17) that led to many deaths, the creation of large number of internal refugees, and large-scale infrastructure destruction. The defeat of ISIS, in late 2017, led to a period of optimism bolstered by increases in both the volume of oil exports – Iraq's primary export – and world oil prices. Violent deaths decreased dramatically.

Now, in 2020, there is once again a resurgence of pessimism. Despite Organization of the Petroleum Exporting Countries (OPEC) efforts to restrict oil supply, oil prices collapsed to the lowest levels in two decades, which has effectively halted reconstruction efforts and the provision of essential services. Protesters filled the streets seeking jobs, access to essential services, and reduced corruption. The response to these protests by the Government of Iraq (GoI) was initially indifference but later turned violent, which led to the deaths of an estimated 600 persons. Thus, political instability has joined with an economic depression.

Are there any reasons for optimism about Iraq's future? There are several. First, as a result of what Fouad Ajami (2006) referred to as the "foreigner's gift," Iraq has a nascent democracy. Second, currently none of the major social, ethnic, tribal, political, or religious groups seek to break up the country. After the debacle of the 2017 independence referendum, even the Kurdistan Regional Government (KRG) is reconciled to staying part of Iraq. Finally, most importantly, Iraq is a very young country with 40 percent of the population less than 15 years of age. Generalizations about Iraq are always dangerous, but many of these young men and women are open to new ways of dealing with the country's many serious challenges. During college seminars or over coffee at entrepreneurship incubators, these young people say that they believe Iraq has the potential to achieve a bright future for its people.

Whether that potential will be realized depends on the ability of the Iraqi people and their leadership. For, in choosing the best route for their country to travel upstream into the future, they must not only steer clear of the wreckage of previous voyages but also deal with new threats only faintly seen in the morning fog.

However, Iraq's response to these old and new challenges must reflect the changed political, economic, and social reality of Iraq at the beginning of the twenty-first century. This chapter provides a basic introduction to Iraq's geography, history, political structure, and economy as well as a discussion of the major trends and challenges. It is intended as a foundation for the rest of this book, but readers who already possess knowledge of Iraq are invited to skim the rest of this chapter or skip it entirely.

Mesopotamia – a common name for Iraq – means "the land between the two rivers": the Euphrates and Tigris. Both rivers originate in the mountains of Turkey. The Euphrates detours through Syria before entering Iraq from the west. The Tigris enters Iraq from the north, and the two rivers flow roughly parallel through Iraq before they merge into the Shatt al-Arab waterway just north of the city of Basrah, before flowing into the Persian Gulf. Most of Iraq's estimated 40 million people live near the Euphrates or Tigris rivers, with the three of the four largest Iraqi cities – the capital of Baghdad, the major port city of Basrah in the south, and Mosul in the north – located on the Tigris. The fourth largest city, Arbil, is in the mountainous region of northeast Iraq. These two rivers have provided transportation, irrigation for agriculture, and drinking water for almost a dozen great civilizations for ten millennia.

Iraq has three climate regions. The southwest of the country is a flat, very arid desert. At the other extreme, the northeast is mountainous with sufficient precipitation for rain-fed agriculture. In between is a band of semi-arid climate that stretches from the Syrian and Turkish borders in the northwest to the Persian Gulf in the southeast. The climate of central Iraq has hot summers, cool winters, and an agreeable spring and autumn.

Politically, Iraq is divided into 18 provinces (Figure 1.1), three of which – Dahuk, Arbil, and Sulaymaniyah – have joined together into a regional government, the Kurdistan Regional Government (KRG). While Iraq is a federal system according to its 2005 constitution, only in the KRG are sub-national entities active. For the rest of the country, the national government in Baghdad makes almost all decisions.

Following the violent civil conflicts of 2005–07 and 2014–17, there is a tendency for outsiders to overly emphasize ethnic and religious differences in Iraq. These differences are important, but there are other factors that may have as much or more influence on the future of the country. With respect to ethnicity and religion, the population of southeast Iraq is mostly Arab and Shi'a Muslim; the northeast are generally Kurds and Sunni Muslims; while the west is largely Arab and Sunni Muslim. However, there are members of every ethnic and religious group in every province of Iraq, and there are cities and provinces where it is not clear whether the majority of the population is Shi'a or Sunni Muslim, Arab or Kurd.

As is well known, Iraq was almost torn apart by conflict following the 2003 invasion by the US-led coalition. From about 7300 civilian deaths in 2003, violence increased sharply, especially after the 2006 insurgent bombing of the Golden Mosque. The worst year was 2006, when an estimated 34 500 Iraqi civilians were killed, almost 100 a day. As the killing accelerated it led to large number of Iraqis becoming refugees, either inside their own country or fleeing to Syria, Jordan, and other neighboring countries. Less noted is how concentrated the killing was. Most of the civilian fatalities were in five provinces, while at the other extreme there were an equal number of provinces with few violent civilian deaths even during the worst periods. And to the surprise of many commentators both inside and outside Iraq, the killing began to slow in 2007, and by 2011 the number of civilian violence-related fatalities was less than 5 percent of those in 2006. This drop in civilian fatalities was not the result of one group obtaining a complete victory, nor was it imposed from outside. Apparently, the Iraqi people looked into the abyss of an unending civil war and backed away.

TURKEY

DAHUK

Dahuk

ARBIL

SYRIAN
ARAB
REPUBLIC   NINAWA   *Mosul*   *Arbil*

Euphrates River   *Sulaymaniyah*

Kirkuk

KIRKUK   SULAYMANIYAH

Bayji

SALAH AD DIN
Samarra   DIYALA

Baqubah

IRAN

Ramadi   *Baghdad*

BAGHDAD

ANBAR   WASIT

Karbala   *Kut*

JORDAN   BABIL

KARBALA   *Hillah*

Najaf   *Diwaniyah*   *Amarah*

QADISIYAH   MAYSAN

Samawah   DHI QAR

Nasiriyah

SAUDI
ARABIA   NAJAF   *Ur (Historic)*

Basrah

BASRAH

MUTHANNA

KUWAIT

0   100   200   300 km

0   100   200 mi

IRAQ

●   Major City (Governorate Capital)

------- Major River

☐   Governorate Border

Source: Map by Professional Maps of Houston, Texas, 2012.

Figure 1.1   Political map of Iraq

Tragically, the rise of ISIS led to another serious conflict in 2014–17. At its widest extent, ISIS occupied almost 40 percent of the country's territory which was home to about 30 percent of the population. Many Iraqis were forced to flee, resulting in an estimated 4 million internally displaced persons. Violent deaths increased to almost 20 000 in 2014 and continued at high levels until the final defeat of ISIS in Iraq. The destruction of homes, farms, factories, and infrastructure was extensive as a result of both conflict and deliberate acts by ISIS to destroy what they could not control. After the defeat of ISIS, violent deaths dropped to the lowest levels in two decades. The diminution of violence has shifted emphasis back to economic development; which in Iraq means oil.

Iraq is an island floating on a sea of oil. It is generally considered to have the fourth-largest petroleum reserves in the world, but ongoing surveys may boost it to first place in front of its neighbor Saudi Arabia. Oil export earnings are the source of almost all government revenues and account for over two-thirds of the country's GDP. However, the Iraqi oil industry employs only a small fraction of the country's labor force. One of the continuing challenges facing the Iraqi government is dealing with oil price volatility. Oil production and export volumes tend to change gradually, but oil prices are subject to rapid changes. As a result, government revenues are on a rollercoaster. For example, these revenues fell rose by 36 percent in 2018, before decreasing by 50 percent two years later.

For four decades, Iraq's oil legacy was squandered by conflict, corruption, and mismanagement. In fact, it was only in 2015 that oil export volumes approached the levels previously achieved in 1979. But living standards for the average Iraqi are much lower than three decades ago because of the high rate of population growth. As a result of having one of the highest fertility rates in the world, Iraq has a young, rapidly growing population. Combined with the destruction of the recent conflicts, rapid population growth has severely challenged the ability of the government to provide universal elementary education and basic essential services. In addition, every year hundreds of thousands of Iraqis join the labor force and start looking for work in an economy characterized by high unemployment and underemployment. If these young people are unable to find useful work, it will be politically destabilizing. Meeting these challenges is made more difficult by the widespread corruption and mismanagement of a government in transition.

Now that Saddam's socialist central planning model has been abandoned, it is not clear which form of capitalism will be most effective at raising living standards rapidly. In 2020, Iraq might best be described as having partially evolved from central planning to state-guided capitalism in which the government tries to guide the market by supporting particular industries that it expects to become "winners" or that are important sources of employment. But if current political and economic trends continue, then there is a real danger that

Iraq's state-guided capitalism is only a way station on the route to becoming an oligarchic capitalistic state – like most of the other countries in the Arab Middle East – in which the bulk of the power and wealth is held by a small group of individuals and families. Despite wealth from natural resources, such oligarchic capitalistic states tend to have great income inequality, sluggish growth, large informal or underground sectors, and massive corruption.

Iraq is facing many of the same types of political, economic, and social challenges that it failed to overcome six decades ago; now Iraq has to face these same challenges with a larger population, a less innovative economy, more hostile relations with its regional neighbors, a recent history of sectarian violence, and probably greater cynicism about the competency and honesty of the nation's leaders. While its petroleum wealth is a cause for optimism, the major determinant of Iraq's future will be the choices that Iraqis make over the next decade. If the Iraqi people and their leadership choose once again to take counsel of their fears and pursue short-term political and economic advantage as opposed to establishing a framework for long-term balanced prosperity and a solid democracy, then Iraq will probably fail again. To paraphrase Toynbee (1972, p. 161): countries generally die from suicide, not by murder.

However, there is another route. If Iraq could restore balance in its political economy by combining reduced economic dependency on oil with reduced governmental corruption and bureaucracy in a more responsive political system, then the future could be very different. Its huge oil wealth could finance a more sustainable economic future with a better life not only for the current population of Iraq but also for generations to come. The many difficulties of following this alternative route are the subjects of the rest of this book.

2. Population and key macroeconomic variables

> It has been suggested that during the period of Abbasid rule (A.D. 754–1258), when Baghdad was the centre of the civilized world, the present state of Iraq may have supported a population of some 20 million people ... [However] in the middle of the nineteenth century it has been estimated that the population numbered only 1.28 million.
>
> (R.I. Lawless 1972, p. 97)

DATA QUALITY

There is no shortage of data on Iraq's macroeconomy. The problem is that the data available is often inconsistent. When it comes to the size of the country's population, gross domestic product (GDP), unemployment, and inflation, not only do different sources provide very different numbers, but also the trend of a single variable from a single source is often unreliable. The latter problem is often caused by unannounced changes in the composition of a variable or the method of its estimation.

The four decades of internal and external conflict have made it difficult and often dangerous to gather data. For example, during post-1991 periods of severe internal conflict, markets in many areas of Iraq were split; it was dangerous or impossible for workers, customers, or products to travel even short distances. Iraq, for periods of time, ceased to be a national market, rendering data on trade, manufacturing, agriculture, employment, and so on, difficult to gather and interpret. Also, since 2003, the Government of Iraq (GoI) has been engaged in a difficult transition from socialist accounting to that of a modern market economy. As shown in Eastern Europe after the collapse of the Soviet Union, this transition can lead to serious gaps in data as definitions and collection methods change. In addition, the GoI continues attempts to conceal unfavorable data in order to put its operations in the best possible light.

In addition, as will be discussed in Chapter 4, Iraq suffers from serious corruption including large-scale smuggling across its borders with Iran, Syria, and Turkey. This distorts its international trade and investment data. Also, corruption combined with an extremely hostile regulatory environment has resulted in a large and growing informal (underground) economy. Production, consumption, and employment in the informal economy is only partially reported in the official statistics, resulting in a wide gap between economic activity reported in official statistics and the reality on the ground. In recognition of severe problems with the GoI's statistical methodology, the World Bank in 2018 ranks Iraq's Statistical Capacity Index at 51.1 (out of 100), almost tied with Afghanistan. The Middle East and North Africa (MENA) average was 59.9, while the world average was 74.9 (World Bank 2018a).

There are also policy implications. Low-quality data makes it more difficult to avoid waste. Major reconstruction decisions must either be delayed until better data is available, often at the cost of continued suffering among the population, or these decisions must be based on educated guesses based on inadequate or contradictory information. It is ironic that once these educated guesses are used to make policy, they take on a life of their own and are treated as being more credible than they actually deserve to be.

In Baghdad in 2006, I was instructed to provide a provincial breakdown of a certain political economic variable. Having wrestled with this data previously, I immediately responded that none of the existing data sources for this variable covered all 18 provinces and, in the few provinces where more than one source was available, they contradicted each other. I immediately received a response that Washington really did not care about my concerns about data quality; they wanted the provincial breakdown submitted within 48 hours. After making a large number of plausible – at least to me – assumptions, I finally produced the data and assuaged my conscience by attaching a note detailing the weaknesses. Of course, this caveat was stripped out before Washington apparently used the data to make a small revision in US reconstruction policy. Maybe my numbers were actually fairly correct, but I still cringe when I run across references to this data as being from an authoritative US government source.

POPULATION

Population data should be much more reliable than that for GDP, inflation, unemployment, and so on. It is easier to count people than to estimate these other variables. However, in Iraq even population data is very uncertain. The last nationwide census was in 1987; the 1997 census covered only the 15 non-Kurdish provinces and was probably contaminated by political considerations. Iraq was under United Nations (UN) sanctions and it was important to Saddam that the census revealed not only the high humanitarian cost of the sanctions but also robust growth among groups that supported the Saddam regime. An example of the latter was the importance of showing a growing Arab population in Kirkuk, compared to earlier censuses which showed that the city and province were mostly Kurdish. Further complicating estimates of population are the devastating effects of the post-1980 violence, which has also resulted in large numbers of internal and external refugees.

During the 1980–88 Iran–Iraq War – the deadliest war ever waged between developing nations – Iraq suffered an estimated 250 000–500 000 people killed or wounded. The Iraq invasion of Kuwait in 1990 and Saddam's reprisals against his own people followed. The true number of dead or wounded from this period will probably never be known but it is estimated at about 300 000. Finally, there were the casualties suffered following the invasion of Iraq by the US-led coalition in 2003. Estimates of deaths during this period range from 50 000 to the controversial figure of 650 000 deaths by *The Lancet*. Probably the most reliable estimate is the 160 000 deaths of the Iraq Body Count organization. Not only were a large number of Iraqis killed over the last four decades, but also the violence was accompanied by large-scale population displacement.

The breakdown of government authority during the worst of the insurgent violence in 2005–07 and again in 2014–17 raises questions about the accuracy of official estimates of the number of refugees. However, by mid-2015, it was thought that Iraq had one of the largest populations of internally displaced persons in the world: an estimated 10 million persons, or roughly 30 percent of the country's total population. In addition, an estimated 250 000 Syrian refugees fled to Iraq to escape their own civil war (IMF 2015, Box 1, p. 6). With the beginning of the decrease in violence in 2018, there is evidence that many internal refugees are beginning to return to their former homes. But with the destruction of towns and cities, especially in Iraq's northwest, combined with the on-again, off-again national budget crisis which limits reconstruction and resettlement aid, it will probably be a decade before most of the internally displaced persons can return to their former homes.

In the absence of a recent census, current population estimates – regardless of the source – are based on multi-decade extrapolations of old census data adjusted by estimates of deaths, international refugees, the results of a variety of smaller surveys, and rules of thumb. Due to the uncertainty about birth, death (especially violent deaths), and migration rates during a period of severe internal and external conflict, combined with deliberate distortions introduced to data so as to favor a particular ethnic or religious group, estimates of Iraq's current population should therefore be considered as no more than educated guesses. This has resulted in a divergence of estimates: while the US Central Intelligence Agency estimated that the country's population was 38.9 million in 2020, the World Bank estimated a population of 40.2 million in the same year. It is probably better to think of estimates of Iraq's population not as a point, say 40 million persons in 2020, but as a range: maybe 40 million persons plus or minus 5 percent. In other words, if a professional census had been performed in that year, it would not have been terribly surprising if the census results showed Iraq's population to be as small as 38 million or as great as 42 million.

Uncertainty about the actual population of Iraq reduces the usefulness of most other data for Iraq. It is unlikely that more accurate population data will be available in the near future. The 2005 Constitution called for a periodic census but, as in previous years, the 2019 GoI budget provided no funding for planning or executing a census. One reason for the delay, as discussed in Chapter 6, is that the results of a census may cause a major political shift in the oil-rich Kirkuk province. If a census were to show a majority Kurdish population in Kirkuk, this would strengthen the movement to have Kirkuk rejoin the Kurdish Regional Government (KRG).

Another controversy concerns the proportion of Iraqis who are adherents of the Shi'a and Sunni denominations of Islam. It is thought that Shi'a are about two-thirds of the population, with the Sunnis about one-third, and Christians about 2 percent. This belief about population is used to justify a disproportionate amount of government expenditures and jobs being reserved for Shi'a. However, there are Sunnis who think that the proportions of Shi'a and Sunni are nearer to being equal, and therefore more expenditures and jobs should be reserved for their community. As a result, the GoI may prefer to live with population uncertainty rather than permit a census that might weaken the national government's authority over Kirkuk or fuel religious disputes.

Although all of the population data is based on educated guesses, there are at least four population characteristics which are so prominent that one can be reasonably certain that the data reflects reality and not statistical artifacts.

First, Iraq is a land of large households. Although it appears that the average household size has decreased over the last several decades, the average household size was still six persons in 2015. Among the 31 percent of the population that live in rural areas, the typical Iraqi family is larger, with four or five children and two or three adults in each household. Among older rural families, polygamous households (families with two or three co-wives) are not uncommon. With the exception of agricultural work, relatively few women or elderly people are in the labor force; the typical Iraqi household has a single male wage earner supporting four to six dependents. Women, mostly widows, head an estimated 11 percent of all households.

Second, Iraq – as a result of a high fertility rate – is a very youthful country. Iraqis aged less than 15 years account for almost 40 percent of the population. This is a substantially greater proportion of young people in Iraq than in neighboring countries: Saudi Arabia (25 percent of the population is younger than 15 years), Turkey (25 percent), Iran (24 percent), and Kuwait (21 percent) (World Bank 2019a, Table 2.1).

Third, providing productive employment for each cohort of young persons is the most important challenge facing Iraq. Each year about 900 000 Iraqis become old enough to work. Adjusting for each year's retirements and deaths among the working population, as well as the very low labor force participation rate among women, means that the nation must create about 340 000 additional jobs each year. In other words, in any year when the country fails to create a third of a million net new jobs, there will be an increase in the pool of mostly male, mostly uneducated, unemployed. This growing pool of unemployed, discouraged young men without any expectation of finding a good job – and therefore supporting a family – is a major source of instability.

Finally, Iraq's population is increasing at a rapid rate by international standards. With about 2.7 percent annual population growth, the Iraqi population will double in about 25 years. Contrast this to the average 2.0 percent population growth rate among other MENA countries (their populations will double in about 35 years) and the 1.6 percent population growth among other lower-middle-income countries (LMICs) (their populations will double in about 44 years). Thus, any year in which real GDP growth is less than 2.7 percent will be a year of decreasing per capita incomes.

GROSS DOMESTIC PRODUCT

The pattern of real per capita income shown in Figure 2.1 is based on data from several sources and depends upon some fairly controversial assumptions. However, it provides a rough guide to changes in inflation-adjusted Iraqi per capita income. While real GDP per capita in 2020 is much greater than immediately following Saddam's 1990 invasion of Kuwait, it is about half of what it was in 1979. Over the last 40 years, year-to-year changes in per capita income were primarily driven by three variables: the value of oil exports, conflict, and population growth.

From 1950 through 1979, there was a steady rise in per capita incomes as a result of increased oil export volume combined with higher world oil prices. The latter year was distinguished not only by the highest living standards ever achieved by the Iraqi people, but also by Saddam seizing absolute power. The decade-long collapse in per capita income that followed was caused both by an almost 65 percent drop in real oil prices as well as the incredible waste of blood and treasure of the eight-year-long war that followed Saddam's September 1980 invasion of Iran. A gradual recovery in oil export revenues following the end of that war was aborted by Saddam's August 1990 invasion of Kuwait, followed by his military defeat, UN sanctions, and the invasion of Iraq by a US-led coalition in March 2003. However, in 2003, world oil prices began to march upward until they reached almost the same real level in 2008 as during the Organization of the Petroleum Exporting Countries' (OPEC) glory year of 1979.

Source: Map by Professional Maps of Houston, Texas, 2012, in Maddison (2003), adjusted
and extended by author.

Figure 2.1   Iraq real per capita GDP (PPP)

If oil prices were as high in 2008 as in the halcyon days of 1979, then why
were Iraqi per capita income and living standards so much lower in 2008? Two
reasons explain this. First, as a result of conflict including deliberate attacks
against the country's oil infrastructure combined with serious mismanagement, the volume of oil exports was about 1.8 million barrels per day (mbpd)
in 2008, almost 40 percent less than in 1979. But the second reason is more
important: rapid population growth. Iraq's population was less than 13 million
in 1979, compared to an estimated 31 million in 2008; a 140 percent increase
(COSIT 2012, Table 1/2). Even if the volumes of oil exports and world oil
prices were to return to 1979 levels, the average Iraqi would only have about
40 percent of the real income and living standards that their parents had three
decades before.

As shown in Figure 7.2 in Chapter 7, in recent years, lower oil prices have
more than offset higher oil export volumes. Combined with continuing rapid
population growth, this has resulted in roughly stagnant real per capita GDP
from 2016 through 2020.

As shown in Table 2.1, 2020 per capita GDP was about $4515 (converting
Iraqi dinars (ID) to dollars at an exchange rate of 1182 dinars/dollar). However,

Table 2.1   *Key macroeconomic variables*

	2013	2014	2015	2016	2017	2018	2019^e	2020^p
Real GDP growth	7.6%	0.7%	2.5%	15.2%	−2.5%	−0.6%	4.4%	−9.5%
Non-oil real growth	12.4%	−3.9%	−14.4%	1.3%	−0.6%	1.2%	4.9%	−5.1%
Per capita	$7021	$6517	$5047	$4843	$5058	$5641	$5841	$4515
Per capita (PPP)	$15 500	$14 200	$10 300	$9800	$10 900	$10 800	$11 300	$8900
Oil production	3.0	3.1	3.7	4.6	4.5	4.6	4.8	4.0
Oil exports	2.4	2.6	3.4	3.8	3.8	3.9	3.8	3.0
Oil price	$102.90	$96.50	$45.90	$35.60	$49.10	$65.50	$61.10	$38.30
Inflation	1.9%	2.2%	1.4%	0.5%	0.2%	0.4%	−0.2%	0.1%
Exchange rate	1166	1166	1166	1180	1182	1182	1182	1182

Note: Superscript "e" indicates data is estimated; superscript "p" indicates data is projected.
Source: World Bank (2020b, Table 1, pp. 16–17), International Monetary Fund (2019a, Table 1, p. 27).

the average Iraqi's share of GDP provides little insight into the quality of life of the average Iraqi because it fails to consider the cost of living. Since Iraq is a relatively low-cost country, especially for services, the $4515 adjusted for the cost of living (purchasing power parity, PPP) has an estimated value of about $8900. In other words, an Iraqi who earned the average income in their country would have approximately the same living standard as a resident of the USA whose total spending for all purposes – food, rent, clothing, medical care, and so on – equaled $8900. Multiplied by the country's estimated population, this results in a 2020 GDP (PPP) of $356 billion. According to this measure, the Iraq economy is roughly the same size as Denmark with its 5.8 million population (World Bank 2020c, Table WV.1).

Iraq is an extremely natural resource-dependent economy, although this dependency has begun to decrease. From 51 percent in 2011, natural resources earnings fell to about 38 percent of Iraq's GDP in 2017. Among other major oil exporting nations, natural resources in 2017 contributed 39 percent of Libya's GDP, 37 percent in Kuwait, 24 percent of Saudi Arabia's GDP, 18 percent in Iran, and about 14 percent in the United Arab Emirates (UAE) and in Venezuela (World Bank 2019a, Table 3.14). As a result of this natural resource dependency, in the short run, the health of Iraq's economy is determined to a great extent by the value of its petroleum exports.

The percentages for non-petroleum sectors of GDP tend to change substantially year-to-year even though the actual production of these sectors is relatively stable. For example, oil prices fell from $91.50 per barrel (pb) in 2008 to $55.60 pb in 2009. Despite a small increase in the volume of oil exports, the

40 percent drop in oil prices led to sharp declines in oil export earnings and therefore in the proportion of GDP accounted for by petroleum.

It might be more useful to think of Iraq as two economies: oil and non-oil. In the non-oil economy, services including financial services (mostly provided by the national government) account for about 39 percent; trade, transport, and communication account for 36 percent; construction 17 percent; electricity, water, and manufacturing account for 5 percent; and agricultural accounts for about 5 percent (GoI 2018, Section 3.2.1). The non-oil economy accounts for about 98 percent of the labor force, with only about 2 percent of the labor force working in the oil economy. Growth in the non-oil economy is relatively stable.

Government investment is the "shock absorber" of the Iraqi economy. When oil prices fall, the GoI tends to adjust by cutting government investment while leaving wages and pensions and other current expenditures untouched. For example, in 2014, before the collapse in oil prices, public sector investment in fixed capital amounted to 41 percent of total expenditures, but this fell to 16 percent in 2018, and an estimated 2 percent in 2020 (MoF 2020; IMF 2019a, Table 2, p. 28). These large year-to-year funding changes in government investment in multi-year infrastructure projects are a source of great ineffi-ciency. When a project's funds are exhausted, construction sites are essentially abandoned until new funding allows projects to be restarted. This abandon-ment permits degradation of materials, if not outright theft. Often, workers continue to receive their salaries even though there is no work to be done. As a result, the total cost of each interrupted project is substantially increased.

UNEMPLOYMENT

There are various surveys that show that Iraqis consider unemployment the most important issue for the GoI to address. In fact, to the average Iraqi, unem-ployment is a more serious problem than lack of essential services, corruption, or security (SIGIR 2012, Figure 4.12, p. 84). As found when looking at data for populations and GDP, there are no shortages of estimates of unemployment in Iraq, with current estimates ranging from 13 percent to 40 percent. But again, one should have little confidence in the accuracy of any given estimate.

There are several possible explanations for this wide divergence in unem-ployment estimates. First, in both official reports and news stories, there is confusion about the different employment concepts or a failure to make it clear exactly which concept is under discussion. Reports of unemployment (no job, but currently looking for work) will often include estimated underemployment (working, but in a job that does not use a person's skills, education; or working part-time when full-time employment is desired) or discouraged workers (wants a job, but not working and no longer looking for a job).

Since, of course, the unemployment rate is the percentage of the labor force – not the population – that is unemployed, estimates of unemployment rates require estimates of a nation's labor force. The labor force is defined as people aged between 15 and 65 years who are either employed or unemployed but actively seeking work. Iraq's labor force in 2019 was an estimated 10.1 million persons, plus or minus 10 percent. Iraq's labor force is less than 25 percent of the population and is below the average of the typical MENA country. The low proportion of the population in the labor force is caused by the fact that 40 percent of the population is less than 15 years old, and the fact that relatively few females either work or are looking for work. The female labor force participation rate is an estimated 11.4 percent, compared to 74.2 percent for males (World Bank 2020c, Table 2.2).

A second reason for the wide range of unemployment estimates is the difficulty of including employment in the Iraq informal (underground) economy in the national employment estimates. Since the Iraq government is bureaucratic and corrupt, private businesses find it difficult and expensive to legally hire a worker. As a result, many businesses in the private sector hire workers "off the books" which excludes them from the official employment statistics. Some of the higher unemployment estimates by Iraqi ministries appear to assume that if a person is neither working for the government nor a farmer, then that person must be unemployed.

Finally, "ghost workers" and unreported immigrant workers may also distort unemployment and employment statistics. Many state-owned enterprises (SOEs) and ministries have workers who receive pay each month from the government for working full-time but, in reality, do little work and may not even show up at the workplace except on payday. For some SOEs, the proportion of these "ghost workers" is believed to exceed 25 percent of total employment. A proportion of these ghosts have jobs elsewhere, maybe even with another SOE. As a result, the number of employed in the economy is exaggerated. Another bias in the data is from the employment of large numbers of undocumented workers from Egypt and other poor countries. To make up for the large proportion of the male population that was drafted into the army during the 1980–88 war with Iran, Saddam encouraged large numbers of foreign workers to come to Iraq to work in its factories and on its farms. Most of these economic immigrants were forced to leave Iraq during the sanctions period. But since 2003, there has been increasing anecdotal evidence that undocumented workers are returning to Iraq in large numbers, especially to the southern provinces. Employers claim not only that hiring Iraqis to do certain jobs costs more, but also that Iraqi workers are more difficult to manage and to motivate.

All of the above distort the official data; like the data for population and real GDP, the best that can be hoped for is an educated guess at the true pattern of employment, underemployment, and unemployment in Iraq. The following discussion is an educated guess based primarily on data from the International Labour Organization (ILO 2019), extrapolations of a 2007 COSIT survey (COSIT 2008) modified by a variety of other sources, and the author's prejudices.

In 2017, the unemployment rate was reported as 13 percent. However, the reported rates were higher for women at 31 percent and young Iraqis at 26 percent. And not only did young Iraqis (15–19 years old) experience a higher unemployment rate, but they were also more likely to be seasonally or irregularly employed. It is estimated that only 20 percent of employed young men worked full-time for 30 hours or more. The rest were employed part-time, engaged in seasonal or irregular work, or listed themselves as "self-employed".

The data for self-employment conceals more than it reveals. As in most developing countries, there is a bimodal distribution of the self-employed. At one extreme of the self-employed, there are skilled workers (carpenters, plumbers, and so on) or educated individuals (merchants, medical personnel, and so on) who are relatively well compensated, tend to have employees, and are self-employed by choice. However, at the other extreme of the self-employed are persons engaged in small-scale service activities, who work alone, and are self-employed as a matter of survival until they can obtain any other job.

Due to missing or defective data, estimates of underemployment are even more uncertain than those of unemployment. One rule of thumb says that underemployment equals the sum of part-time workers and seasonal and irregular workers plus half of those listed as self-employed, all divided by the labor force (underemployed = (part-time + seasonal and irregular + half of self-employed)/labor force). Using this rule, maybe 33 percent of the general population and 55 percent of the young are underemployed.

Despite the crude nature of the employment data, there are two important conclusions. First, underemployment is a serious problem in Iraq. The combined unemployment and underemployment rate for the general population was almost half of the labor force. Second, the young (aged 15–19 years) are experiencing much higher rates of both unemployment and underemployment, with a combined rate of almost 80 percent. A major cause of the much higher rates of unemployment and underemployment of the young is the GoI's reliance on the public sector to provide employment for young people just entering the job market. In other words, until the oil price fall of 2014, the government was the "employer of first resort".

What are the sources of funds used to pay Iraq's estimated 8.3 million employed persons? Since oil export earnings account for approximately two-thirds of GDP and over 90 percent of GoI revenues, one would expect that oil export earnings fund a large proportion of the jobs in Iraq, but it is interesting to delineate the routes that these funds took. While roughly 50 percent of the Iraqi labor force is paid with oil money, actual employment in the oil industry is small: maybe 2 percent of the labor force.

The largest numbers of recipients are ministerial employees that account for about 2.9 million or over one-third of the total employed (GoI 2020b, Figure 4, p. 8). This reflects not only the government workers involved in the provision of a large range of public and merit goods, but also the government's role as the "employer of first resort". As will be discussed further in Chapter 10, the SOEs (even those that are only an empty shell with zero production) are encouraged to maintain their employment levels. These accounted for another estimated 600 000 (GoI 2020b, p. 10). Therefore, these two forms of direct government employment account for 3.5 million persons or about 42 percent of the country's employed.

There is a more controversial adjustment to government employment. Agricultural employment, about 1.7 million persons, can be roughly divided into two categories. Farmers who grow grain are government employees in everything but name. As will be discussed in Chapter 8, the GoI owns much of the land, supplies most of the inputs – seed, fertilizer, fuel, and so on – and sets the prices at which the grain crops can be sold. On the other hand, farmers of non-grain crops, although subject to detailed and arbitrary government regulations and policies, might be more accurately described as self-employed. Including half of the agricultural labor force brings government employment up to about 4.4 million, or about 52 percent of all employed.

About 17–20 percent of the nation's labor force are employed in the nation's small formal economy, by the relatively rare Iraqi-owned private firms, or foreign investors especially in the construction and hotel industries, or by foreign government, international organizations, and non-governmental organizations. Another 17–20 percent of the labor force is employed in the informal or underground economy. Iraq has an extremely hostile regulatory environment (see Chapter 11) that discourages small businesses from operating openly. They would rather avoid government regulation and demands for bribes by operating in the informal economy despite the associated inefficiencies. Some of these activities are clearly illegal, such as prostitution or serving alcohol, but others are activities that would be legal if the small business owner was willing to pay the necessary fees and bribes. It is not uncommon for officials to count Iraqis working in the informal economy as unemployed.

Female and Male Employment

In 2019, while an estimated 74.2 percent of Iraqi males 15 years or older were in the labor force, the estimate for females was about 11.4 percent (World Bank 2020c, Table 2.2). Iraq's female labor force participation is low compared to the average MENA country that has a female rate of 20 percent, while the rate among upper-middle-income countries (UMICs) is 55 percent. Female participation in Iraq's labor force continues to be severely constrained by religious and cultural strictures. Saddam's Arab Ba'athist Socialist Party strongly supported female equality in government employment, although there was substantial backsliding from this ideal during the sanctions period. However, females still fill a higher percentage of government jobs compared to their participation in the private sector, mostly agriculture employment.

Of course, the unusually low labor force participation rate of women in Iraq tends to slow the country's economic growth. In 1920, Weber wrote of the importance of the "protestant work ethic" as a major cause of economic development in the West (Weber [1920] 2002). A more recent study concluded that Muslims ranked even higher than Protestants with respect to pride in work and willingness to sacrifice to make a better future (Zulfikar 2012). Therefore, it is ironic that the great benefit of this strong religion-based "work ethic" is more than offset by an unwillingness to allow one-half of the population to fully participate in building a better Iraq.

Data collected before the rise of ISIS showed not only that there is a wide gap in the labor force participation rates of men and women, but also that they tend to be employed in different professions. Most working, educated women are engaged in teaching at the primary or secondary levels (33 percent of all female employment), while the major source of employment for uneducated women is agricultural work (31 percent). There is a big gap until the next most important profession: about 5 percent of women are employed in the financial sector. Male employment is much more diverse. To reach two-thirds of male employment, one must include retail and wholesale trade (19 percent), building and construction (14 percent), farming (13 percent), communication and transport (12 percent), as well as public administration and military (11 percent) (COSIT 2008, Table 5-20A, p. 324).

As expected, not only do Iraqi women have a lower labor force participation rate combined with fewer employment options for those women who do seek employment; but also the average Iraqi woman earns a much lower wage than her male counterpart. In fact, among agricultural workers, 85 percent of women aged 15–19 years and 34 percent of those aged 20–29 years reported zero wages. The corresponding non-wage employment for men in agricultural employment was 20 percent and 8 percent, respectively (COSIT 2008, Table 5-17A, p. 316). Many rural women are expected to work on the family farm as part of their family duties.

The "Five Iraqs": Primary Sources of Provincial Employment

Each of the major categories of employment is represented in every one of the 18 provinces, although the range is wide. To take agriculture, for example, at one extreme an estimated 39 percent of the employed in Salah ad Din province are farmers, while at the other extreme only 4 percent of the residents of Basrah province till the soil. However, based on the major sources of employment in each province, one can roughly divide the country into "five Iraqs", each defined by its dominant source of employment and income. (Data is from COSIT 2008, Table 5-20A, pp. 324–5, but the division is by the author.) The "Five Iraqs" are:

1. Oil Iraq: Basrah and Kirkuk. While oil and gas reserves have been discovered in almost every Iraqi province, Basrah and Kirkuk accounted for almost 94 percent of the nation's production. As a result, while the oil industry accounts for only 2 percent of national employment, it is the most important category of employment in these provinces.
2. Farming Iraq: Salah ad Din, Babil, Anbar, and Qadisiyah. These provinces tend to be more agricultural, as either the Euphrates or the Tigris river flows through each of these provinces, providing water for irrigation. While only 31 percent of the total Iraqi population lives in rural areas, in these four provinces a majority of the population lives in rural areas (51 percent).
3. Business Iraq: Ninawa, Baghdad, Karbala, and Najaf (and Basrah again). Since the days of the Ottoman Empire, the three most important trading and manufacturing cities in Iraq have been Mosul (Ninawa province), Baghdad, and Basrah. However, ISIS devastated Ninawa province and especially its provincial capital of Mosul. As will be discussed in Chapter 13, several million religious tourists annually visit Najaf and Karbala provinces. Tourism-related construction and services for these mostly Iranian tourists are major businesses.
4. Government job Iraq: Dahuk, Arbil, and Diyala. Dahuk and Arbil (along with Sulaymaniyah) comprise the KRG in northern Iraq. Diyala is south of the KRG on the Iranian border. Over 30 percent of the labor forces in these three provinces work for the government in public administration, military service, or education. The average for the rest of the country is 19 percent.
5. Balanced Iraq: Sulaymaniyah, Wasit, Muthanna, Dhi Qar, and Maysan. In these provinces there is no single dominant employment sector.

The existence of the "Five Iraqs" is a reminder that policy initiatives or unexpected shocks will probably have very different impacts on the various provinces. For example, if religious tourism increases sharply post-Covid-19

while agriculture stagnates, then not only will there be a net overall effect –
positive or negative – on Iraq's GDP and unemployment, but also there will
be some provinces that will boom while others decline. It is also unlikely that
large-scale labor migration will substantially mitigate these differences. For
cultural and historical reasons, it is unlikely that large numbers of the unemployed or underemployed from a province with slower real growth will perma-
nently move to a province with a better economy. For example, according to
the GoI, Dhi Qar province had the highest unemployment rate in 2011, while
Sulaymaniyah province had one of the lowest. However, it is unlikely that an
Arab Shi'a from Dhi Qar province would move his family to Sulaymaniyah
province that, being mostly Kurdish and Sunni, has a different language and
follows a different form of Islam.

INFLATION

Inflation – according to official sources – has experienced large year-to-year
changes: from about 32 percent in 2004 and 2005, the rate of price increases
almost doubled to almost 65 percent in 2006, before falling to between 5 and
7 percent during the period 2007–08, and 4 percent deflation in 2009. Since
then, Iraq's inflation has stabilized. As can be seen in Table 2.1, inflation has
tended in the last half-decade to be less than 1 percent. Considering the political and economic turmoil of the last five years, this apparent price stability is
an incredible achievement.

But is this price stability real? Like unemployment, estimates of Iraqi inflation are subject to severe definitional as well as data problems. These problems
are so severe that I think the reported pre-2008 inflation figures are essentially
meaningless. And the data from 2008 and later are questionable. They may
provide a rough guide to the direction of inflation change – for example, "inflation is getting worse" – but the published data does not provide a reliable guide
to actual price changes. There are at least five reasons for price uncertainty,
which provide insight into the difficulties of developing sound macroeconomic
policies in Iraq.

First, inflation – a rise in the average price level – is measured by estimating
changes in the number of Iraqi dinars that it costs to buy a "basket" of goods
and services. As long as these goods and services are sold in markets, the
biggest challenges in estimating inflation are determining which items should
be in the basket and periodically gathering price data for each of the goods and
services. But markets in the formal sector do not provide many of the goods
and services consumed by the average Iraqi. Instead, these goods and services
are either provided by the GoI or are produced in the informal (black market)
sector. This makes determining the "price" of a particular good or service an
arbitrary and, often, a political exercise. The GoI is the sole or dominant sup-

plier of essential services such as electricity, water, fuel, food rations, and so on. The GoI generally provides these essential items at zero or very low cost, substantially less than any market clearing price for these items. As expected, provision at below market prices leads to shortages and rationing. But what is the relevant price of a rationed good?

Electricity is a good example of the problem. It is rare for a family or business in Iraq to have 24 hours a day access to grid electricity. When the grid is "dark" a family or business must purchase electricity from an entrepreneur operating a community generator, buy black market fuel to operate a private generator, or wait through the heat and the dark until the grid is on again. (See Table 12.1 in Chapter 12 for the use of backup sources of electrical power.)

Under these conditions, what is the true price of electricity when the grid is dark? One might argue that it is the price of electricity from the alternative – probably black market – source. Obviously, reliable pricing data on such a source is difficult for government officials to obtain. What if a family is forced to sit in the heat and dark and wait until the grid is back on because an alternative source is not available? It seems deceptive to argue that the relevant price is, say, 50 ID per kWh if one cannot buy electricity at any price. In theory, one might try to estimate the highest price that the family would be willing to pay for electricity if it was available, but as can be imagined, any such estimate would be extremely arbitrary. The GoI does not seem to use either of these options, rather it uses the official price for government-supplied goods and services, even when these goods and services are not available.

This leads to a severe distortion of inflation data when the GoI raises official prices. One of the drivers of the sharp rise in inflation in 2005 and 2006 was the increase in official fuel prices. In order to reduce smuggling and under pressure from the International Monetary Fund (IMF), the GoI raised the official price of diesel fuel from about 1 US cent per liter in 2004 to 6 cents per liter in 2005, and 11 cents in 2006; over a 1000 percent increase in a two-year period in the price of a basic consumer item. But, in reality, did fuel prices actually increase? In 2004, there was a severe shortage of diesel fuel at official prices due, primarily, to the illegal diversion of such fuel into the black market. Consumers and businesses were forced to buy diesel fuel in the black market at a price that was 10–20 times the official price. When official fuel prices started to be raised in 2005, consumers were able to obtain more fuel from legal sources at the official price because of a decline in the profitability of illegally diverting fuel into the black market. Thus, actual fuel prices paid by consumers and private firms generally grew much more slowly than was reported in the official inflation statistics. In some provinces, actual fuel prices may have declined. By focusing on official prices, the official inflation statistics exaggerated actual price increases.

Second, one of the continuing characteristics of the Iraq economy is the large-scale smuggling of consumer products from neighboring countries. This smuggling of everything from frozen chickens to fuel to refrigerators both distorts Iraq's trade data and further complicates the GoI's attempts to estimate changes in domestic prices. In addition to shortages brought about by artificially low official prices, smuggling was also encouraged by the GoI's complex and expensive trade regulations. As will be discussed in Chapter 13, although the GoI has simplified its tariff structure, Iraq has some of the most restrictive trade regulations among the MENA countries. Therefore, for most products, it is more profitable for a merchant to pay a bribe to the appropriate official to look the other way when an item is smuggled into the country than to legally import it.

Third, from 1980 through at least the end of 2017, Iraq suffered from severe internal and external conflict. As a result, through much of this period Iraq was not a single market. For example, during 2014–15, ISIS controlled an estimated 40 percent of Iraq's territory including one of the country's largest cities, Mosul. This ISIS-controlled territory had been the home of 25 to 35 percent of the country's population (Gunter 2015). This national fragmentation meant that the same products sold for different prices (differences much greater than the costs of transportation) not only in different provinces but also in different towns within the same province. The gathering and analysis of this data proved extremely difficult.

Fourth, there was a strong incentive for elements of the GoI to exaggerate inflation. The real wages of government employees had declined substantially during the sanctions period. As a result, during the initial years of the US-led coalition occupation of Iraq there were continuing requests that the coalition permit substantial increases in government employee wages in order to restore the living standards of government employees to reasonable levels. In particular, the 65 percent official inflation rate in 2006 was used to justify a substantial increase in both the salary and pension benefits of government employees. In retrospect, the adjustments were probably excessive, since the average government employee is now believed to earn about 50 percent more than someone working in the private sector with similar skills.

Finally, the ubiquitous corruption in Iraq that is the subject of Chapter 4 means that, for many goods and services, a price has two parts. The consumer or firm must not only pay for the product itself, but also pay a bribe to a government official to allow the purchase. This two-part purchase both raises the true price of obtaining a product and – since there is no reason to think that the amount of the bribe is either constant or even closely related to the price of the purchase – the failure to report the actual amount paid in the bribe is another wedge between actual price changes and the official inflation rates.

Raising government salaries and benefits was not the only policy response to the perceived acceleration of inflation. The Central Bank of Iraq (CBI) began a deliberate appreciation of the ID that resulted in the exchange rate increasing from 1467 ID/$ in 2006 to 1170 ID/$ in 2009; about a 20 percent appreciation of the ID. This was intended to reduce the rate of inflation by decreasing the prices of imported goods. Of course, as will be discussed in Chapter 13, this appreciation also reduced the competitiveness of non-oil exports from Iraq; the "Dutch Disease". In December 2020, in an effort to reduce the drain of dollars from the CBI, the Iraqi dinar was devalued by about 23 percent to 1450 ID/$. This is expected to lead to a substantial increase in 2021 inflation.

IMPROVING THE QUALITY OF MACROECONOMIC DATA

The primary statistical authority of Iraq, the Central Statistical Organization (CSO), is an invaluable source of data not only on the Iraqi economy but also on its society. This book makes extensive use of CSO periodic surveys and studies. As an example of the careful preparation of these publications, in Volume III of the Iraq Household Socio-Economic Survey: 2007, COSIT (2008) lists not only standard errors for the data (ibid., pp. 780–795) but also the exact questions asked during the survey (ibid., pp. 799–986). In addition to the data and analysis of the CSO, the World Bank, and the IMF, the author has benefited greatly from the research of the Al-Bayan Center for Planning and Studies, a Baghdad think tank. However, there are at least three challenges facing the publication of timely accurate data concerning Iraq by the CSO and other Iraqi organizations with statistical responsibilities.

First, CSO staffing is inadequate, and many employees lack the necessary training. It is important that the CSO be adequately funded to allow it to hire and train the best people for this important job (IMF 2019b, Annex I, p. 36).

Second, in most cases, the CSO must rely upon the various government ministries to submit reasonably accurate data on a timely basis. But the degree of cooperation differs greatly among ministries. For example, the Ministry of Finance and the Ministry of Industry and Mineral Resources generally provide quality data on a timely basis. At the other extreme, the Ministries of Health and Trade are considered to be particularly opaque. It is apparent that some ministries consider the provision of data to be of little importance, or seek to hide unfavorable data. The prime example is the unwillingness of some ministries to cooperate in the ongoing census of government employees. This census, which was supposed to have been completed in 2006, is still in progress almost 15 years later. The census results are expected to support accusations of either mismanagement or corruption, especially with respect to ghost workers, and some ministries have gone beyond failing to cooperate and have actually physically threatened census takers. In addition, CSO access to KRG data is also limited.

The third, and potentially the most serious, challenge to data collection that is both timely and useful is the universal temptation of all government departments to restrict or distort data that puts the government in a bad light or that might have adverse political consequences. This of course is another example of "Goodhart's Law", which states that whenever a formerly reliable variable becomes important for policy then it ceases to be reliable.

This summary of Iraq's recent GDP, employment, and inflation data is intended to provide a framework for the rest of this book. It should be kept in mind that there is no shortage of data about this economy. The challenge is determining which data most accurately describes what is actually happening in Iraq. While data challenges exist in almost all developing countries, Iraq's combination of recent conflict, socialist market transition, and massive corruption make its economic data more questionable than most.

3. Health, poverty, education, and gender issues

Health conditions in Iraq are considered some of the worst in the region. Indicators over the past twenty years show that the health of the population has seriously deteriorated.

(GoI 2010, p. 119)

The ISIS education system was designed to incite violence and extremism, and prompted parents to refrain from sending their children to school. This had two major effects: (1) it undermined parent's trust in the Iraqi educational system, and (2) it planted the seeds of violence and radicalism in the minds of children and adolescents.

(World Bank 2018b, p. 28)

Good health, low poverty levels, and access to education are important determinants of the quality of life. Prior to the initiation of 40 years of conflict in 1980, Iraq had made dramatic progress in all three areas. In fact, in the late 1970s, Iraq was considered by many to be the leading Arab country with respect to equitable social development. However, the physical and moral destruction of the wars initiated by Saddam with Iran and Kuwait, followed by defeat by the US-led coalition, the United Nations (UN) imposed sanctions, Saddam's overthrow in 2003 by a second US-led coalition, widespread corruption, and the vicious fights with the al Qaeda and ISIS insurgencies that brought the country to the edge of civil war, tore the fabric of Iraqi society apart.

Most recently, the 2014–17 battle against the ISIS insurgency affected the social structure of Iraq in three ways. First, the destruction of two of Iraq's cities, Mosul and Ramadi, as well as the devastation of as much as 40 percent of the countryside, led to the widespread destruction of hospitals, schools, and other critical infrastructure. Second, as was discussed in the previous chapter, the fight with ISIS led to a wave of refugees, internally displaced persons (IDPs). In mid-2020, many of the IDPs are still in refugee camps with limited medical and educational services. Finally, the ISIS insurgency and its aftermath have led to a substantial decline in the quality of data concerning the Iraqi people and the economy. This has complicated the development of policy to ameliorate the damage done.

It is therefore surprising that, with the exception of variables related to gender equality, Iraq's key social indicators of development are reasonably close to

Table 3.1   *Social indicators of development*

	Iraq	MENA	UMIC	Data
Per capita income (PPP)	$11 300	$17 400	$17 200	2019
Life expectancy (male/ female)	68/73	72/76	73/78	2018
Malnutrition (less than 5 years old)	13%	14%	6%	2011–19
Child mortality (per 1000 births)	26	22	13	2019
Maternal mortality (per 100 000 births)	79	57	43	2017
Births to women aged 15–19 (per 1000)	72	40	33	2018
Contraception prevalence (among married women)	36%	N/A	71%	2013–18
Tuberculosis (per 100 000 persons)	42	31	69	2018
Universal primary education (male/female)	89%/86%	N/A	96%/96%	Iraq 2012, UMIC 2018
Youth literacy (male/ female)	95%/92%	92%/88%	99%/99%	2010–19

Source: World Bank (2020c, Table WV.1: Per capita income; Table WV.2: Malnutrition, child mortality, maternal mortality, TB; Table WV.3: Improved sanitation; Table WV.5: Life expectancy; Table 2.8 and CSO Iraqi Woman and Man in Statistics: Primary education; Table 2.10: Literacy; Table 2.14: Contraception; Per capita income WDI).

the average of the Middle East and North Africa (MENA) countries as well as that of the upper-middle-income countries (UMICs). In other words, Iraq lost its position of primacy in social development, but it did not fall to the bottom; it is now close to the average of countries which have roughly the same level of per capita income.

The UN 2000 Millennium Development Goals program provided development goals in eight areas to be achieved by the year 2015. Iraq's progress towards these goals is monitored both by Iraq's Central Statistical Organization (CSO) and the World Bank. Unfortunately, for some of the same reasons discussed in the previous chapter, the CSO and the World Bank often report different data for Iraq for the same indicators. However, the trends are similar. Table 3.1 provides the World Bank's most recent data for Iraq with respect to the key social indicators of development

Iraq's results are compared to those of MENA countries as well as those of the UMICs. The UMICs have a gross national income (GNI) of between about

$4000 and $12 500 (2019 GNI per capita). Iraq's per capita GNI in 2019 was $5840.

ABSOLUTE AND RELATIVE POVERTY

By international standards, absolute poverty is high in Iraq, but relative poverty is low. Absolute poverty measures whether persons are able to obtain the essentials of life, while relative poverty measures the gap between the poor and the well-off in a country. Levels of absolute poverty range greatly across Iraq. In every province, rural regions tend to have higher levels of absolute poverty than urban areas. Among provinces, those in the south and those that were occupied by ISIS have higher levels of poverty than Baghdad, the Kurdish Regional Government (KRG), and Kirkuk. As expected, families and individuals displaced by violence tend to experience greater poverty. Finally, households with a female head are more likely to be poor compared to those with a male head.

Between 2007 and 2012, evidence points to a decrease in absolute poverty. However, these gains were almost completely reversed as a result of the drop in oil prices and the ISIS insurgency. By 2014, poverty had returned to 2007 levels (World Bank 2018b, p. 27).

The usual measure of absolute poverty is the proportion of the population that earns less than $1.90 or $3.20 per day, adjusting for differences in the cost of living (at purchasing power parity, PPP). Trying to live on less than $1.90 (PPP) a day probably means insufficient food, little access to education or medical care, and a shortened lifespan. Earning more than $1.90 but less than $3.20 (PPP) is not as bad, in that one has the potential of maintaining long-term health. Of course, if an individual is alone and very young, pregnant, or very old, then life is difficult on even $3.20 (PPP) a day. It is a precarious life, in that a single error or accident has the potential to plunge an individual or family into desperate – sub-$1.90 – poverty.

Even before the devastation caused by the ISIS insurgency, absolute poverty was high in Iraq. In 2012, about 2.5 percent of the Iraqi population were living on less than $1.90 a day, while almost 18 percent of the population were living on less than $3.20 a day (World Bank 2020a, Table 1).

The major determinant of absolute poverty is personal income. Iraq's per capita gross domestic product (PPP) was an estimated $10 830 in 2018, $11 280 in 2019 and, if early real growth estimates are correct, may fall to $8900 in 2020 (World Bank 2020a, Table 1, p. 16). Combined with the ISIS destruction, the decline in incomes can be expected to lead to an increase in absolute poverty.

Relative poverty looks at the distribution of income or consumption across a country's population. A Gini index usually measures this distribution. If everyone in a country has the same income, the Gini index would be zero. At the other extreme, if one person received all of the income in the country and everyone else received nothing, then the Gini index would be 100.

Surprisingly, relative poverty in Iraq is low. Iraq's 2012 Gini index was an estimated 30, about the same as Austria and Poland. This is a more even distribution of income than most of Iraq's neighbors. For example, Iran had a Gini Index of 41, while Jordan's was 34. In fact, of the 164 countries whose Gini indices are reported by the World Bank, almost 90 percent had a more uneven distribution of income than Iraq (World Bank 2020c, Table 1.3). It is possible that the Gini estimate for Iraq is distorted by hidden earnings from corruption or the underground economy. However, if accurate, the low relative poverty in Iraq is driven, in part, by the large proportion of the population that are government employees, where the range of salaries tends to be less than in the private sector. This is not to deny that there are large income differentials in Iraq. The bottom 20 percent of the population received only about 9 percent of the national income, while the top 20 percent of the population received almost 39 percent of the national income (World Bank 2020c, Table 1.3).

Will Iraq's low Gini index – relatively low income inequality – continue? The "Kuznets curve" predicts that the trend of income inequality is U-shaped. Initially, economic growth leads to a worsening of income inequality. It is only when the country achieves high levels of per capita income that income inequality begins to fall again. In other words, low- and high-income countries tend to have lower Gini indices than middle-income countries. Note that an increase in income inequality does not necessarily mean that incomes of the bottom 20 percent of the population are actually declining. All that is required is that the incomes of the bottom 20 percent of the population grow at a slower rate than incomes at the top of the income pyramid.

Like many developing countries, Iraq faces a poverty choice. If the GoI continues to play a dominant role in the economy, then Iraq will probably experience slower real growth and therefore higher absolute but lower relative poverty. On the other hand, more rapid growth from a liberalized economy will probably result in lower absolute but higher relative poverty. Lower absolute poverty is usually accompanied by better health.

HEALTH

Life Expectancy

A primary measure of health is life expectancy. As can be seen in Table 3.1, the average life expectancy from birth in Iraq is worse than the average in both the UMICs and the MENA countries. According to the last Iraqi census of 1997 – which excluded the KRG – more boys are born each year than girls.

Up to approximately age ten, girls actually die at a more rapid rate than boys. However, beginning at age ten, the death rate of males substantially surpasses that of females, so that by the age 15–19 cohort the number of females exceeds that of males. This trend continues, so that among Iraqis older than 65 years, the number of females exceeds that of males by 27 percent (COSIT 2012, Table 2/7).

While the violent death of countless males is a major contributing factor to the gap in life expectancies, there are also differences between the genders in life choices. Many young men in Iraq are addicted to painkillers, that are readily available without a prescription. Despite the injunctions contained in the Qur'an, alcohol abuse is not uncommon. Possibly associated with substance abuse among young males, there is a disproportionate rate of auto-mobile accident fatalities. According to the latest data, males accounted for 80 percent of automobile-related fatalities. Fatalities and serious injuries are common in workplaces and, since the labor force is mostly male, this also has a disproportionate impact on life expectancies. Finally, about 42 percent of men are smokers, compared to less than 7 percent of women (COSIT 2012, Tables 6/11 and 19/22).

Malnutrition and Child Mortality

Malnutrition is measured as the percentage of children less than five years of age whose weight is less than two standard deviations below the average weight for that age. It is estimated that, in 2011–19, about 13 percent of Iraqi children aged 0–5 years were malnourished, compared to 14 percent in MENA countries and 6 percent in UMICs (World Bank 2020c, Table WV.2). Malnourishment at this age often leads to stunting and lifelong health prob-lems. Since this sample included the pre-ISIS period, it is likely that malnour-ishment has worsened. Even if this data is reasonably accurate, considering the social and economic disruptions of the last few decades one would expect more cases of severe malnutrition among the Iraqi young. There are three explanations generally given to describe why Iraq has less malnutrition than expected. First, the official data for Iraq severely underestimates actual cases of malnutrition. Second, the Public Distribution System (PDS), discussed in detail in Chapter 8, succeeds in providing a basket of food to over 99 percent of the poor in Iraq (World Bank 2018b, p. 28). While the baskets are often short by one or more items (COSIT 2012, Table 15/71), and it can be difficult to add an additional person to a family's allocation, these baskets do supply food essentials – at very low cost – that can provide a foundation for a reasonable diet. Third, although there are not enough quality orphanages and there is a continuing tragedy of abandoned children living on the streets of major cities, many orphaned children are not abandoned, but rather end up being cared for by someone in their extended family.

Comparisons of child mortality rates among different countries are almost meaningless, since countries use different standards for a "live" birth. Many countries require at least a 26-week gestation. If a child has not been carried until at least the end of the second trimester of pregnancy, then if the child dies it is not counted as a death in the nation's child mortality data. Other nations require a certain minimum length or weight to be considered a live birth. Some governments do not count any child's death that occurs within the first 24 hours after birth, while at the other extreme, the USA counts it as a live birth if the child shows any signs of life regardless of gestation, length, weight, or time of death. These different definitions greatly complicate cross-country comparisons of child mortality. It is more meaningful to focus not on the level of a country's child mortality, but on the trend.

Between 1990 and 1999, according to GoI statistics, Iraq experienced a doubling of infant mortality to 101 deaths per 1000 live births. This sharp deterioration is generally blamed on shortages of food and medicine as a result of the UN sanctions. However, as revealed by investigations into the Oil-for-Food Programme scandal, even when funds were available and imports of food and medicine were allowed, Saddam diverted a large portion to his family, the army, and his supporters. It is also likely that Saddam exaggerated the official rate of infant mortality in his attempt to make a moral argument that the UN was engaged in a crime against humanity. Since the fall of Saddam in 2003, infant mortality decreased rapidly to about 26 deaths per 1000 live births in 2019 (COSIT 2012, Table 19/7; World Bank 2020c, Table WV.2). Iraq failed to meet its 2015 Millennium Development Goal for infant mortality of 17 deaths per 1000 live births. However, if current trends continue then Iraq should achieve this goal in the next 8–10 years.

Maternal Mortality

Maternal mortality followed a similar path as infant mortality. From 117 maternal deaths per 100 000 live births in 1990, these deaths rose 150 percent to 291 maternal deaths in 1999. Following the invasion of Iraq, maternal deaths fell to an estimated 79 maternal deaths in 2017 (COSIT 2012, Table 19/9; World Bank 2020c, WV.2). As Table 3.1 shows, this compares unfavorably to the averages in both MENA countries and lower-middle-income countries (LMICs). The 2015 Millennium Development Goal goal for Iraq was 29 maternal deaths per 100 000 live births. It is unlikely that Iraq will achieve this in the foreseeable future.

One factor in the failure to rapidly reduce maternal mortality is a shortage of female doctors and midwives, especially in rural areas. For cultural reasons, many Iraqis consider it unacceptable for a male doctor to examine or attend a woman during childbirth. While an estimated 96 percent of births are assisted by a health professional, it is believed that many women still fail to receive optimal pre-natal care by a skilled health professional (GoI 2018, p. 204). The GoI is attempting to increase the number of female health professionals, but the more difficult challenge may be to provide incentives for such profession-als to settle in rural areas that tend to impose restrictions on how single women should live.

A contributing factor to maternal mortality, if there are complications, is the shortage of ambulances, hospitals, and clinics. As will be discussed below, these shortages have worsened since the rise of ISIS.

Covid-19

In late December 2020, reported Covid-19 deaths in Iraq were 12 808, which is equal to 333 deaths per million of population. This is about the same number of deaths per million as Israel and Latvia. The number of cases of Covid-19 infection were an estimated 595 000. However, Iraq's estimates of Covid-19 infections and fatalities are probably too low, for several reasons. There is a shortage of both testing kits and trained personnel to both administer and interpret them. Also, Iran's neighbor to the East, Iran, has reported 694 Covid-19 deaths per million, although international public health authorities believe the actual rate in Iran in much greater, maybe double the officially reported figure. While the border between Iraq and Iran is officially closed, there are unofficial border crossing points, leading to a high probability of the movement of infected persons across the border. Finally, the large number of Iraqis who work in the informal economy do not have the option of avoiding contact. If they do not work, they do not eat (UN OCHA 2020).

To reduce the rate of infection, the government imposed severe restrictions. Restaurants could only operate on a delivery basis. Schools, mosques, cafes, and malls were closed, as were government offices. Gatherings of more than three persons were forbidden. In addition, the government imposed strin-gent curfews and movement restrictions. Combined with the collapse in oil prices, these restrictions resulted in a sharp economic decline. The impact on internally displaced persons (IDPs) was particularly severe, as many IDPs in refugee camps were unable to access cash assets. While food supplies were generally adequate, several of the camps had limited access to clean water and sanitation services.

Hopefully, by the time this chapter is read, the initial Covid-19 crisis will have passed. However, there can be expected to be substantial long-term effects both for IDPs and the general population. In particular, medical resources were concentrated on the Covid-19 effort, which resulted in denial or delay of non-emergency medical care. This can be expected to lead to an increase in infant and maternal mortalities as well as reduced life expectancy for those unable to access preventative or emergency care. School closures are expected to lead to decreased educational achievements as young people fail to return to school after the initial crisis is over. Finally, the economic decline is expected to lead to a significant increase in the percentage of the population below the poverty line (World Bank 2020b, Fall, p. 22).

Tuberculosis (TB), Human Immunodeficiency Virus (HIV), and Sanitation

TB is a common infectious disease that attacks the lungs and is often fatal. It is spread through the air when a person with active TB coughs or sneezes. About one-third of the world's population is thought to have TB. While data on TB in Iraq is difficult to obtain, it is estimated that 42 out of every 100 000 Iraqis – about 17 000 persons in total – had the disease in 2018. But only about one-third of these people were receiving medical care (COSIT 2012, Tables 19/1 and 19/12; World Bank 2020c, Table WV.2).

Antibiotics are used to kill the TB bacteria, but these antibiotics in pill form must be taken daily for six months according to a very strict regimen. I tested positive for TB and was instructed that if I missed taking my pill for one day then I could double up on the next, but that if I missed two days in a row – even in the final week – then I must restart the six-month regimen. And no alcohol for the entire six months. Not only will failing to follow the strict TB antibiotic regimen reduce the probability of curing the infection, but it also increases the likelihood of developing a strain of TB resistant to standard antibiotics. As a result, the preferred treatment option requires that a TB patient take their daily pill under the direct observation of a medical professional. About 92 percent of Iraq's TB patients are currently taking medication under direct observation, which is less than the country's 2015 Millennium Development Goal of 100 percent directly observed treatment.

Persons with HIV are particularly vulnerable to TB since HIV weakens resistance to disease. HIV first came to Iraq in a shipment of contaminated blood in 1985. Under Saddam, those infected were forcibly segregated. Since Saddam's fall, the GoI has opened over two dozen clinics nationwide to provide free HIV testing and treatment. However, there is widespread ignorance and fear of HIV. Families often abandon infected members and there are rumors that religious fanatics are murdering those believed to be HIV positive on the grounds that the infection is proof of sin. While the GoI reported in 2009 that there were only 44 infected Iraqis, it is believed that many cases are not reported because of fear that victims will be ostracized (IRIN 2009). Although there is no reliable data, it has been estimated that the actual number of HIV-infected Iraqis might be as great as 500. While the GoI has begun a program to educate the population about HIV, they are proceeding tentatively for fear of sparking a panic and possible reprisals against the infected. In 2009, only 79 percent of males aged 15–30 and 70 percent of females in the same age cohort knew that HIV could be sexually transmitted (COSIT 2012, Table 11/12). Of course, as an UMIC with substantial poverty that has just passed through a near civil war, the Iraqi people are at risk for a large variety of health problems in addition to TB and HIV. But research over the last several decades has shown that as long as the average person can obtain sufficient amounts of potable water and has access to improved sanitation facilities, they can either avoid most other diseases or at least have a reasonable chance of recovery if they become sick.

According to the World Health Organization (WHO) (World Bank 2020c, Table 2.13), only an estimated 59 percent of Iraq's population had access to sufficient water from an improved source in 2017. Access is defined as at least 20 liters (5.3 gallons) per person per day from a source within 1 kilometer (0.6 miles) of the person's dwelling. About one-third of rural households obtained their water from a household connection, the remainder obtained their water from a public tap, protected well, rainwater, or water tanker truck. Almost all urban residents obtained their water from a household connection. This access is not dependable, with 22 percent of rural residents and 32 percent of urban residents reporting that their water supply was interrupted daily. With respect to the quality of water, 14 percent of Iraqis report that they boil, filter, or chemically treat their water before drinking (COSIT 2008, Table 2-00, p. 56).

How does access to potable water in Iraq compare to other countries? Access in Iraq is much less than the 78 percent reported for the MENA states. And, due to the destruction caused by ISIS, the GoI reports little progress over the last few years in increasing either rural or urban access to potable water. It is unlikely that the country will reach its 2015 Millennium Development Goal of 100 percent urban and 89 percent rural access any time in the next decade.

Increasing access to sufficient amounts of potable water is less a problem of inadequate infrastructure than of political will. Despite the fact that published water tariffs are quite low, fees for water usage are rarely collected, due in part to a lack of metering. As a result, almost all users currently pay a zero price for their water and there is little incentive for consumers to conserve, or for the Ministry of Municipalities and Public Works to aggressively strive to ensure that every Iraqi has access to potable water. Despite these inefficiencies, raising water tariffs or even collecting the existing low tariffs is extremely unpopular and is expected to reduce support for any politician that proposes it. The impact of such perverse pricing of essential services will be discussed at greater length in Chapter 12.

Improved access to potable water and access to improved sanitation are closely related, since ad hoc sanitation measures often lead to water contamination. If improved sanitation means that the population has ready access to facilities that prevent human, animal, or insect contact with excreta, then only about 41 percent of Iraq's population have access to such facilities. While the Iraqi population has better access to improved sanitation facilities than the 35 percent access reported by the average MENA state, access in Iraq is much less than that of the average UMIC, 60 percent (World Bank 2020c, Table 2.13).

Hospitals and Clinics

Iraq was once considered to have among the best medical facilities in the region. However, decades of conflict and mismanagement have reduced the quality and quantity of such facilities. The 2014–17 ISIS insurgency was particularly destructive. Of the 56 hospitals in the provinces occupied by ISIS (Anbar, Diyala, Salah ad Din, and Ninawa), 19 were completely destroyed, while 24 were partially damaged. Similarly, of the 97 health centers in these provinces, 42 were either completely or partially destroyed (World Bank 2018c, Table 6, p. 21). As a result, Iraq currently has only 1.2 hospital beds per 1000 people, which is relatively low compared to other countries at a similar level of economic development.

There is also a severe shortage of doctors and of nurses and midwives. There are only 0.7 doctors per 1000 persons in Iraq, and only 2.0 nurses and midwives per 1000 persons. The average MENA country has doctor and nurse ratios of 1.3 and 2.5, respectively; while UMICs have ratios of 2.0 doctors and 3.5 nurses and midwives (World Bank 2020c, Table 2.12).

The shortage of skilled medical professionals was caused primarily by large-scale emigration since 2003. Some of these emigrants were attracted by better economic opportunities abroad (known as "brain drain"), while others fled the sharp rise in violence in the middle of the decade, especially since some insurgent groups deliberately targeted medical and other professionals. Exacerbating the impact of the loss of medical professionals was a deterioration in the quality of medical education, and a shortage of medical supplies. Beginning in the early 1990s with the UN-imposed sanctions and continuing until at least the reduction of al Qaeda violence in 2007, medical training in Iraq suffered from both shortages of laboratory and other teaching materials as well as increased corruption. As a result, poorly trained physicians and nurses were able to obtain certification. While the GoI is executing a complex plan to increase the proportion of the population that has ready access to medical care, ensuring the quality of that care by improving medical education will be very difficult. Educating future doctors and nurses is not the only educational challenge in the country.

EDUCATION: CURRENT STATUS

In 2016–17, the enrollment rates were 93 percent in primary (6–11 years old), 55 percent in intermediate (12–14 years), and 30 percent in secondary education (15–17 years). The rise in enrollment has substantially exceeded the increase in instructional personnel. As a result, the student–teacher ratio average is 32, with some districts in Baghdad having 50 students in a classroom (GoI 2018, Figure 32, pp. 194–5). While the quality of primary and secondary schools varies greatly, it is generally poor. Teachers are often unskilled, unmotivated, or absent. The curriculum is generally outdated, and there is often a shortage of textbooks and other educational material. Facilities are limited, with an estimated nationwide shortage of 3000 schools. As a result, some schools operate two or even three shifts, and long commutes by even young children are required. Many school buildings are run down, lacking adequate lighting, running water, and sanitation facilities. In fact, an estimated 15 percent of all schools are in danger of collapse (GoI 2010, p. 115). While the GoI and international organizations have built many schools in the post-Saddam era, it is not uncommon to find the new buildings not in use, looted, or confiscated by a government or non-government organization for non-educational use.

While progress was made prior to 2014, the illiteracy rate in Iraq among persons who are 15 years or older was still high: about 22 percent. With the school destruction and the sharp rise in IDPs during the ISIS insurgency, levels of illiteracy are believed to have risen. This high rate of illiteracy not only has an adverse economic impact, but also reduces the ability to effectively participate in commercial and political activities. Corruption also tends to thrive in societies with high levels of illiteracy. If you cannot read, you are an easy target for every government official who can wave papers in your face.

As can be seen in Table 3.2, more than twice as many females as males were considered illiterate by International monetary Fund (IMF) standards in 2013. The IMF definition of literacy is being able to both read and write, with understanding, a short statement about one's daily life. Illiteracy is much worse in rural areas, where not only is the overall level of illiteracy higher but there is also a greater gap between male and female levels of illiteracy. Also, before ISIS, there was a strong correlation of illiteracy with age. For example, while only about 16 percent of women aged 10–14 were unable to read and write, this proportion rose to almost 70 percent among women who were 50 years old and older (COSIT 2008, Table 3-1, p. 222). However, with the closure and destruction of schools by ISIS, combined with the limited education opportunities in displaced persons camps, there is rising illiteracy among young Iraqis. Why are so many Iraqis – especially females – either illiterate or have only limited ability to read and write? Table 3.3 gives the results of a survey that

Table 3.2   *Education and gender*

Maximum education (2013) for persons 15 years or older	Male (%)	Female (%)
Illiterate	14.1	31.0
Read only	2.6	3.2
Read and write	14.5	15.0
Primary (6–11 years)	30.9	24.4
Intermediate (12–14 years)	15.5	11.7
Secondary (15–17 years)	9.0	6.0
Vo Tech diploma	6.1	4.5
Bachelor (college)	6.7	3.9
Master and PhD (university)	0.5	0.2

Source: CSO (2014, Figure 11, p. 24).

Table 3.3   *Reasons for never attending school (%)*

	Family not interested	No school	Disability or disease	Not interested	Cannot afford	Other reasons
Primary	32	14	9	9	5	37
Intermediate	39	18	12	9	8	18
Secondary	41	13	11	13	6	18

Source: COSIT (2008, Table 3.8, p. 233).

asked families across Iraq for the reasons why their children (both male and female) had never attended primary, intermediate, or secondary school. Lack of family or individual interest was the most common answer. It is difficult to deconstruct this answer, but it may be a function of both culture and economic opportunities. Many Iraqi families think that it is inappropriate to have a daughter instructed by a male teacher at any age, but especially after she reaches puberty. Thus, if there are an insufficient number of female teachers at the primary and especially at the intermediate level, then parents will withdraw their daughters from school. The shortage of female teachers in rural areas is especially acute, since these generally more religious communities can impose restrictions on the freedom of women to fully participate in community life or, in some cases, even to move about without being accompanied by a male relative. As a result, it is difficult to persuade educated women to accept teaching assignments in rural areas. If they are unmarried, then their lives might be severely constrained. If they are married, there is the difficulty of finding employment for their spouse in a rural community that probably does not provide a great variety of employment opportunities.

According to the survey, 5 percent of Iraqi families say that they either keep their children (both male and female) out of primary school or curtail their attendance because of the expense. While education is technically free, there are three types of costs that must be paid by students or their families. First, the average annual out-of-pocket costs range from an estimated 93 000 ID ($74) to attend a primary school to 300 000 ID ($240) for secondary school (COSIT 2008, Table 3-16, p. 244). If a family has two or three children of school age, the out-of-pocket costs alone might equal as much as 25 percent of household income. Often families choose to continue the education of their male children while saving money by curtailing female schooling.

Second, like most public institutions in Iraq, there is corruption in the educational system. Families must be willing to pay a succession of small bribes to ensure that their children actually receive the promised instruction.

Finally, the opportunity cost of having a child in school might be high, especially for families engaged in farming or small commercial activities where children can contribute productively even at a very young age. For example, a family might decide that their ten-year-old might be better engaged in herding the family's goats or doing chores in the family shop rather than in a classroom. This may be especially true for females, where higher education levels are associated with higher unemployment. The unemployment rate for females who complete intermediate schooling is almost three times greater than for those who are illiterate. As mentioned in the previous chapter, female employment in Iraq is limited, with almost two-thirds employed in either farming or teaching. The traditional labor-intensive farming that is practiced throughout much of Iraq tends to reward experience more than education. In addition, although the legal minimum age for marriage is 15 years, there are cases of girls as young as age 12 getting married. Girls and women almost invariably leave school upon marriage.

EDUCATION: LESSONS LEARNED

Research on education in Iraq is in its infancy and it will probably be a decade before it can provide a solid foundation for an Iraq-specific education policy. One could look to other countries' experiences with educational policy, but it may not be applicable. One of the challenges of developing a sound educational policy is that the same policy will often have different results in developing countries – such as Iraq – than in developed countries. However, among developing countries, research has discovered four initiatives that will probably improve education in Iraq.

First, returns on education decline with the level of schooling. In other words, investment in primary education tends to produce greater returns to a society than investment in secondary education, which produces greater returns than investment in university education (Psacharopoulos and Patrinos 2002, Tables 1 and 5, pp. 12, 15). Therefore, if human and financial resources are constrained, then the GoI should concentrate on primary and secondary education before funding university education. While it is difficult to obtain and interpret the cost of education data in Iraq, it is estimated that the cost per student of the GoI providing one year of elementary or secondary education was at least 500 000 ID ($420), while a university education cost at least ten times that amount. This multiple of ten seems very low by international standards, and better data on overhead costs of university education will probably increase the multiple.

Second, research shows that the more general the curriculum, the higher the returns to education. This is a surprising result, since it is often thought that vocational training is more practical and useful than general education. However, teaching a general curriculum costs less than vocational training. Classes tend to be larger, and there is less of a requirement for specialized equipment. In addition, whether or not they have a higher unemployment rate, general curriculum graduates tend to spend less time unemployed, possibly because they have more job flexibility (Psacharopoulos 1991, pp. 190–91). One of the many challenges facing vocational training in Iraq is the inflexibility of the educational bureaucracy. For example, rather than teach students to maintain and repair the modern imported tractors that Iraqi farmers prefer, teachers will instead instruct them on obsolete Iraqi tractors because that is what they know.

Third, there are many initiatives that are being sold to improve educational quality, but only some have been shown to be successful. The successes include reducing home–school distance and increasing the availability of textbooks. About 11 percent of Iraqi families state that excessive distance to school is why their children are not attending (COSIT 2008, Table 3-13, p. 240). This non-attendance could be reduced either by constructing and staffing more local schools or by improving transportation to existing schools. Greater availability of textbooks tends to significantly reduce the number of students held over for another year. One estimate was that every $1 spent on textbooks resulted in a $12 saving in education expense by speeding students through school. Iraq should seek to have at least one textbook for every five students for every subject. It should also be noted that at the secondary level, low-quality teachers significantly impair learning. Countries that require teachers to specialize at the secondary level tend to have better results; for example, having mathematics majors rather than education majors teaching mathematics classes.

However, popular (and expensive) ideas such as decreasing class size and providing computers to students do not often lead to significant improvements in education quality. While smaller classes with high-quality instruction are better than larger classes with the same quality of instruction, this is not the usual choice. Instead, because of a shortage of skilled teachers, developing countries such as Iraq often have to choose between large classes with high-quality instruction or smaller classes with lower-quality teachers. The former tends to lead to better education (Hanushek 2005, pp. 201–5).

Finally, it is an error to attempt to use education to solve problems that are not really educational in nature. In Iraq, there is a continuing push to increase the number of university seats in order to train doctors and other skilled professionals to replace those that are leaving the country ("brain drain"). This is a common error in developing countries, exemplified by India which trains some of its brightest people at great expense to be outstanding medical professionals in Canadian and US hospitals. It is more efficient to improve the pay and working conditions of Iraqi physicians and other skilled professions so as to decrease their motivation to migrate, than to train additional new doctors to replace them for several years until they migrate as well.

Excessive Post-Secondary Education

The unemployment rates are roughly the same for Iraqis whose final education was at the intermediate level and those that received more advanced degrees. This is evidence that Iraq either has an excess of highly educated persons or is providing education in the wrong fields. This has resulted in an estimated 40 percent of new college graduates in Iraq being unemployed or underemployed upon graduation. Further evidence is found in the constant pressure on the GoI to hire college graduates to do jobs that do not really require advanced education, and the high rate of "brain drain" in technical fields. These results support a controversial policy of decreasing government support for higher education until a falling unemployment rate for college-educated Iraqis shows that there is no longer an excess of highly educated Iraqis. However, shifting educational priorities away from universities is an uncomfortable recommendation for the Iraqi elite, who do not want to reduce the chances of their own children obtaining a university education as a stepping stone to a high-pay, high-status job. In other words, the private benefit of higher education to the student or the student's family is greater than the benefit of one more college-educated person to the entire Iraqi society. Not least is that having an advanced degree improves the chances of a young man making an advantageous marriage.

YOUTH CHALLENGES

Due to high fertility rates, Iraq's population is relatively young. In 2019, an estimated 38 percent of the population was younger than 15 years. As previously discussed, educational opportunities are limited and, since the arrival of low oil prices in 2014, many young Iraqis expect to be unemployed or underemployed for long periods. Lack of steady well-paid employment has resulted in an increase in proportion of indebted young Iraqis to 59 percent, the fourth-highest rate in the MENA. The lack of opportunity in Iraq has also contributed to increased interest in emigrating. One survey showed that 65 percent of young Iraqis are considering emigrating, the fourth-highest rate in the MENA. The United Arab Emirates and the United States are among the most desired destinations (Arab Youth Survey 2020, p. 14).

There are other survey results that point to an increasing disconnect between young Iraqis and their society. About 76 percent of young Iraqis think that the country is corrupt (Arab Youth Survey 2020, p. 26), which is second only to Yemen. Religion appears to have a decreasing influence, with only 44 percent saying that religion is important to their personal identity (ibid., p. 42). As a result of this growing disconnect, young Iraqis are increasingly engaged in protests demanding that the GoI improve the quality of life. In fact, 89 percent of young Iraqis (ibid., p. 19) support the ongoing protests, higher than in Lebanon or Sudan.

MARRIAGE CULTURE

One of the most controversial issues involving the Iraqi population is the culture of marriage. It includes religious, historical, and tribal aspects of marriage that are widely accepted – if not widely practiced – among some groups in Iraq, and yet harshly disparaged by members of other groups. As a result, these aspects often cause confusion among foreigners trying to understand the country and its people. These controversial aspects include early marriage, cousin marriage, wife discipline, fertility decisions, polygyny, temporary marriages, and honor killings. It is true that Islam generally provided greater rights for women in the seventh century AD than they possessed in Christian lands during the same period. It is also widely accepted that women's rights are stated in the Qur'an, and that women control the private space in Islamic life (Allawi 2009, p. 204). But these truths are insufficient to justify the harsh or unfair treatment inflicted on married and unmarried women in Iraq.

Early Marriage

Although the current legal minimum age of marriage in Iraq is 18, exceptions can be made to allow marriage as early as 15 years with the approval of a judge and the minor's guardian. Before this approval is given, the judge must evaluate the minor's physical capacity for marriage as well as overall fitness. And if the judge decides that an objection by the minor's guardian is unreasonable then the judge can overrule the objection and allow the marriage to take place without the guardian's approval. About 17 percent of all Iraqi women are married before age 18. Despite the legal minimum age of 15 – with a judge's approval – in about 3 percent of all marriages the bride is younger than 15 years old (UNICEF 2012). In these cases, since tribal or religious traditions allow earlier marriages, the national law is ignored. A number of the girls who married below the age of 15 were found to have given birth to one or more children before they turned 15 years of age (COSIT 2008, Table 4-13, p. 276). As expected, there are often substantial long-term adverse health effects from young girls experiencing intercourse and childbirth. Even among young women between 15 and 19 years of age, fertility rates are high. In 2018, the fertility rate among women in Iraq aged 15–19 years was 72 births per 1000 women of that age, compared to the MENA average of 40, or the UMIC average of 33 (World Bank 2020c, Table 2.14).

Cousin Marriage

Iraq also has one of the highest levels of cousin marriage in the world. Marrying a first or second cousin is not only permitted but, among some groups in Iraq, actually encouraged. It is estimated that one-third of all marriages in Iraq are between cousins, while in certain rural areas the ratio rises to about half of all marriages. Cousin marriage increases the importance of family over other loyalties; after all, your uncle is your father-in-law. As discussed in the next chapter, this belief that nothing is more important than one's family encourages nepotism and other activities generally considered to be corruption. In addition, some women rights advocates think that cousin marriage tends to facilitate wife abuse, since an abused woman cannot flee to a female relative for sanctuary. It is likely that an abused woman's sister is not only married to the brother of the abusive husband but also lives next door. With respect to the health consequences, surveys show that only about one-third of women and men aged 12–30 are concerned that marriage to a close relative might affect their child's intelligence, spread genetic diseases, or lead to deformities (COSIT 2012, Tables 11/13 and 11/14).

It is expected that the ongoing decline in fertility will reduce cousin marriage in the future by reducing the number of potential marriage partners. In 1990, when the Iraqi fertility rate was 5.9, the average Iraqi had about 30 cousins, including 15 of the opposite sex who were possible marriage partners. However, with the 2018 fertility rate of 3.7, the average Iraqi only has 12 cousins, which reduces the number of potential marriage partners to only six.

Wife Discipline

Physical, sexual, or psychological violence within marriages is unfortunately common. In some provinces, over 60 percent of married women were reported to have been abused by their husbands. There is a rough geographic distribution, with the highest proportion of cases of reported marital violence against women occurring in the Maysan, Anbar, Qadisiyah, Karbala, and Baghdad provinces; while in the KRG, reported cases of husbands' violence against their wives is substantially lower (CSO 2014, Table 26, p. 34).

According to a 2006 GoI survey, a husband's right to physically discipline his wife was widely accepted among Iraqi women. Almost 60 percent of all Iraqi women believe that a husband may physically discipline his wife for a variety of reasons. The most commonly-stated justifications are: if a wife leaves her home without telling her husband; if the wife is neglecting the children; if she argues with her husband; if she abstains from sexual relations; or burns a family meal. The right of a husband to beat his wife is most accepted among rural, married, and uneducated Iraqi women. But, even among urban, unmarried, and educated women, at least 46 percent of women believe that a husband is within his rights to strike his wife under certain circumstances (COSIT 2012, Table 19/26).

This widespread acceptance of an Iraqi husband physically disciplining his wife is based, at least in part, on a statement in the Qur'an. Whether the Qur'an can be understood in translation rather than in the original Arabic is often debated. However, a standard translation states:

> Men are in charge of women, because Allah hath made the one of them to excel the other, and because they spend of their property (for the support of women). So good women are the obedient, guarding in secret that which Allah has guarded. As for those from whom ye fear rebellion, admonish them and banish them to beds apart, and scourge them. (Pickthall [1930] 1992, Sura 4:34, p. 97)

In another widely read translation, the critical words are given as: "beat them (lightly)" (Ali 1985, p. 220). The text in parentheses within this quote represents clarifying remarks by Ali that other translators do not include. Some commentators argue that the "beating" should be primarily symbolic rather than a severe infliction of pain or injury, possibly a husband slapping or spanking his wife. Others say that the critical words in question can also mean: "go away from them". Regardless of the disagreements among translators, commentators, and religious thinkers, it is widely believed in Iraq that a husband may physically discipline his wife.

Although Article 29 of the Constitution prohibits "All forms of violence and abuse in the family, school, and society", the penal code does not list the physical discipline of a wife by her husband as a crime. An Anti-Domestic Violence Law to make such violence illegal was rejected by the Council of Representatives (CoR) in 2015. Religious members argued that the law as then written was inconsistent with Islamic values. An amended version is being considered. Until a domestic violence law is approved, it is unlikely that an Iraqi husband will be sanctioned by the courts or even disparaged by his neighbors for beating his wife, unless he kills her or causes serious injury.

Fertility Decisions

One of the most important decisions for a married couple is how many children to have. This is usually measured as the total fertility rate: the number of children that would be born to the average woman through her entire childbearing years. If a country has a fertility rate of about 2.1 children per woman, then its long-run population will stabilize. If the fertility rate is below 2.1 then the country's population will continue to grow for a period of time (population momentum), but the average age will increase and, eventually, the population will begin to decline in size. Based on its total fertility rate, Iraq is an outlier, with a substantially higher fertility rate than its regional neighbors. In 2018, Iraq's fertility rate was 3.7, compared to a 2.8 average for MENA countries and 1.9 for the UMICs.

The high fertility rate and the accompanying rapid population growth in Iraq is expected to continue to have adverse effect on the country's economic development and public finance primarily because of the necessity of providing ever more essential services for a growing population (Hamilton 2020b, p. 27).

Iraq's fertility rate has declined over the last three decades due, in part, to increased use of contraceptives by married couples. As shown in Table 3.1, about one-third of married women use some form of modern contraception method. While Iraq's fertility rate is falling, it is decreasing at a much slower rate than among its regional neighbors. For example, Iraq's fertility rate fell from 5.9 children per woman in 1990, to 3.7 in 2018, a decrease of 2.2; while during the same period, Iran's fertility rate fell from 4.7 children per woman to 2.1, a decrease of 2.6 (World Bank 2020c, Table 2.14).

Will the decline in Iraqi fertility continue? What might be called the Goldman hypothesis states that fertility falls when women become educated and they cease to be religious (Goldman 2011, Chapter 13, 191–204). That younger women in Iraq tend to be more educated than older women is generally true. And there is some evidence of a decrease in the importance of religious beliefs among women. If the Goldman hypothesis is correct, these changes should lead to a further decrease in fertility and population growth. In addition, the GoI has a voluntary family planning strategy pending approval by the CoR. One facet of this strategy is to increase access to contraceptive methods among married women from 36 percent to 45 percent (Hamilton 2020b, pp. 26–27). In addition, aside from the moral issues, a national strategy to reduce pregnancies among girls younger than 15 years would also contribute both to slower population growth as well as to healthier children and mothers.

Polygyny

Among the more controversial aspects of Islam are the possibility of marrying up to four wives simultaneously, the divorce procedure, and temporary marriage. Although accepted in some groups in Iraq, all three are atypical of the whole nation. The average Iraqi marries one woman, does not divorce her, and does not engage in temporary marriages. However, the Qur'an permits a man to have up to four wives simultaneously – polygyny – but a woman is forbidden to have more than one husband at the same time. Since the Qur'an also requires that a husband with multiple wives must be willing to divide his time and wealth equally among his wives, polygyny is not common, except among tribal leaders in rural areas. A husband can divorce his wife simply by publicly stating three times: "I divorce you". The wife may not divorce her husband in this way.

Temporary Marriage

The Shi'a branch of Islam accepts temporary marriage where a man and woman agree to be married for a set period of time, possibly as short as one night. The temporary marriage contract sets out the conditions of the marriage, including any cash gift that the man will provide the woman. It is possible that the contract may even forbid any physical contact between the temporary husband and wife. There is no divorce; the marriage ends on the contracted date. Temporary marriages can be negotiated for a variety of reasons. While in Iraq, I heard that four school teachers had temporarily married their van driver. The marriage contract forbade physical contact and required the four teachers to periodically pay their "husband" for driving and maintaining the van. The marriage had two advantages for the teachers. First, it freed them from following the strict clothing requirements, the "hijab", in the evenings when they shared a house with their driver. Second, and more importantly, the presence of their "husband" reduced the restrictions on activities faced by unmarried women in more traditional communities. At the other extreme, temporary marriage is condemned as a form of legal prostitution, especially when very young women or girls receive cash to temporarily marry older strangers in Basrah and some other major cities.

Honor Killings

Honor killing is the murder of a family member, usually a girl or woman, for having brought dishonor on the family. The dishonor is usually sexual in nature. A girl or woman might be murdered by her family for being a rape victim, engaging in adultery, or even talking to a boy in public in the absence of proper supervision. There have also been honor killings on the basis that a family member engaged in homosexual activity. It is difficult for a girl or woman to defend herself against accusations of a sexual offense, since it is unlikely that she will receive an opportunity to defend herself in any kind of open forum. Even if a religious court decided to hear her case, her testimony is worth half that of a man (Ali 1985, Sura 2:282, pp. 128–30; Pickthall [1930] 1992, p. 63). While there is no reliable data, there are probably at least 200 honor killings each year in Iraq. While such murders are against the law, sentences are often lenient. During one year in Anbar province, the typical sentence for a man convicted of an honor killing of his daughter, wife, or sister was six months in prison.

FEMALE POLITICAL PARTICIPATION

As required by the 2005 Constitution, 25 percent of the members in the Iraqi Council of Representatives are women. The 2015 Millennium Development Goal is that women will hold 50 percent of all seats in Iraq's national parliament (COSIT 2012, Table 19/1). It is extremely unlikely that Iraq will achieve this goal even within a generation. At the provincial level, there is substantial female representation among the upper- and middle-management members of provincial governments. In 2010, 22 percent of these positions were held by women, but with a wide range among provinces. In the more religiously conservative provinces female representation is low: only 9 percent of the upper and middle management positions in Najaf province are held by women, while in Anbar province the proportion is 13 percent. At the other extreme, provinces with large Kurdish populations tend to have more females in senior positions. In Kirkuk province women hold 30 percent of the senior positions, while they hold 35 percent of these positions in Sulaymaniyah province (COSIT 2012, Tables 19/23 and 19/28).

Some research on developing countries shows that female elected representatives tend to have different budgetary priorities than males. Female representatives are more concerned with spending for essential services, especially drinking water and better roads, and less interested in funding large-scale economic development projects or – surprisingly – education (Chattopadhyay and Duflo 2005, pp. 290–291). But there is little evidence that female government officials are less corrupt than their male counterparts. Will an increase in female representation in national and provincial governments lead to greater opportunities for women in Iraqi society? The evidence is mixed. In some MENA countries, the women who are chosen by the various parties to run for parliamentary seats tend to be very traditional concerning gender issues.

WHAT SHOULD BE DONE?

In its National Development Plan: 2018–2022, the GoI (2018) provides a long list of detailed recommendations for improving health, education, and narrowing the gaps between genders (ibid., Chapter 9, pp. 192–220) as well as another chapter devoted solely to poverty alleviation (ibid., Chapter 7 of the plan). But of these many valuable recommendations, which is the most important? Ensuring that 100 percent of the population receives at least an elementary school education is not only a desirable goal in itself, but also can be expected to have favorable impacts on other important social goals.

Currently, almost 7 percent of Iraqi children aged 6–11 years are not receiving an elementary education. In addition, there are a large number of Iraqis – especially girls and women – older than 11 years of age who have not received an elementary education, and in fact many are illiterate. There are strong positive externalities for literacy and primary education: we all benefit when more of our neighbors can read, write, and count. In addition, compared to the illiterate, men and women with a primary education not only are more economically productive but also tend to have better health, raise healthier families, and live longer. Politically, illiterate or uneducated voters are dangerous to democracy. They tend to be more ignorant of the issues and more easily led by demagogues.

Ensuring that 100 percent of those girls and boys aged 6–11 years are enrolled in elementary schools will require not only more local schools but also – more importantly – changing incentives for teachers and potential teachers so that they are willing both to locate to where the students are and to provide a quality education. Educational circumstances differ dramatically across Iraq. For example, rural families who are engaged in agriculture may require their school-age children to work in the fields during planting and harvest; the periods of planting and harvest are different for different crops. Urban families might require their children to work in a family business during key events such as religious holidays, and so on. The difficult challenge of achieving a quality elementary education without excessively burdening families so that they withdraw their children might be best achieved through a division of responsibility. The national Ministry of Education could set national standards, such as for subject matter and number of required hours of education per year, while delegating the actual operation of schools to provincial or even local governments. One possibility would be for the GoI to annually transfer to each province a sum to educate all of the province's school-age population. Provinces would then have the flexibility to spend these funds to achieve the goal of 100 percent quality elementary education.

More difficult will be eliminating illiteracy among older Iraqis. Adult illiteracy has an adverse impact on economic development, reduces democratic stability, and impairs progress towards an improved status for women. Yet, reducing this illiteracy raises difficult challenges of organization to ensure adequate incentives both for instructors to teach older Iraqis and for illiterate Iraqis to participate. Again, there is no template or blueprint for reducing adult illiteracy that can confidently be applied nationwide. But if Iraq can achieve 100 percent primary education among the young, and 100 percent literacy among all Iraqis, then it should have a strong favorable effect on health, political stability, and the general quality of life.

4. Corruption

> Corruption benefits the few at the expense of the many; it delays and distorts economic development, pre-empts basic rights and due process, and diverts resources from basic services, international aid, and whole economies. Particularly where state institutions are weak, it is often linked to violence.
>
> (Johnston 2006, as quoted by Allawi 2020)

Like sand after a desert storm, corruption permeates every corner of Iraqi society. According to Transparency International (2020), Iraq is not the most corrupt country on earth – that dubious honor belongs to Somalia – but Iraq is in the bottom 12 percent, ranking 162nd out of the 183 countries evaluated. Corruption in Iraq extends from the ministries in Baghdad to police stations and food distribution centers in every small town. (For excellent overviews of the range and challenges of corruption in Iraq, see Allawi 2020 and Looney 2008). While academics may argue that small amounts of corruption act as a "lubricant" for government activities, the large scale of corruption in Iraq undermines private and public attempts to achieve a better life for the average Iraqi. The former Iraqi Minister of Finance and the Governor of the Central Bank stated that the deleterious impact of corruption was worse than that of the insurgency (Ministry of Finance 2005). This is consistent with the Knack and Keefer study (Knack and Keefer 1995, Table 3) which showed that corruption has a greater adverse impact on economic growth than political violence.

Corruption is the abuse of public power for private benefit. Corruption occurs if a government official has the power to grant or withhold something of value and – contrary to laws or publicized procedures – trades this something of value for a gift or reward. Corruption is a form of rent-seeking. Among corrupt acts, bribery gets the most attention, but corruption can also include nepotism, official theft, fraud, certain patron–client relationships, or extortion (Bardhan 1997, pp. 1320–1322; Gunter 2008b). Private corruption, such as insider trading, is not considered in this chapter.

One of the challenges of studying corruption in Iraq is that the word "corruption" is both descriptive and pejorative. Societies that have long cultural traditions of patron–client relationships or giving "gifts" to officials tend to object strongly to describing such behavior as corruption. It is often argued that calling such cultural traditions corruption is a distortion, an attempt to apply Western standards to non-Western societies. It would be useful if there were a separate word to describe corrupt behavior in a neutral fashion and another to refer to such behavior critically, such as the distinction between "killing" and "murder".

In this chapter, corruption is intended to describe a particular form of behavior, not to make a pejorative statement. However, an adverse connotation may be deserved, since corruption tends to adversely affect economic development regardless of culture. The correlation between corruption and per capita income adjusted for cost of living is strongly negative. Regardless of culture or geographic location, there are no (very) corrupt rich countries and few honest poor countries.

To maintain a reasonable size, this chapter focuses on the corruption of Iraqi officials and institutions. Therefore, it excludes the rapidly increasing number of studies on the corruption of international institutions such as the United Nations during the sanctions period, as well as evidence of corrupt acts on the part of representatives of other nations, such as the US, that have played major roles in Iraq's recent history.

CORRUPTION UNDER SADDAM

Under Saddam (1979–2003), corruption was controlled from the top in a classic case of "state capture". With Arab Ba'athist Socialist Party control, almost all of the Iraqi economy was nationalized. Positions of power in the economy were assigned to members of Saddam's family and his loyal supporters. As was expected, this elite received a dominant share of the benefits from any economic activity. This pattern is consistent with that of most socialist developing countries. In fact, according to Gordon Tullock (2005b, p. 29), "what is called socialism in much of the backward world is simply an elaborate mechanism for transferring rents to friends and close supporters of the dictator".

But corruption under Saddam's dictatorship differed in two fundamental ways from the corruption that exists during the current Iraqi transition to full democracy. First, Saddam's family and his immediate ring of supporters captured a large proportion of the bribes and other gains from corruption (Allawi 2020, pp. 9–10; Marr and al Marashi 2017, p. 130). Consequently, those at the base of the governmental pyramid captured relatively small amounts of the gains from corruption. In contrast, current Iraqi corruption is more "democratic", with corrupt gains more widely distributed.

Second, under Saddam corruption was more "honest": honest in the ironic sense that one refers to an "honest judge" as one who, once bribed, stays bought. Since corruption under Saddam tended to be more structured, when bribes were paid to a public official there was a high degree of confidence that the promised favor would be rendered. Current Iraqi corruption is more entrepreneurial. Government officials are engaged in sometimes cooperative, sometimes competitive efforts to extract the maximum rents not only from private citizens but also from other branches of the state bureaucracy. As a result, it is difficult to discover the proper person to bribe in order to obtain a specific favor, and there is less confidence that that favor will be provided even if the bribe is paid. (This is a not uncommon phenomenon; see Rose-Ackermann 1999, p. 32).

It is likely that although Saddam, his family, and supporters were able to capture a larger proportion of the nation's economic income through their well-organized corruption, current entrepreneurial decentralized corruption imposes a more serious burden on Iraq because of increased uncertainty (Bardhan 1997, p. 1324).

EXTENT OF CORRUPTION AFTER SADDAM

Estimating the amount of corruption in Iraq is difficult, for several reasons. First, this offense is often perceived as lacking a victim. For example, Iraqi private citizens may find themselves excluded from business opportunities because of the length of time, expense, or complex procedures required to pursue the opportunity legally. If, in order to speed up the bureaucratic process, citizens either offer bribes or agree to public officials' demands, then the citizens often see the officials as doing favors, not imposing burdens. Even if Iraqi bribe-paying citizens feel victimized, they hesitate to report corruption for fear of retaliation or legal sanction. Second, conflict with al Qaeda and ISIS both increased the opportunities for corruption and made it easier to conceal. This symbiotic relationship between corruption and the insurgency is discussed in greater detail below.

Victims in Iraq rarely report the crime of corruption, so almost all information on corruption is obtained from investigative reporting, including publicized corruption investigations or surveys. In Iraq as elsewhere, publicized investigation reports tend to grossly underestimate actual levels of corruption, because only a fraction of corruption cases are investigated and the results of some investigations are often not released to the public. Further complicating the analysis is the fact that decisions to initiate corruption investigations in Iraq are often political in nature. However, reported investigations help to reveal the scale of the corruption problem in Iraq.

Corruption in Iraq extends from the top to the bottom of official Iraq. Allawi (2020, pp. 13–15; 2007, pp. 348–68) in his excellent discussions of Iraqi corruption went as far as to state that corruption had turned the Government of Iraq (GoI) into a "Potemkin State". Ministers responsible for Defense, Trade, Electricity, Oil, and Interior (the police) have been investigated for corruption and several have fled the country with hundreds of millions of dollars. Former provincial governors have been arrested and charged with large-scale theft of public funds. It has been estimated that one-third or more of some agencies' budgets are lost to corruption (see Al-Rahdi 2007; Rubin 2008; Sattar 2012; Shilani 2020). Entire initiatives such as the construction associated with Baghdad's 2013 celebration as a "Capital of Arab Culture" have been defeated by corruption. For example, almost 25 billion dinars was spent on a new opera house, but little construction actually occurred (Saqr 2019).

One of the major examples of corruption at the highest levels of Iraq's government is the *muhasasa* system. *Muhasasa* is an organized sectarian power sharing among Shi'a, Sunni, and Kurds justified as a means of preventing one group from dominating Iraq. It has devolved into a division of spoils where each group obtains control of various ministries. A group's control of a ministry provides not only jobs for the group's members but also the possibility of diverting ministerial spending to favor the group (Tabaqchali 2020c). As an example, the formation of the government in early 2020 resulted in Sunni groups gaining or maintaining control of six ministries – Planning, Defense, Trade, Industry, Education, and Sports – nine commissions, and more than 60 special grade positions. Other ministries will be controlled by Shi'a and Kurdish groups, while even the Turkmen and Christians demanded a ministry each (Haddad 2020, p. 7).

At the other extreme, there is evidence that the official village grain merchants who are responsible for distributing the monthly food baskets (under the Public Distribution System (PDS)) are substituting lower-quality items in the baskets and selling the higher-quality products. That corruption is ubiquitous is exemplified by the findings of the Inspector General of the Higher Education Ministry that as many as 4000 of the almost 14 000 candidates in the January 2009 elections had forged university degrees (Al-Jawari 2009).

At the individual level, corrupt acts are inequitable. They allow some to avoid laws, regulations, and practices that others must follow. Thus, corruption undermines the average Iraqi's confidence that success results from individual effort, rather than from bribery or political connections. As shown in Table 4.1, most of the key institutions of Iraq are perceived to be corrupt. Political parties are perceived to be the most corrupt institutions, with 47 percent of Iraqis stating that they are corrupt. At the other extreme, the same survey showed that the judiciary is perceived to be least corrupt institution at only 26 percent. With respect to the percentage of Iraqis who had paid bribes within 12 months of the survey, 39 percent paid bribes to land services and 35 percent to the police.

Corruption is good for insurgencies. Or as Iraqi President Barham Salih stated, corruption is the political economy of conflict (al Monitor 2019). Terrorist groups in Iraq finance their operations, in part, with the proceeds from corruption. Some state factories, including an oil refinery, were taken over by insurgent groups or by groups willing to pay the insurgents for security. Organizations and smuggling routes ("ratlines") that handle smuggled

Table 4.1   *Corrupt institutions*

Iraqi institutions	Perceived corruption (%)	Paid bribe (last 12 months) (%)
Political parties	47	
Legislative bodies	34	
Civil servants	32	
Land services		39
Registry and permits		27
Tax services		17
Utilities		14
Police	30	35
Medical services	28	23
Judiciary	26	22

Source: Transparency International (2013b).

or other black-market goods provide terrorists with routes into and out of the country as well as safe houses for the terrorists, their weapons, and the making of improvised explosive devices (IEDs) (Gunter 2007; Looney 2006). Not only is corruption good for the insurgency, but also the insurgency facilitates corruption because it justifies bypassing accounting and regulatory procedures. It also increases the urgency of getting things done regardless of the cost, and provides an acceptable excuse for corruption-related losses. Corruption also directly undermined the GoI's anti-insurgency efforts against al Qaeda and ISIS. One of the key elements of a successful counterinsurgency endeavor is to immediately begin reconstruction, especially the restoration of essential services, in order to build confidence and support for the government and counterinsurgent forces (Gunter 2007; US Army 2006, p. 272). However, since 2003, massive corruption has restricted or delayed reconstruction and the provision of food, electricity, water, medical care, and so on, to the Iraqi population.

In addition to its impact on the GoI's anti-insurgency campaign, corruption adversely affects the Iraqi economy. In fact, there are two economies in Iraq: an oil-funded public sector economy (about 50 percent of Iraq's labor force; author's estimate) and a non-oil-funded economy with a large informal sector. Both sectors are burdened by corruption, although employees in the public sector often have the opportunity to accept bribes as well as pay them. The weak performance of the non-oil Iraqi economy is one factor in the growing number of unemployed and underemployed. This further undermines support for the government among the Iraqi public.

In Iraq, the burden on private businesses is substantial. When asked about the impact of corruption on business, a 2011 survey of about 900 business owners in nine Iraqi provinces showed that it is prevalent in such "basic business transactions as business registration, banking and even garbage collection" (CIPE 2011, p. 29). In Basrah, a study stated that 100 percent of government construction contracts required a bribe (World Bank 2011b). With respect to the costs of corruption compared to the overall costs of doing business, 51 percent of businesses surveyed thought that corruption increased costs by more than 20 percent. In addition to increased costs for businesses, Iraqi markets for goods and services tend to be inefficient because of the uncertainty and risk associated with corrupt activities. Firms deliberately stay inefficiently small and organize their activities in complex manners to avoid coming to the attention of those in authority who will seek payoffs. The amount of directly unproductive economic activity – labor and resources used not to produce or trade but to "get around" artificial barriers to production and trade – is large. Also, since the Iraqi banking system rarely lends to firms in the informal economy, less efficient high-interest money lenders or more prosperous family members are asked to provide financing.

Not only does corruption in Iraq lead to slower real growth, but it also worsens the distribution of income: the poor in Iraq must pay bribes, but rarely receive them. The impact of corruption also leads to an expansion of the budget deficit, reduced services for the money spent, and a waste of needed investment spending. Because of corruption, the Iraqi national government has to spend much to get little.

For example, an estimated 10–25 percent of Iraqi government workers are "ghosts" who receive a paycheck but rarely show up for work. Instead, these "ghosts" give a portion of their pay to their supervisor to ignore their absence, while the ghosts stay home or work at another government or private sector job. Other government workers use their government office, equipment, staff, and so on, for private businesses. The completion of a biometric census of all government employees would allow the identification of ghost workers, but this census has been delayed for over 15 years. Census takers have been bribed, threatened, and beaten in government offices to prevent them from providing an accurate list of government workers.

Of those employees who do show up for work, many may not actually be qualified for the job they are employed to do. According to the 2010 investigation by the Iraqi Commission of Integrity and the Ministry of Justice, between 20 000 and 50 000 government employees obtained their positions with forged education documents (Niqash 2011a).

Finally, the widespread perception of corruption tends to discourage legitimate foreign entities from trading, lending, or investing in Iraq. Corruption was one of the major reasons that the $53 billion Common Seawater Supply Project to pump seawater into oil fields to maintain pressure has stalled (Watkins 2019). In addition, outside of the oil sector, corruption and associated pathologies tend to discourage much-needed foreign investment in

Table 4.2   *Long-term institutional and environmental causes of*
                   corruption

	References	Applicable to Iraq
Low levels of literacy	Glaeser et al. (2004)	Yes – literacy: 74%
Inhospitable climate	Acemoglu et al. (2001)	Yes
History of French or socialist legal system	La Porta et al. (1999) and Djankov et al. (2002)	Yes – socialist
Catholic or Muslim	Treisman (2000) and Landes (1998)	Yes – 97% Muslim
Cousin marriage		Yes – 33% or more

Iraq (Cheung 1996; Gunter et al. 2020; Thede and Gustafson 2010; US–Iraq Business Council 2020).

Corruption in Iraq is not static; in the absence of an effective anti-corruption strategy, it tends to worsen over time. Corrupt officials are motivated to increase the inclusiveness and complexity of laws, maintain monopolies, and otherwise restrict legal, economic, or social activities in order to be able to extract even larger bribes or favors in the future. Perhaps the most damaging aspect of corruption is that it increases the level of uncertainty and forces individuals and organizations to expend a great deal of effort in attempts to reduce this uncertainty. Investors have to worry about not only changing market conditions, but also whether various unknown officials in Baghdad will seek to block their investment in order to extract additional bribes. Reducing corruption is critical to Iraq's long-term economic development. But in order to develop an effective anti-corruption strategy, it is necessary to first determine the causes of corruption in Iraq.

WHY IS IRAQ CORRUPT?

The determinants of corruption in Iraq can be divided into two general categories. Long-term demographic, environmental, and cultural aspects may be important causes of corruption, but are difficult to change. On the other hand, there may be economic or political incentives for corruption in Iraq that, optimistically, may be subject to short-term policy solutions. Dependency on a single natural resource – petroleum in the case of Iraq – does not fit neatly into either category and, somewhat arbitrarily, will be discussed under policy. There is an extensive literature on the long-term causes of corruption. In Table 4.2, five of these determinants are listed along with their applicability to Iraq.

Low levels of literacy provide fertile soil for corruption. An illiterate population is not only vulnerable to exploitation by low-level government officials, but is also unable to effectively monitor its government. While the quality of the data on literacy in Iraq is questionable, it is believed that in 2013 about 86 percent of the male population is at least basically literate, compared to 69 percent of Iraqi females (CSO 2014, Figure 11, p. 24). Not only is this a lower level of literacy than in the average Middle East and North Africa (MENA) country or upper-middle-income country (UMIC), but also the level of literacy is believed to have declined since 2014 as a result of ISIS-related population dislocation and school destruction (World Bank 2018c, pp. 25–30).

Acemoglu et al. (2001) found that countries with climates hostile to Europeans because of heat or disease tended in the colonial era to get lower-quality administrators who left their families at home and lived in the colony as little as possible. There was a tendency for these administrators to have a short-term focus and establish "extractive" institutions in order to make as much as they could and get out while they still had their health (Acemoglu et al. 2001, p. 1370). However, this colonial influence on corruption was limited in Iraq. While commercial interests continued, the British mandate lasted only three decades from about 1918 (with the conquest of Mosul) to about 1948 (the Treaty of Portsmouth) (see Marr and al Marashi 2017, Chapters 2–4).

If, during the colonial or post-colonial period, countries adopted French or socialist legal systems then they tend to be more corrupt (La Porta et al. 1999). Napoleonic code or socialist legal systems tend to be extremely bureaucratic with complex regulations covering every aspect of life. Corruption becomes a way of life in such societies as a way of dealing with the bureaucracy. The Arab Ba'athist Socialist Party that dominated Iraq from 1968 through 2003 imposed an extensive bureaucratic socialism on the country, with most of the important economic decisions made by party members in the ministries in Baghdad.

More controversial is the finding by Treisman (2000) and Landes (1998, Chapter 24) that countries with large Catholic or Muslim populations tend to be more corrupt than Protestant ones. Possibly this is because Protestants tend to be more distrustful of any authority – including the state – and therefore monitor it more aggressively. Also, both Catholics and Muslims may place less emphasis on education. Of course, Iraq is almost 97 percent Muslim. While there is a great deal of respect for religious leaders in Iraq, there are limits. Several analysts state that the significant losses sustained by religious parties in the January 2009 provincial elections were caused in part by the public losing confidence in the administrative competence or honesty of religious parties' officials (Arraf 2009; *The Economist* 2009b).

A demographic characteristic related to Iraqi corruption is cousin marriage. As discussed in Chapter 3, while the marriage of first and second cousins is not uncommon in Arab society, in Iraq it is estimated that about one-third of all marriages were formerly within this degree of consanguinity, with the rate

Table 4.3   Policy-related causes of corruption

	References	Applicable to Iraq
Dominant natural resource	Ades and Di Tella (1999), Leite and Weidemann (1999)	Yes – petroleum is two-thirds of GDP
Lack of market competition	Ades and Di Tella (1999), Bliss and Di Tella (1997), Djankov et al. (2002)	Yes
Weak free press	Brunetti and Weder (2003)	Yes – 1979 to 2003 No – 2004 to present
Lack of political competition	Persson and Tabellini (2004)	Yes – 1979 to 2005 Partial – 2006 to 2008 No – 2009 to present
Large-scale subsidies		Yes
Lack of legal sanctions		Yes
Inadequate public sector salaries	Krueger (1974), Mookherjee and Png (1995), Rose-Ackerman (1999)	No – but problems with excessive public sector salaries

of cousin marriage rising to about 50 percent in rural areas (Bobroff-Hajal 2006; COSIT 2008). In fact, Iraq may have had the third-highest rates of cousin marriage in the world after Pakistan and Nigeria. Such marriages tend to strengthen the influence of families and clans, because not only do they multiply the relationships between any two members (your father-in-law is also your uncle), but also they reduce interactions between different clans. With widespread cousin marriage, it is clear why nepotism is not seen as an act of corruption but rather as a positive virtue: caring for a member of your tight-knit family and clan.

Short-Term Determinates of Corruption

While a society can possibly break its ties to its socialist past, it is difficult or impossible to change a country's physical environment, history, or culture. However, there are other determinants of corruption that are more of a function of policy error and therefore can potentially be improved in a reasonable period of time. Table 4.3 lists seven policy-related causes of corruption and their applicability to Iraq.

Oil is the curse of Iraq

In Leite and Weidemann's (1999, pp. 22, 24) study, it was shown that export dominance by a single natural resource such as petroleum tends to be associated with greater corruption. Natural resources tend to have high economic rents:

a large gap between the cost of production and the export price. This results in large incentives for corrupt behavior to capture these rents (Rose-Ackerman 1999, p. 19). Oil accounts for almost 99 percent of Iraq's export revenues and almost two-thirds of its gross domestic product (GDP). In fact, Iraq has the highest level of natural resource dependence in the world (World Bank 2019a, Table 5, p. 31; World Bank 2012b, pp. 204–7).

In Iraq, the Baghdad ministries control the massive revenues from oil exports. Since the revenues from oil exports account for over 90 percent of total government revenues in Iraq, the ministries are not dependent on tax-payer voters. The Baghdad ministries use the oil funds primarily to maintain the salary and benefits of their employees, including both those who are directly employed by the ministries as well as those employed by state-owned enterprises (SOEs) associated with many of the ministries. The perverse but predictable priorities of the ministries were dramatically revealed during the unexpected revenue shortfalls in 2006, 2009, 2014, and 2020. Maintenance and capital expenditures were slashed, while government employment actu-ally increased. The ministries and the associated SOEs are, to a great extent, welfare programs providing generous salaries and benefits in return for little work.

In addition, the independence that the Baghdad ministries enjoy as a result of oil export revenues allows them to follow economic and social policies that block or undermine the economic liberalization goals set forth in the National Development Plan or Strategy (GoI 2005b, 2007, 2010, 2018).

Lack of market competition
Lack of market competition is associated with corruption for at least two reasons. First, the possibility that the government may allow the creation or maintenance of monopolies tends to provide strong incentives for corruption. Individuals or groups would be willing to pay bribes or otherwise favor certain government officials in order to capture monopoly profits. Second, Iraqi firms will try to influence officials to reduce or eliminate competition from imports (Ades and Di Tella 1999; Djankov et al. 2002). Under Saddam, detailed plan-ning was done in Baghdad by dozens of ministries. These plans determined in exhaustive detail almost all economic activities, from fertilizer consumption by farmers to setting the price for imported automobiles. Almost all of the manufacturing entities were combined into around 200 SOEs. Although these SOEs were generally low-quality, high-cost producers, they served at least three purposes. First, they reinforced Saddam's Arab Ba'athist Socialist Party control of the political levers of power. Or to be more precise, political and economic power was joined. Second, control of the ministries and the SOEs provided a means for Saddam to reward his supporters and, by exclusion,

Table 4.4   *Relative ease of doing business in Iraq, MENA, and the USA*

2020 relative rank (190 countries)	Iraq	MENA range	USA
Starting a business	154th	31st to 169th	55th
Getting credit	186th*	20th to 186th	4th
Trading across borders	181st	41st to 189th	39th
Closing a business	168th**	63rd to 168th	2nd

Note: *Iraq's ranking is worst in MENA. **Iraq's ranking is tied for worst in world.
Source: World Bank (2020d).

punish those who were less than enthusiastic. Finally, the SOEs provided multiple opportunities for government officials to extract bribes and divert funding for their personal benefit.

Since 2003, there has been little discernable progress in reducing the dominance of the SOEs in the domestic economy. As will be discussed in Chapter 10, this dominance has been maintained both by continuing to provide large direct and indirect subsidies as well as by discouraging the private business sector. The burdens placed on private businesses in Iraq are onerous even by regional standards.

Table 4.4 compares the relative regulatory burden on Iraqi businesses. To provide context, similar rankings are provided for the other MENA countries and the USA. As can be seen in the table, of the 190 countries surveyed, Iraq ranks 154th in the ease of starting a business. Not only does Iraq possess one of the most hostile regulatory environments for private business in the world, but also there has been little progress over the last 12 years. While there has been some improvement in reducing the regulatory burden involved in the categories of "Getting Electricity" and "Dealing with Construction Permits", most other categories have deteriorated.

Legally starting a business, obtaining credit, engaging in foreign or domestic trade, or going bankrupt are extremely complex and expensive processes in Iraq. The bureaucratic complexities that tie Iraqi businesses in knots are not random or unloved artifacts of earlier days. Government ministries continue to expend a great deal of influence to preserve these complex procedures, for a very simple reason: the more complex, illogical, and time-consuming the procedures are, the greater the number and size of bribes that can be extorted from businessmen (Cheung 1996, p. 3; Tanzi 1998).

Weak free press
A free press reduces the potential for corruption both by increasing the likelihood that corrupt acts will be uncovered and by providing a mechanism through which public opinion against corruption can be marshaled and

expressed (Brunetti and Weder 2003). Iraq has made a rapid transition from an extremely restrictive media environment before 2003 to a media free for all, with a sharp rise in the number of media outlets (Brookings Institution 2009, p. 45). This increase in outlets was accompanied by an increased variety of views and opinions. Stories about corruption are increasingly common. A few are careful investigative pieces, but most make undocumented claims of responsibility combined with calls for action. Based on a sample of stories, letters to the editor, and the call-ins to radio shows in the Arabic media, there appears to be little patience with the notion that corruption is part of Arab or Iraqi culture. The more common view is outrage at corrupt government officials.

Unfortunately, there is strong opposition to the newly freed media. The government offers cash bonuses, subsidized apartments, and in some cases free land, to a select list of journalists. The government states that these benefits recognize the courage of these journalists and are not an attempt to purchase favorable coverage. Other media outlets have been constrained by a flood of lawsuits requiring, in some cases, that editors spend one-third or more of each month in court defending their coverage of the news (Al-Ansary 2011).

In some cases, media outlets have been shut down for aggressive corruption coverage. Alhurra, a popular Arabic-language, US government-funded broadcaster had its license temporarily suspended as a result of its investigative report alleging corruption within *waqfs*, religious endowments (Reuters 2019).

In addition to corruption, the authorities have tended to crack down on any reporting unfavorable to the government concerning the protests that began in late 2019 and the government's reaction in 2020 to the Covid-19 virus. At least ten news organizations have been forced to suspend coverage of these issues, and violence against news agencies or their employees is not rare. In the last four months of 2019, five journalists died under suspicious circumstances. According to the Reporters Without Borders (2021) World Press Freedom Index, Iraq has dropped from 153rd to 162nd out of the 190 countries evaluated.

Lack of political competition
Political competition serves a similar role as a free press in reducing the prevalence of corruption (Persson and Tabellini 2004). From 1979 through 2004, Iraqis did not have a realistic chance to express themselves politically, but after the approval of the Constitution in October 2005 there have been three national elections, with the most recent in October 2018. Over the same period there were three provincial legislative elections, in 2009, 2013, and 2019. The election results showed that there is a realistic opportunity for opposition parties to gain power through elections (Freedom House 2021).

News stories in both the Arabic- and English-language press emphasized that perceived corruption was an important determinant in the repudiation of previously dominant parties (see Black 2009; Crisis Group 2009, p. 27; Parker and Redha 2009). However, there appears to be little difference in the willingness of various political parties or coalitions to engage in corruption. All parties divert funds from the ministries they control and take bribes for government contracts.

Large-scale subsidies

The greater the value of the good or service controlled by the government official, the greater the value and possibly the number of bribes that they can extract. In other words, the existence of large government subsidies leads to increased corruption, since the maximum bribe – the difference between the official and market prices – is greater.

Under Saddam, gasoline, kerosene, diesel, and other fuels were almost free, and the post-Saddam governments continued this policy although the size of the subsidy was reduced beginning in December 2005. For example, while the average price of premium gasoline in the Gulf in September 2005 was $1.06 per gallon (28 cents/liter), in Iraq the same fuel sold for 13 cents per gallon (3 cents/liter). This gap provided tremendous profits for corruption. If a single tanker truck of premium gasoline was diverted from the official market and its cargo sold in the Iraqi black market or smuggled to a neighboring country, a profit of $6000–$7000 was possible. A poorly paid customs official or border guard could easily double their annual income by accepting a cash bribe to turn a blind eye to such diversions. Surprisingly, in the face of political opposition and street protests, the GoI raised the official fuel prices to approximately equal to the prices in neighboring countries. However, this left the official price below the market clearing price, and there are still large profits to be earned from diverting fuel from official channels into the black market. For example, in one month in 2011, 120 fuel tankers that were headed to Samarra and 260 Baghdad-bound tankers went missing and were believed to have delivered their fuel to the black market (Faruqi 2011).

Similar situations exist with other essential goods and services, such as water, electricity, and food. There is widespread divergence from official channels into the black markets. For example, food is heavily subsidized through the system of monthly food baskets for every Iraqi family. One survey showed that in some provinces almost 82 percent of the food rations (PDS) were illegally diverted between the seaports (most of the subsidized food is imported) and the village or town food distribution centers (*Khaleej Times Online* 2011).

Weak legal sanctions

There are weak disincentives to accepting a bribe. In Iraq, although laws may call for severe punishment for bribery, the chances of being caught and convicted are practically zero. Over a 15-month period in 2017–18, 1366 arrest warrants for corruption were issued (out of almost 3 million government employees) but only about 28 percent of these warrants were executed. And of those cases that reached trial, only about 56 percent resulted in a conviction (Commission of Integrity 2018a, pp. 5–7). While the number of investigations has increased substantially, it is still extremely unlikely that a corrupt act will be discovered, investigated, prosecuted, adjudicated, and punished. In fact, despite the massive insider theft of ministerial funds, "not a single senior official has been indicted, tried and then jailed for corrupt practices" (Allawi 2020, p. 12).

Inadequate public sector salaries

A more controversial possible cause of Iraqi corruption is that inadequate public sector salaries motivate government employees to seek and accept bribes. In theory, raising the compensation of government employees will tend to discourage corruption, although the pattern of incentives can be complex (Mookherjee and Png 1995, p. 154). Examining whether public sector compensation in Iraq is adequate is especially complicated. Immediately after the 2003 US-led invasion, civil servants were seriously underpaid (Allawi 2020, p. 11). In addition to motivating government employees to seek bribes in order to support their families, it also led to severe distortions. For example, a skilled Iraqi doctor quit a high-status but poorly compensated position at a Baghdad hospital to become a low-level translator for the US military. However, as a result of a series of salary increases, the typical government employee is now better compensated than a worker in the private sector. Combined with benefits, protection against dismissal, and a less intense – some would say, relaxed – pace of work, government employment is eagerly sought after. There are multiple applicants for each government service opening.

However, once someone has obtained a government position, the financial rewards for advancement in Iraq tend to be meager. For example, a senior enlisted member of the US Army would earn almost 300 percent more than a Private even before adjusting for time-in-service raises. This is not true in the Iraqi Army, Iraqi Police, or in almost any other Iraqi institution. A senior enlisted member of the Iraqi Army earns only about 20 percent more than a Private.

The existence of generous compensation for entry-level positions combined with meager raises for seniority provide strong incentives for accepting bribes. New employees are expected to "purchase" their entry into government employment by bribing senior officials (classic rent-seeking; see Krueger

1974). The payment is usually some combination of an initial bribe – possibly financed by borrowing – and a monthly cash "contribution" to one's supervisor. Of course, each level of management is expected to make a "contribution" to the next most senior level of management. As a result, the actual paycheck received by many senior members of the bureaucracy may account for only a small fraction of their compensation. Many seek bureaucratic advancement primarily to increase their ability to extract larger bribes.

REDUCING CORRUPTION

Over the last 50 years, many countries have attempted to eradicate corruption in their economies. However, not only have there been no successful eradications of corruption, but also most attempts to reduce it to tolerable levels have failed. Some of the failures were not surprising. Anti-corruption policies that are comprised entirely of exhortations to virtue and a spurt of well-publicized investigations tend to have little long-term effect. In addition, various political factions will often usurp anti-corruption campaigns to settle scores with their opponents and, as a result, anti-corruption efforts often collapse in mutual recriminations.

There has been some progress in Iraq's war against corruption. The Commission of Integrity reports increased prosecutions and substantial recoveries of misappropriated funds (Commission of Integrity 2020). This progress has been reflected in a gradual improvement in Iraq's Transparency International corruption rankings. Realistically, however, Iraq has only improved from being "extremely corrupt" to being "very corrupt". The remaining high level of corruption continues to undermine the Iraq economy, polity, and society. What Iraq needs is an integrated anti-corruption strategy and the political will to execute it.

Successful anti-corruption campaigns, such as Hong Kong's (Speville 1997), must take into account each country's cultural, social, political, historical, and economic situation. These campaigns include institutional changes to reduce the economic incentives for corruption, combined with improved governance, transparency, and an aggressive effort to communicate the purpose and progress of the campaign to the public. Successful anti-corruption campaigns must have widespread support to enable them to move forward in the face of tenacious covert opposition. Finally, for lasting results, there must be a serious effort to change the culture of corruption. In summary, successful anti-corruption campaigns tend to be complicated, difficult, politically risky, and expensive (Klitgaard 1988, Chapter 8; Rose-Ackerman 1999, Chapter 1).

In view of the costs of a possibly ineffective Iraqi anti-corruption campaign, is the game worth the candle? In view of the other security, political, health, and economic challenges facing the Iraqi government, should an anti-corruption campaign be put aside until a certain degree of stability is achieved? There are two contrary arguments.

First, the current cost of corruption is probably unsustainable. Corruption finances the remaining insurgent activity. It slows economic growth and worsens income distribution. Corruption hurts Iraq's budget situation from both sides. Oil revenues and other government revenues are lower because of corruption, while as much as one-third of government expenditures are diverted from their assigned purposes (Latif 2020). Second, in the absence of an aggressive anti-corruption campaign, corruption tends to worsen.

The government of Iraq has long been aware of the criticality of the corruption fight, although its emphasis has shifted. In Iraq's first National Development Strategy: 2005–2007 (GoI 2005b), one of the primary goals was to eradicate corruption. This theme is continued in the revised National Development Strategy: 2007–2010 that states: "Corruption – the abuse of public office for private gain – is arguably the most critical component of governance in a natural resource rich country like Iraq" (GoI 2007, pp. vi, 92). However, the National Development Plan: 2018–2020 reduced statements on fighting corruption to two paragraphs (GoI 2018, pp. 70, 72).

Despite public statements of the importance of the anti-corruption effort in Iraq, follow-through on these statements has been weak and confused. The GoI did join the Extractive Industry Transparency Initiative in 2010 and, in March 2008, signed the United Nations Convention Against Corruption (UNCAC) with its more than 160 provisions. However, the required implementing legislation for both of these agreements is progressing slowly. In Iraq's defense, the UNCAC is a very complex document that combines broad coverage – corruption in the private sector is included (Article 12, pp. 14–15) – as well as strange lacunae: it lacks a specific definition of corruption. The UNCAC concentrates on improved governance and rule of law (United Nations 2004).

It is challenging to find quality studies on countries that have most of the same characteristics as Iraq, including being a developing, post-conflict, natural resource-dependent country that is transitioning from socialism and is divided by religious and ethnic animosities. However, the 2020 paper by Ali A. Allawi, the 2015 study by Transparency International, Chêne (2015), as well as Jon S.T. Quah's (2013, pp. 219–55) work, have identified at least five lessons for countries fighting corruption that would appear to be applicable to Iraq. However, I argue that, in the case of Iraq, there is a missing component, a sixth lesson, for developing an effective anti-corruption strategy. In addition to the five lessons commonly discussed, Iraq must also reduce the economic incentives for corruption.

Lesson 1: Political Will and Good Governance are Needed for Effective Corruption Control

It is difficult to measure the political will to reduce corruption in Iraq. Iraqi politicians of all parties regularly speak against corruption and, periodically, there are well publicized arrests of senior government officials. However, the Iraqi public is cynical about these speeches and arrests, believing that many of the latter are driven by political disputes, not extraordinary levels of corruption.

That this public cynicism about government anti-corruption efforts may be justified is illustrated by the August 2015 announcement of an anti-corruption program by Prime Minister Haider al-Abadi. While the Iraqi Parliament initially passed these reforms by a unanimous vote, less than three months later the Parliament revoked the Prime Minister's mandate to implement his program. Instead, the Prime Minister was forced to renegotiate every specific anti-corruption step with various political groups in the Parliament. Needless to say, progress in these reforms ground to a stop (Oldfield 2017, p. 7).

One simple way that officials can demonstrate commitment to fighting corruption is by personally abiding by anti-corruption regulations such as accurately reporting personal financial assets. However, a substantial proportion of Iraqi officials covered by reporting regulations have failed to submit their reports. Reporting rates were high for the executive and judicial branches, but poor for the legislative branch. According to the Iraq Commission of Integrity, the worst offenders were members of the national legislature – only 39 percent submitted their financial statements – and the members of the provincial counsels (non-KRG) at 50 percent. And audits of a small fraction of these reports have revealed that about 5 percent are very inaccurate (Commission of Integrity 2020).

With respect to improved governance, since 2003 the US government, the International Monetary Fund, and several international non-governmental organizations have emphasized rationalizing governance mechanisms, both to improve the operation of the Iraqi government and to reduce corruption. These initiatives include developing a professional civil service, a single integrated open budget in the place of the multiple secret budgets of the Saddam era, and moving from paper to electronic records. Results have been meager. Government employees have resisted procedural changes in part because increased efficiency may reduce the need for their jobs. However, even if these efforts at improving the quality of governance had succeeded, they probably would not have resulted in a substantial reduction in corruption. Improving governance may be necessary, but it is not sufficient.

A more radical initiative would be to shift many of responsibilities currently handled by the Baghdad ministries to regional or provincial governments. This might include most of the current functions of the departments of Housing,

Labor and Social Affairs, Trade, Health, Policy, and Education (Allawi 2020, p. 16, Notes 6–8). This option, which will be discussed at greater length in Chapter 6, is consistent with Articles 105 and 114 of the 2005 Iraq Constitution.

Lesson 2: Rely on a Single Anti-Corruption Agency (ACA) Instead of Many ACAs for Effective Corruption Control

Why have an independent ACA? In most corrupt countries including Iraq, the police and the courts are extremely corrupt. As shown in Table 4.1, one survey revealed that 35 percent of the Iraqi population had paid a bribe to a police officer within the last 12 months. And 22 percent had paid a bribe to a judge or member of the court (Transparency International 2013a). Other countries have found that when there is widespread corruption among the police and judiciary, making the national police forces responsible for rooting out corruption in the rest of the government usually fails. Often, after the exposure of a particularly egregious case of police or court corruption, there will be calls to establish an independent ACA. However, a poor or badly managed ACA may prove ineffective or may actually damage anti-corruption efforts (Meagher 2005).

There are multiple reasons why a country tends to make more progress against corruption with a single ACA. First, if there are multiple ACAs, then responsibility for fighting corruption is diffused (IMF 2019b, p. 59). Incompetent or corrupt ACAs can attempt to shift the blame for lack of progress to another ACA. Second, several ACAs with overlapping areas of responsibility can lead to bureaucratic infighting that can result in the spoiling of an anti-corruption investigations as one ACA withholds information from another. Attempts to prevent such overlap by delineating different areas of responsibility for each ACA intentionally or unintentionally can lead to gaps or "corruption free-fire zones," where no ACA has statutory responsibility. Third, with multiple ACAs, there is a greater likelihood that one ACA will be "captured" by corrupt senior politicians and/or senior officials in the gov-ernment. The captured ACA then can be used to divert investigations of these senior officials. This is often the result when an ACA reports directly to the Prime Minister or Parliament. Fourth, a single ACA provides the public with a single unambiguous source of information on anti-corruption efforts as well as a single point of contact for reporting suspected corrupt acts. Countries with several ACAs often suffer from confusing or contradictory anti-corruption messaging that confuses the public. And multiple ACAs increase the likeli-hood that those citizens who muster the courage to report a corrupt official will report it to the wrong ACA, resulting in their statement either being ignored or possibly being released to the official involved. Finally, multiple ACAs compete for personnel and funding as anti-corruption resources are split multiple ways.

Iraq violates this guidance to have a single ACA. There are at least eight agencies with anti-corruption responsibilities in Iraq (Pring 2015, pp. 7–10). The Joint Anticorruption Council (JACC) was established in 2007 and is composed of senior government officials. The JACC focuses on policy issues and is perceived to have little impact.

The Federal Board of Supreme Audit (FBSA), founded in the same year as independent Iraq, performs audits of government organizations and oversees public contracts in order to expose fraud, waste, and abuse. When the FBSA uncovers violations, it refers them to either the Commission of Public Integrity or the Inspector Generals. Performing these audits can be dangerous, and the FBSA at times has had to send auditors out of the country to protect them from physical attacks by thugs hired by persons at audited agencies. While progress has been made, the FBSA suffers from a shortage of quality staff and has been plagued by outdated technology. As a result, many of its audits are not completed and released in a timely manner and some fail to meet international standards (Saeed 2016, pp. 91–100). There is a separate organization, the Kurdistan Board of Supreme Audit, with similar responsibilities in the Kurdish Regional Government (KRG).

There are about 30 Inspector Generals (IGs) who are supposed to unearth corruption within assigned ministries. This agency is generally considered to be the least effective of the major anti-corruption agencies. It is not simply that many of the IGs are unqualified or corrupt themselves, but rather that each ministry's IG reports to the Ministry Director, not to an independent organization. As a result, while the IGs have exposed low-level corruption, there have been very few cases where an IG has exposed corruption involving a minister or one of their senior staff members (Allawi 2020, p. 16, note 10). For these reasons, in March 2019, the Council of Representatives (CoR) voted on a non-binding resolution to phase out the country's Inspector Generals (Sattar 2019).

The Iraqi ACA that is perceived as being most effective is the Commission of Integrity (CoI). It not only investigates cases of corruption, but also seeks to educate the public on transparency-related issues. Unlike other Iraqi ACAs, the CoI publishes a quarterly report providing data on investigations and arrests as well as whether government officials are meeting their personal income and asset reporting requirements (Commission of Integrity 2020). The CoI faces two serious challenges. First, the CoI, the Iraqi equivalent of the US Federal Bureau of Investigation (FBI), has limited authority and must rely on the police to make most arrests. But the police are often corrupt. There have been multiple cases where the police have either demanded bribes to release persons freed by the courts, or accepted bribes to release persons that the courts have ordered to be held. As a result, a large proportion of senior Iraqi officials charged with corruption are tried *in absentia* since they were able to leave the country before being arrested. A second challenge is interference from the highest levels of the Iraqi government. One former Director of the CoI resigned because of such interference.

There are four other ACAs that are believed to be relatively ineffective. There is an Office of the Ombudsman that seems to be limited to publicizing corruption complaints. The Parliamentary Committee on Integrity is subject to substantial political pressure. The Central Bank of Iraq Monetary Laundering Reporting Office has had only limited success in reducing money laundering and the associated capital flight. Finally, the Financial Intelligence Unit launched in 2015 combines a very broad mandate with very limited power and resources.

As discussed above, Iraq's multiple ACAs diffuse responsibility and resources and therefore retard the fight against corruption. One possible solution would be to reduce anti-corruption efforts to three organizations. Because of their important institutional continuity and international roles, preserve an independent FBSA and Central Bank Monetary Laundering Reporting Office. Close the JACC and the Parliamentary Committee of Integrity as hopelessly compromised. Make all of the other ACAs part of the CoI with all of their budget and personnel decisions controlled by the CoI's Director. Carrying out such a change will require substantial restructuring of lines of responsibility. For example, if IGs are not phased out then they should no longer be assigned and report to their respective ministries, but rather they would be assigned by and report to the CoI.

Lesson 3: Importance of Cultural Values in Minimizing Corruption

Hong Kong and Singapore are among the few entities that have changed their peoples' attitudes towards corruption successfully within a fairly short period of time. Their experiences appear to show that culture changing efforts such as billboards with anti-corruption messages and mandatory corruption classes for government employees are necessary but insufficient (Allawi 2020, pp. 15–16). If this is the limit of attempts to change the culture of corruption, then the anti-corruption strategy will fail. As expected, changing the culture of corruption is extremely difficult. The few successes have utilized a top-to-bottom approach with careful coordination of the different facets of creating an anti-corruption culture.

There is no single template for successfully creating such a culture. Many of the relevant country characteristics that will determine success or failure are hidden. Every country has both formal rules and regulations as well as informal norms. And changing the formal rules is much easier than changing the country's informal norms. During the 2003 invasion and the fights against al Qaeda and ISIS, there was a withdrawal of the state, a reduction in government authority, in Iraq. This withdrawal reduced the effects of formal rules and motivated greater reliance on the informal norms of community, tribe, political, or religious groups. These groups were not only providers of security and other public goods but also established standards governing political, social, and economic relationships. Changing the culture of corruption in Iraq will require more than changing formal rules and regulations. It will also require identifying and motivating changes to the country's informal norms.

In the case where formal and informal rules or norms differ, there are several decisions that must precede proclaiming anti-corruption cultural initiatives. Otherwise, contradiction and confusion will weaken the effort. First, there must be a clear definition of corruption, or rather the type or types of corruption that are targeted. Is the intent to reduce or eliminate grand corruption, also known as state capture? An example would be when an oligarch uses political influence to gain wealth "legally". Or illegally. There have been reports that some ministries with potentially large numbers of jobs available are actually for sale (China News Asia 2020). An alternative is to target petty corruption. While state capture may do more damage to the national politics and economy of Iraq, Iraqi citizens may experience greater frustration at the daily demand for small bribes by a hoard of low-level officials.

Second, should private corruption such as insider trading be included, or only corrupt acts involving officials? Third, should there be a lower limit to acts of corruption that will be investigated? In a society with a long tradition of gift-giving to persons in authority, should small bribes – a cup of tea for a policeman – be ignored or should the standard be an unambiguous "not one dinar"?

The final question to be answered before an effort to change the corruption culture is launched is especially controversial. Should there be an amnesty for corrupt acts that occurred before a certain date? On the one hand, an amnesty weakens the anti-corruption message by saying that the same act that would be prosecuted as a serious offense if it occurred in, say, 2019 would be forgiven and forgotten if it occurred earlier. However, in Iraq, it is a safe bet that almost every current Iraqi official has been engaged in corruption sometime in their career. Without an amnesty, there would be little official support for, and maybe active opposition to, anti-corruption enforcement. It should be noted that Hong Kong initially decided against an amnesty before adding one in order to gain police support for the anti-corruption effort.

Once the tough strategic decisions discussed above have been made about the scope of Iraq's anti-corruption strategy, it will be important to have many influential individuals and groups openly support the effort. Which individuals and groups will have the most influence on public attitudes towards corruption varies from place to place in Iraq. However, previous efforts in Iraq show that politicians and sports figures seem to have little credibility in the anti-corruption fight. On the other hand, religious leaders, writers, and – for some reason – soap opera stars seem to have a strong impact on the average Iraqi's attitude towards corruption. The Iraqi government has made efforts to encourage religious leaders to add an anti-corruption message to their Friday sermons (Commission of Integrity 2018a). It also may be possible to encourage writers and performers to add an anti-corruption element to their works in print and on radio and television. However, progress has been uneven. The Ministry of Culture and Cultural Heritage has denounced the Iraqi television drama *Al Funduq* (*The Hotel*) for openly discussing prostitution, drugs, and corruption. *Al Fundaq*'s station, Al Sharqiya, was forced to apologize (*Arab Weekly* 2019).

Just as important is adding strong anti-corruption modules to school instruction starting in the first grades and continuing through secondary and tertiary education. The intent is not only to make corrupt acts unacceptable to the young, but also to encourage them to object to such acts by their parents and wider society. Combined with a well-designed anti-corruption media campaign, the support of well-regarded individuals and groups and integrating anti-corruption into education curricula gradually could change Iraq's culture of corruption.

Among the Iraqi ACAs, only the Commission of Integrity appears to be making a serious attempt to change the culture of corruption. It has begun distributing anti-corruption messages to Iraqi newspapers, television, and radio programs. In addition, anti-corruption material has been prepared for children in kindergarten, primary, and secondary schools (Commission of Integrity 2018a). However, these efforts are in their infancy, and there are no published studies of the effectiveness of these efforts or even the coverage; what percent of the Iraqi public is aware of these efforts.

Lesson 4: Adequate Salaries are Necessary, but Insufficient for Effective Corruption Control

It long has been recognized that inadequate salaries for police and other government officials tend to lead to corruption (Krueger 1974; Rose Ackerman 1999). However, as discussed above, the application of this general rule to Iraq may be counterintuitive. In theory, raising the compensation of government employees will tend to discourage corruption (Mookerjee and Pring 1995). Whether public sector compensation in Iraq is adequate is especially complicated. If one looks at average salaries alone, then the typical government employee is better compensated than a worker in the private sector. However, as discussed above, once anyone has obtained a government position, the financial rewards for advancement in Iraq tend to be meager.

What can be done about corruption encouraged by excessive government wages and benefits in Iraq? It is probably political suicide for the government to attempt to reduce public sector wages to the level of those in the small private sector with similar skills and responsibilities. A more politically realistic approach would be to hold the line on public sector wages and benefits and wait for inflation – about 1 percent in early 2020 – to reduce gradually the real value of government employment to levels competitive to those in the private sector.

Lesson 5: Constant Vigilance is Needed for Sustained Success in Corruption Control

The fight against corruption is never over. Especially if there are great financial incentives for corruption, each "generation" of government officials will find it difficult to resist the temptation. Since 2003, three successive Iraqi Ministers of Defense have been charged with corruption. Each took office promising to "clean house", but the opportunities to amass $100 million or more in illegal funds could not be resisted. In summary, if a society is not constantly fighting corruption in accordance with a well-planned strategy, then corruption will get worse. And every failed anti-corruption policy in Iraq since 2003 has led to increased public cynicism, making eventual success more difficult.

The missing component: reducing economic incentives for corruption

A serious lacuna in Iraq's current anti-corruption strategy is the absence of meaningful efforts to reduce the economic incentives for corruption. This is somewhat surprising, because the first post-2003 National Development Strategy: 2005–2007 contained a substantial discussion of reducing government subsidies as part of the anti-corruption effort. However, reducing economic incentives for corruption disappeared from later five-year plans and was not included in either the Iraqi government anti-corruption plan of 2008 or the US government recommended anti-corruption strategy of 2009.

Iraq's National Anti-Corruption Strategy 2010–2014 (Joint Anti-Corruption Council 2010) contained a detailed action plan for combating corruption with 201 elements, while the Commission of Integrity's 2018 Anti-Corruption Roadmap (Commission of Integrity 2018b) discusses 43 critical initiatives; however, there are only a few ambiguous references in either document to reducing economic incentives. In economic terms, Iraq's current anti-corruption strategy focuses primarily on reducing the supply of corrupt acts by government officials, but with little attention paid to the demand for corrupt acts by the Iraqi public. Such an unbalanced strategy is doomed to fail. Investigating, prosecuting, convicting, and punishing corrupt officials has little long-term effect if the replacement officials face the same strong economic incentives to accept bribes. Currently, Iraq provides strong economic incentives for corruption in two overlapping areas: the Government of Iraq provides generous subsidies and has created a regulatory environment which stifles the non-energy private sector.

Iraq has an extensive system of subsidies and price controls. The International Monetary Fund estimates that direct subsidies of food, electricity, and fuels amounted to almost 13 percent of total government expenditures (IMF 2017, p. 4). Including indirect subsidies, the cost may reach 12 percent of Iraq's GDP. These generous subsidies result in low or zero official prices for food, electricity, fuels, water, medical care, and so on. As a result, at the official price, the amounts demanded of these goods or services are substantially greater than the amounts supplied. In effect, Iraqi families are competing with each other to obtain these subsidized items. Each must try to persuade some government official to favor their family over another family in the distribution of scarce subsidized goods and services. Families may use political pull, tribal membership, or religious connections to persuade officials to favor their requests for scarce items. However, the most common means of persuading an official to part with a subsidized good or service is to offer a bribe.

Lesson 6: Changing the Regulatory Environment

The need for relaxing regulatory hostility towards private business long has been recognized in Iraq. It was a major topic in all five post-invasion economic development plans. But little has been done, for two overlapping reasons.

First, strict government regulation of private business is part of the government's DNA following four and a half decades of dominance by the Arab Socialist Ba'th Party. The Ba'thist Party imposed complete government control of the private sector both because the party thought that doing so would increase economic efficiency – it was wrong – as well as a means to strengthen the dictatorship's control of the country.

Second, the government bureaucracy carefully maintains the existing "rats' nest" of extremely complex regulations in order to motivate private businesses to pay bribes. For example, if a businessman is unable or unwilling to wait an average of 167 days to get government permission to build a warehouse, then he can speed up the process by paying several $100 bills under the table. Also, research has shown that the impact of regulatory burden in motivating corrupt acts can be exacerbated if the regulations are implemented in a bureaucratic way (Duvanova 2014). This problem is especially severe in Iraq where almost every restructuring of a ministry's organization increases the complexity of the bureaucratic process. One new minister noted that for his ministry to grant a relatively minor approval to a private firm required 75 signatures of various officials in the ministry, including his own signature five times.

Some progress is being made. Over the last decade, Iraq has combined multiple registration procedures for starting a business, launched a new credit agency, allowed simultaneous processing of building permits, and enforced tighter deadlines on electricity connections (World Bank 2018d, p. 65). However, there are still requirements that most businesses in the modern global economy would consider primitive. For example, the GoI still requires "wet" signed and "live" rubber-stamped documents rather than accepting electronic signatures (US–Iraq Business Council 2020, p. 26). Optimistically, it will take only another decade to complete a total overhaul of the country's commercial and labor codes. Unfortunately, with the expectation of a decade or more of low oil prices, Iraq cannot afford to wait another decade before freeing the private sector.

As discussed in Chapter 11, there are several options that could be adopted to speed up the process of rationalizing Iraq's incredibly complicated commercial code. First, instead of changing regulations one at a time, Iraq could adopt the best practices commercial code that has been developed by the World Bank. However, it is possible that the formal institutions and practices in this code are inconsistent with Islamic or Arabic cultures, or that the changes may conflict with informal institutions and practices that are widely accepted by the Iraqi people. If such inconsistencies exist, then imposing such a code regardless of its theoretical excellence will probably fail.

A second option would be for the Iraqi government rapidly to rationalize its regulations by substituting the existing commercial code of an Islamic Arab neighbor such as the United Arab Emirates (UAE). While far from perfect, the regulatory environment of the UAE is much friendlier to its private sector. For example, Iraq ranks 154th in the ease of establishing a new business, since there are nine procedures taking 27 days to complete at a cost equal to 43 percent of the average Iraqi's income. The Emirates rank 51st in establishing a new business, with five procedures taking nine days to complete at a cost of 13 percent of average Emirati income (World Bank 2019a).

A third option that has received less attention is to revitalize the "free zones", a form of special economic zone (SEZ). An SEZ is a geographic area within a country where the rules of business are different from those that exist in the rest of the country (Farole and Akinci 2011, p. 3). Many developing nations have experimented with SEZs, with two primary motivations. In some countries, SEZs are established primarily to encourage foreign direct investment in labor-intensive production of goods for export. These SEZs have often been criticized as not only leading to the exploitation of labor, but also resulting in an industrial enclave that has little interaction with the rest of the country's economy. In others, such as the Shenzhen "miracle" SEZ in China, the primary motivation was to experiment with economic reforms before imposing these reforms nationwide (*The Economist* 2015).

Most SEZs fail to achieve their stated goals. This failure often occurs when the government provides generous tax reductions, while failing to provide the physical and human infrastructure to connect the SEZ to the rest of the country. Also, it is difficult to achieve the necessary political oversight without permitting unnecessary bureaucratic interference. However, in view of the great uncertainty about the appropriate level of private sector regulation in Iraq, an SEZ might provide a useful laboratory to experiment with a different level of regulation. And if this experiment is successful, then the revised regulatory environment could be extended to the whole country.

In the Free Zone Law of 1998, Iraq authorized such zones, but this law is inconsistent with modern SEZ management practices. However, a revised Free Zone Law might provide a viable transition route to a more rational regulatory environment in Iraq. This topic is discussed at greater length in Chapter 11.

The key point of this chapter is that corruption is the binding constraint to the development of the political economy of Iraq. Without significant progress in reducing corruption, all other efforts to diversify the economy, provide adequate essential services, and reduce unemployment will fail.

5.   Preventing al Qaeda 3.0

A dynasty rarely establishes itself firmly in lands with many different tribes and groups.
(Ibn Khaldun [1384] 1967)

Twice in the last two decades, Iraq has come close to breaking up. Al Qaeda drove the country to the edge of civil war in 2006–07, while beginning in 2014, ISIS (al Qaeda 2.0) conquered almost 40 percent of the territory containing almost one-third of the population. Baghdad itself was in danger of falling to ISIS. As can be seen from Figure 5.1, the defeat of the ISIS "state" in December 2017 led to a sharp drop in civilian deaths by violence. But will this period of relative peace be disrupted by al Qaeda 3.0? How can Iraq reduce the likelihood of such a resurgence? To answer these questions requires not only analysis of the origin of these pathologies but also the adoption of the appropriate policies. Otherwise, Iraq's chances of creating a peaceful multi-region, multi-religion, multi-ethnic state are low.

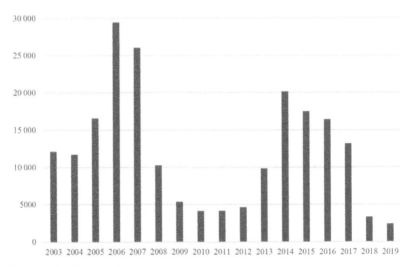

Source: Iraq body count.

Figure 5.1   Civilian deaths from violence

These political and social challenges have been the subject of several excellent books (see Barkey et al. 2011; Haddad 2011; Visser and Stansfield 2008) as well as a wide range of studies by various governmental, non-governmental, and international organizations (for example, see Cordesman and Khazai 2012; Haddad 2020). In view of the complexities involved, it is impossible to do complete justice to them in a small space. Therefore, this chapter will be limited to a few issues related to the relationship between economic development and political stability.

VIOLENCE IN IRAQ

As is well known, Iraq not only has a long history of internal conflict but also has been involved in large-scale wars with several of its neighbors. What is not as well known is that Iraq is an extreme outlier among the nations of the world with respect to conflict. Adam Szirmai (2005, p. 448) argues that Iraq was the most conflict-ridden nation in the world between 1945 and 2003. During this period Iraq experienced 48 years of civil conflict (resulting in an estimated 181 000 Iraqi casualties) and seven years of international wars (resulting in around 116 500 casualties). As shown by Figure 5.1, the violence continued after 2003, with Iraq descending into what many analysts saw as a full-fledged civil war in 2006–07.

It should be noted that even during the worst periods of 2006–07 and 2014–17 violence was concentrated in relatively few provinces. While six provinces had terribly high levels of violence (Ninawa, Baghdad, Babil, Anbar, Salahuddin, and Basrah), six provinces experienced relatively low levels of violence against either police or civilians (Dahuk, Arbil, Sulaymaniyah, Qadisiyah, Maysan, and Muthanna). The differences in the levels of violence among villages and cities were equally dramatic, with atrocities occurring in one village, while on the same day another village only several hours away enjoyed a market day with various groups mixing peacefully.

In view of the complexity of the situation, what can be said about the likelihood of further conflict in Iraq? There is no shortage of literature about the causes and implications of ethnic, religious, or regional differences on political stability in Iraq. Optimistically, there are many anecdotes about military, business, political, and personnel relationships that cut across religious, ethnic, and regional lines. Even during the worst of sectarian violence in 2006–07 and 2014–17 there were examples of heroic members of all groups risking their lives to protect their neighbors who were members of another ethnic or religious group. Of course, at the other extreme, there were the horrible atrocities involving murder and mutilation of both adults and children. So which picture reveals the true Iraq? Some researchers, such as Haddad in his 2011 book *Sectarianism in Iraq: Antagonistic Visions of Unity*, advocate a more nuanced understanding of the origin and possible outcome of Shi'a–Sunni relations (see also, Haddad 2020). While Haddad (2011) focuses on Shi'a–Sunni relations, his analysis also sheds light on the other fault lines in Iraqi society including Arab–Kurd, Muslim–Christian, and regional disputes.

Haddad lists three major points. First, "The social and political relevance of sectarian identity advances and recedes according to wider socioeconomic and political conditions" (Haddad 2011, p. 2). Two examples were the crushing by the Government of Iraq (GoI) of the mostly Shi'a revolt in the southern provinces in 1991, and the routing of the GoI by the US-led coalition in 2003 that reduced the status of many Sunnis. These shocks led both communities to emphasize religious affiliations to a greater degree. The initial efforts by the US-led coalition to establish a post-invasion Iraqi government strengthened sectarian loyalties by making sectarian affiliation synonymous with representation. For example, rather than seeking regional balance, the 25 member Iraqi Governing Council established in 2003 was deliberately composed of 13 Shi'a Arabs, five Sunni Arabs, five Kurds, a Turkman, and a Christian (Haddad 2011, pp. 150–51).

Second, even during the worst periods of conflict, relatively few of the leadership of any of the warring groups sought to break away from the Iraqi state. Their efforts were not to destroy the state so much as to have the state better represent their interests and reflect their ideals. It is ironic that while well-meaning foreigners, based on large-scale maps, were recommending that the country be broken into four states – Sunni, Shi'a, Kurd, and Baghdad – the Iraqi leadership of the disputing groups generally sought to reform rather than eliminate the GoI. The desired reforms were both real and symbolic. As an example of the former, there was a demand that government services and employment be "fairly" distributed across ethnic, religious, and regional lines. Almost as important was the demand that the symbolism of the Iraqi state not be too closely associated with that of any of the disputing parties; that the GoI at least appear to be neutral and inclusive (Haddad 2011, pp. 36–7).

Third, the defeats inflicted during the 1990–91 Persian Gulf War and the 2003 US-led invasion caused a retreat of the Iraqi state both in reality – it was no longer able to provide security and essential services throughout the country – and symbolically: the state was perceived to represent only a fraction of the population. This created a vacuum in post-Saddam Iraq. This vacuum was filled by an expansion of the influence of religious, ethnic, regional, tribal, and other groups that not only tried to provide various social services, but also provided a sense of community with which people could identify during an extended period of social, political, and economic disruptions (Haddad 2011, pp. 87–116, 145–7).

If Haddad's analysis is correct, what are the implications? Optimistically, the longer the GoI exists, especially if there is an increase in its capacity to effectively maintain order and efficiently ensure that essential services are available to the population, there will be a gradual reduction of the intensity of an individual's identification with a religious, ethnic, or regional group (Haddad 2020, p. 3). An Iraqi may still look to their religious leader for spiritual guidance and, maybe, advice on which politician to vote for, but as the state returns to fill the vacuum, Iraqis will no longer rely on non-government groups for security from gangs or access to essential services. Pessimistically, certain groups will resist the GoI attempts to restore the influence of the state.

ANYTHING THAT YOU SAY ABOUT IRAQ IS TRUE SOMEWHERE

Many media and government reports on the potential for instability in Iraq simplify their analysis by focusing on a single characteristic – such as religion – along with an explicit or implicit claim that this characteristic is the only one that matters. This treatment is related to the idea that Iraq is an artificial construct, an artifact of British imperial arrogance. If the Iraqi state is an illegitimate construct imposed by foreigners – the argument goes – then one must look to religious or ethnic affiliations to understand the "real" Iraq. Such a view ignores the strong historical, political, economic, and social ties connecting the major urban areas along the Tigris and Euphrates rivers: Mosul, Baghdad, and Basrah. Therefore, it is important to remember several truths about Iraq.

First, categories such as Arab, Kurd, Chaldo-Assyrian (Christian), Sunni, Shi'a, Southerner, or Baghdadi conceal as much as they reveal. Each ethnic, religious, or regional group in Iraq generally encompasses a wide variety of attitudes and beliefs. For example, in an attempt to understand the politics of Iraq's large Shi'a population, one study divided them into five categories based on how the Shi'a define themselves:

1. Non-sectarian Shi'a, often highly educated.
2. Arabic Shi'a, who define themselves primarily by Arab language and culture.
3. Religious Shi'a, who define themselves primarily by their religion.
4. Pro-Iranian Shi'a, who see Iran as a cultural center or big brother.
5. Nationalist Shi'a, who identify themselves with the Iraqi state and generally oppose closer relations with Iran.

Of course, on some issues such as outrage over the 2006 attack on the Golden Mosque, all Shi'a agree. But on many matters, such as federalism, female education, attitudes towards Iran or Saudi Arabia, budget priorities, and so on, there is not only deep disagreement within the Shi'a community, but also a willingness to make "common cause" on certain issues with members of other groups that have a similar stance on these concerns. One discovers a similar complex reality of differing motivations, beliefs, and policy recommendations in Iraq's Sunni, Christian, Arab, Kurd, regional, and other communities.

Second, over 1000 years there have been extended periods of relatively peaceful economic, political, social, and religious contacts among the various ethnic and religious groups in Iraq. Sometimes these contacts were associated with simple commercial transactions. Others reflect long historical connections. For example, of the five major tribes of Iraq, three have both Shi'a and Sunni branches.

In part this reflects the tendency of people to build closer relationships with people of the same village or region than with co-ethnic or co-religious persons who live further away. This tends to lead to loyalties that are primarily geographic. For example, people of southern Iraq often think of themselves as Southerners or Basrawis (residents of Basrah) rather than Shi'as, Sunnis, Arabs, and so on. Regions might provide a better framework for understanding Iraq, since they have a "historical depth" (Visser 2008, pp. 2–4) that may gradually produce greater loyalty than ethnic or religious affiliations. Another example of forming a common cause is the coming together of various ethnic, religious, and regional groups in opposition to Saddam. (See Al-Bayati 2011, especially Chapter 5, for an insider's account of the difficult efforts to form a united opposition to Saddam). However, like almost everything else is Iraq, the degree of regional identification differs dramatically across the country.

Finally, political instability, short of conflict, is not always a bad thing; it can have a favorable impact on long-term economic development by providing an outlet for social frustration arising from uneven development. Instability short of conflict can take the form of regime change – for example, from an autocratic government to a bureaucratic or parliamentarian one – as well as changes in dominant parties without a change in regime; or there may be changes in party leadership.

One of the difficulties of writing on political instability in Iraq is the rapidity of change in the policies and persons involved. Even if there is an impression of stability, since the same policies and persons continue to resurface again and again, there are sometimes subtle, sometimes dramatic, changes in the likelihood that a particular policy will be executed or that a particular person's political career will be strengthened. Analysis of these changes is the subject of the diplomatic cables that every country's Baghdad representative constantly sends to their national capital. Due to publication schedules and the shifting state of affairs in Iraq, it is almost certain that any statement about the current political situation in Iraq will be seriously outdated by the time this book is read. Therefore, this chapter will concentrate on long-term domestic and foreign trends concerning political instability and their impact on Iraqi economic development.

GREED OR GRIEVANCE

From October 1, 2019 through the Covid-19 shutdown in March 2020, over 600 Iraqis have been killed and thousands injured in protests throughout the country. The accompanying political and social chaos has strengthened efforts by other nations to interfere in Iraq, with deadly effects. What caused this sudden surge of violence? There are a wide range of political, economic, and social problems in Iraq, such as corruption, lack of public services, and mass unemployment. But most observers focus on disputes among the Iraqi leaders or among various political, ethnic, or religious groups as the causes of current civil disorder in Iraq (Slim 2019, pp. 161–86).

Is asking about causes really the right question? All countries have disputes about important issues. And in most countries, these disputes are resolved or at least ameliorated by congressional or parliamentary debate and decision; by the political process. In these countries, non-violent protests may play an important role in getting decision-makers' attention, but violence on the part of the protesters or the government is rare.

A better question concerning the current violence in Iraq is: why do disputes in this country tend to travel in a downward arc from parliamentary debate to non-violent protests to widespread civil disorder and, finally, *in extremis*, to civil war? In part, the current violence in Iraq appears to be part of a global or regional trend away from state-on-state conflict towards civil conflict. As a result, of the 259 wars that have occurred worldwide since 1989, 95 percent have been civil conflict, not state-on-state. These civil wars tend to last four times longer than state-on-state wars. Civil conflicts do more damage to society, polity, economy, and lead to greater losses of life. In the Middle East and North Africa (MENA) states, an estimated 4 million persons have died in civil conflicts since World War II (Salem 2019, pp. 1–31). Currently, there are serious civil conflicts in many MENA states, including Iraq, Libya, Morocco, Syria, Turkey, West Bank/Gaza, and Yemen.

The Narrative of Grievance in Civil Conflicts

Academic studies find the source of civil conflicts in either greed or grievance (Berdal and Malone 2000). One or more groups may engage in civil disorder or conflict to capture some source of income or wealth, such as the nation's earnings from oil exports, or to control a ministry to provide jobs and funds for group members. Alternatively, the cause might be grievance. One or more groups may be seeking justice or revenge for wrongs previously committed by another tribe, ethnic group, race, religious group, or language group: 50 years or 500 years ago, your group burned my group's temple or stole my group's land, and now my group seeks justice.

Further complicating the search for the real causes of a particular civil disorder is the fact that, even if greed is a group's true motivation, this group will still promote a narrative of grievance for several overlapping reasons. First, fighting for justice is a more effective way of inspiring fighters. One can pay young men to attack unarmed civilians or other low-risk targets. But it is difficult to get fighters to run a serious risk of death or injury if their primary motivation is greed. Second, a narrative of grievance is useful to increase the number and enthusiasm of non-fighter supporters, especially if there is little chance that these supporters will directly benefit from a successful insurgency. Finally, a narrative of grievance can be used gain support by swaying domestic or foreign media and governments.

Since a narrative of grievance will be used regardless of the true motivation of a group, it becomes difficult or impossible to determine whether a narrative of grievance is legitimate or fraudulent. A more productive approach to understanding civil disorder in Iraq, and possibly to limiting its adverse effects, is to focus less on the spark that started the fire – whether greed or grievance – and focus more on the fuel that feeds the fire: economic conditions.

The Conditions for Civil Disorder

Economic conditions are important because organizing and maintaining an insurgent group is expensive. Among the expenses are paying fighters, buying supplies, and funding a costly propaganda effort that must be employed in order to maintain enthusiasm among supporters, as well to gain support from the domestic and foreign communities. In a rare study of the costs of an insurgency, Paul Collier estimated that maintaining the Tamil Tigers, a medium-sized insurgency in a low-income developing country, was actually seven times more costly than maintaining the Conservative Party, one of the two largest parties in the United Kingdom (Collier 2009, pp. 133–4). If economic conditions are such that the profits from a successful insurgency are small relative to the costs, disputes may be debated and possibly resolved through parliamentary actions. But if the profits expected from a successful insurgency are large compared to the costs, one can expect the result to be civil disorder. Do current economic conditions in Iraq make civil disorder more likely because returns to disorder are high and costs of insurgency are low?

In a seminal study, Collier analyzed the economic conditions that tend to facilitate civil disorder[1] (Collier 2000; see also Blattman and Miguel 2010, pp. 3–57). Depressingly, almost all of these conditions are currently present in Iraq.

First, dominance of primary commodity exports tends to lead to civil disorder. A country whose economy is dominated by small-scale agriculture and small businesses is difficult and costly to loot. However, if a country's economy is dominated by a single commodity export, such as copper, diamonds, or oil, there is a prize that makes the risk of an insurgency worthwhile. A successful insurgent group can immediately reward its followers by diverting export earnings. Of course, the economy of Iraq is dominated by oil, which accounts for an estimated 63 percent of the state's gross domestic product, 92 percent of its government revenue, and 99 percent of its total exports. Control of Iraq's oil is a prize that will compensate for the costs of even a long and expensive insurgency.

Second, if the population of a country is geographically dispersed, maintaining an insurgency even in the face of a reasonably competent government becomes less costly. In a country such as Singapore, where the population is geographically compact, the government can more easily maintain control of its citizenry and, if necessary, crush an insurgency at relatively low cost. There is nowhere for an insurgent group to retreat to in response to a determined attack by government forces. In this case, the insurgency must be prepared to take on government forces in a direct fight; an expensive, difficult, and desperate alternative. On the other hand, if a state's population is geographically dispersed, crushing an insurgency is more difficult because insurgent forces can retreat in the face of a government effort. Only about 17 percent of Iraq's population lives in the capital, Baghdad, with the rest of the population widely dispersed and the state's second-largest city, Mosul, and its third-largest, Basra, at geographic extremes. Defeating an insurgency in Iraq is also hindered by its porous borders with Iran and Syria, which allow Iraqi opposition groups to retreat to those countries to regroup. Further, the mountains of northeast Iraq have long provided a sanctuary for rebels.

Third, a history of conflict tends to reduce the cost of future civil disorders. As a result of previous conflicts, there are many persons with experience in creating insurgencies and a knowledge of tactics and techniques. Weapons and other equipment can be recovered from caches established at the end of previous disturbances. And it is easier to revive previous sanctuaries and supply routes then to develop them from scratch. This reduction in the cost of creating civil disorder is especially prominent in Iraq, which since World War II has had more years of foreign and domestic conflict than any other country in the world. A history of conflict can also lead to large-scale international migration as persons flee the violence. A large diaspora can encourage future civil disorders by lobbying the governments of their new homes to support an anti-government group or, more directly, the diaspora may provide financial support or weapons to insurgents. One example would be the financial and other support provided to rebels during the decades of the Irish "Troubles" by the Irish diaspora in the United States. There are an estimated 400 000 Iraqis in the United States, and they are deeply concerned with events in their former home.

Fourth, low levels of education reduce the cost of creating civil disorder in several ways. People who are illiterate or have low levels of education tend to have few economic opportunities and therefore are more willing to join an insurgency. In economic terms, these persons have a low opportunity cost to participating in civil disorder. In addition, low levels of literacy make persons more susceptible to a group's propaganda. Finally, if there are few economic opportunities, the chances that a young man will be able to marry are reduced. In many societies, an unmarried, unemployed young man is an embarrassment to his family and the subject of public amusement. Joining an insurgent group that gives him a rifle to carry is one way in which such a man can regain respect.

Education in Iraq is failing in several dimensions. Before ISIS took over almost 40 percent of the country in 2014, 14 percent of Iraqi males and 26 percent of females were illiterate, unable to read and write. And among literate Iraqis, education was limited with only 53 percent of males and 45 percent of females of appropriate ages enrolled in secondary education (CSO 2014). Current percentages of literacy and secondary education are even lower as a result of ISIS closing schools and forcing many families to flee from their homes. Millions of these refugees ended up in camps with limited educational opportunities. Finally, a massive "brain drain" has led to a severe shortage in Iraq of trained teachers at all educational levels. In summary, there has been a substantial decline in both the quantity and quality of education in Iraq.

The fifth economic characteristic that tends to facilitate civil disorder is a high proportion of young men in the population. Due to its high fertility rate – 3.9 children per woman of childbearing age – Iraq is a very young country. Almost 38 percent of the total population is less than 15 years old, while 19 percent is between 15 and 24 years old. As a result, each year, almost 900 000 Iraqis become old enough to work (US Bureau of Census 2020). Even after adjusting for deaths, retirements, and the low labor force participation rate among females, the economy of Iraq must create almost 340 000 new jobs a year just to keep the unemployment and underemployment rates from rising (author's estimate). However, since the 2014 collapse in oil prices the Iraq economy has produced few new jobs, resulting in a steady growth in the pool of unemployed men. As expected, this large "army" of unemployed or underemployed Iraqi young men reduces the opportunity cost of participating in an insurgency.

Finally, an economic decline tends to favor civil disorder by reducing the perceived return to working and investing. If the economic future is grim, people are more willing to consider radical change. The combination of the

2014 oil price drop, the huge costs of the fight against ISIS, and the economic shutdown in response to Covid-19, severely damaged the Iraqi economy. The real economy of Iraq actually shrank in 2017, 2018, and again in 2020 (World Bank 2020b, p. 16), and projections for the next three to five years are ugly.

Surprisingly, ethnic and religious diversity does not appear to increase the likelihood of civil conflict. There are countries with very diverse populations that have had little civil disorder, and more homogeneous states that have had widespread violence. That ethnic and religious diversity is not a cause of conflict is especially controversial in Iraq, because most of the political/militia groups in the country have an ethnic or religious character. One possible explanation is that an individual must make two sequential decisions when they are deciding whether to join an insurgency. First, should they join an insurgency? And, second, which insurgent group should they join? The answer to the first question will depend on how the economic characteristics of the country affect that individual. However, once an individual decides to join an insurgent group, they will generally tend to join a group that matches their ethnicity or religion. Joining a group with members of similar backgrounds tends to facilitate both communication and trust.

The primary advantage in Iraq of focusing on the "fuel" of insurgency that is the country's economic characteristics, rather than attempting to resolve the "spark" that is the expressed grievances, is that resolving grievances is often impossible. In some cases, competing groups have incompatible demands. Two or more groups may want the same symbolic or real result. And, since any offer of compromise is seen as betrayal of the groups' members, both parties may demand a Solomonic judgment that the baby be cut in half.

Further complicating the resolution of grievances is that a group's primary motivation may be greed. If a group's real goal is to "capture" a ministry to provide government jobs for the group's members, discussing or even satisfying the group's expressed grievances will be a wasted effort. However, simply understanding the correlations of various economic occurrences and civil disorder does not provide a useful framework for reducing such disorder. What is needed is a specific model showing the steps that lead from failures in economic development to political instability.

POLITICAL INSTABILITY AND ECONOMIC DEVELOPMENT

Samuel Huntington's (1968) hypothesis of the relationships among modernization, economic development, and political stability has been described as the last great attempt to integrate social, economic, and political causes of instability (Fukuyama 2011, p. ix). There are more elaborate recent models of the relationships between economic and political change, but Huntington's hypothesis provides a solid framework for examining the challenges facing countries at Iraq's stage of political and economic development. (For an alternative model, see Acemoglu and Robinson 2006, pp. 673–92.)

Huntington (1968, p. 41) attempted to explain an oft-observed phenomenon that: "Modernity breeds stability but modernization breeds instability". In other words, high-income countries generally experience less political instability or conflict. However, low-income countries that undergo an acceleration of economic growth tend to experience more political instability or conflict. Huntington sought to explain this counterintuitive result by looking at the rates of change of modernization, economic development, opportunities for mobility, and political institutions. Huntington's explanation is shown in Figure 5.2.

Source: Huntington (1968, p. 55).

Figure 5.2   Economic development and political instability

Urbanization, increases in literacy and education, and increased exposure to media all lead to social mobilization. People who live in low-income countries begin to realize that there are better ways of living in other countries. Thus, social mobilization leads to expanded aspirations. People begin to believe that progress is possible, and they want a better life for themselves and their children. Iraq has a relatively well-educated population by Middle Eastern standards, and following the fall of Saddam's regime there was not only a sharp increase in access to the rest of the world through communication and travel, but also a widespread perception that living standards in Iraq were going to rapidly improve. In Huntington's (1968) terms, since 2003, social mobilization increased rapidly in Iraq, leading to higher aspirations.

Economic development increases a society's capacity to satisfy those aspirations. It does not have to be immediate or complete; for example, farmers may be aware that others have trucks, but if they are able to buy motorbikes then this may partially satisfy their increased aspirations, at least for a while. However, if economic development lags too far behind in fulfilling the aspirations created by social mobilization then social frustration – a growing dissatisfaction with current circumstances – will increase.

Unfortunately, economic development in Iraq has generally proceeded slowly and, in some sectors has stalled completely. This was not caused by a lack of overall economic growth, defined as increases in per capita gross domestic product. As discussed in Chapters 2 and 7, Iraq experienced fairly rapid economic growth prior to 2014 primarily as a result of increased petroleum prices. However, as a result of conflict, government mismanagement, and corruption the average Iraqi has yet to see substantial qualitative improvements in quality of life or employment opportunities. The result of increased social mobilization in Iraq combined with slow economic development is increased social frustration.

Social frustration tends to increase most rapidly in countries that have a youth bulge, where the young – usually defined as 15- to 24-year-olds – are an increasing proportion of the population (Urdal 2012, pp. 121–2, 125). This definition describes Iraq, where persons 15–24 years old account for 20 percent of the population, the highest proportion in the Middle East. One of the major sources of social frustration among the Iraqi young is the inability to obtain a good job. As discussed in Chapter 2, the combined unemployment and underemployment rate may exceed 80 percent for those new to the labor force.

Social frustration can be dissipated if the society allows sufficient opportunities for mobility. This mobility can be geographic (people moving from one province to another, or from rural areas to urban), occupational (moving to a higher-paying or higher-status job or profession), or international (migrating to another country: "brain drain").

Unfortunately, opportunities for economic or social mobility are limited in Iraq. Rural to urban migration has been going on for decades and, as a result, the urban population now accounts for almost two-thirds of Iraq's population. This has resulted in severe shortages of essential services and housing in urban areas.

In addition, there is evidence that increased education is no longer a sure route to higher income or status in Iraq. For both males and females, unemployment rates are higher for educated Iraqis with a bachelor's degree than for persons with only a secondary school diploma. The gap is especially serious for females with a bachelor's degree, who have a greater chance of unemployment than a female with only a primary education (COSIT 2008, Table 5.3). International mobility is also constrained. The worldwide Covid-19 recession that began in 2020 has not only reduced employment opportunities in the more economically developed countries but has also resulted in higher legal barriers to foreign workers. In addition, the perceived decline in the quality of education at an Iraqi university has also reduced the likelihood of successful migration.

If social frustration grows faster than can be dissipated by improved mobility, people will want increased political participation in order to demand that the government either accelerate economic development or increase mobility. People will seek to either join existing political organizations or create new ones in order to give voice to their demands.

Demands for expanded political participation often collide with rigid or closed political institutions that are intended to protect the interests of those currently in power or who have previously been in positions of power. Such institutions restrict incorporating new political participants in any meaningful way. This exclusion tends to lead to a worsening of political instability or, if the situation continues to deteriorate, to conflict (Urdal 2012, p. 123). One of Huntington's surprising results was that the openness of political institutions to new voices is positively related to the age of the institution. In other words, the older the institution, the more accepting it is of new political groups.

The collision between demands for expanded political participation and closed political institutions is probably one of the major causes of the political instability that began in Egypt early in 2011 and has now spread across many nations in the Middle East and North Africa. Democracies tend to be more open to increased political participation than autocratic governments. As a result, democracies tend to have more "micro-instability" with new parties or factions constantly jostling for power, but less "macro-instability" such as regime change. However, even democracies with widespread support tend to have difficulty incorporating a rapidly growing youth population. Urdal (2012, p. 127) reports that, for each one percentage point increase in the proportion of youth, there is a four percentage point increase in the risk of conflict. Of course, these are only probabilities; each country is unique and may adopt policies that increase or decrease the likelihood that political instability will occur.

As Figure 5.2 illustrates, according to the Huntington (1968) hypothesis, there are three ways that Iraq can forestall political instability or conflict: by accelerating economic development; by increasing opportunities for geographic, occupational, or international mobility; or by increasing the flexibility of political institutions when confronted with new demands for political participation. As will be argued in Chapters 7–12, accelerating economic development and increasing mobility will require major structural changes in the political economy of Iraq. However, while making these changes will be both controversial and difficult, the failure to radically restructure Iraq's political economy will increase the likelihood that increasing political instability will lead to a return to civil conflict.

NOTE

1.   Study based on 53 civil wars and 550 episodes that could have led to civil war.

6.   Domestic and international politics

All powers not stipulated in the exclusive powers of the federal government belong
to the authorities of the regions and governorates that are not organized in a region.
With regard to other powers shared between the federal government and the regional
government, priority shall be given to the law of the regions and governorates not
organized in a region in case of dispute.

(GoI 2005a, Article 115)

COMPETITIVE VIOLENCE

"A government is an institution that holds a monopoly on the legitimate use
of violence" (Weber [1918] 2015). If one accepts Weber's statement that
holding a monopoly on the use of violence is the fundamental characteristic of
a government, Iraq is failing. There are many groups in Iraq that use violence
in Iraq with varying degrees of "legitimacy" and government control. As is
true about almost every other characteristic, this competitive violence differs
substantially among provinces and sometimes among districts or cities within
a province.

As an extreme example, consider the armed groups in Ninawa province and
its capital city of Mosul as described in Ala'Aldeen's 2020 study. As noted
in Table 6.1, Ninawa was the scene of vicious attacks by both al Qaeda in
2005–07 and ISIS in 2014–17. Even three years after the defeat of ISIS, there
were dozens of state and non-state groups which periodically made use of
violence to achieve their ends. Among the formal security forces there were
the Iraqi Army, the Ninawa Provincial Police, the feared National Security
Service, the Counterterrorism Service, the Iraqi Border Guards on the Syrian
border, the Federal Police, and the Ministry of Interior's Emergency Response
Division.

In addition, there were numerous armed groups that are considered part
of the Popular Mobilization Forces (PMF). These PMF include Shi'a groups
from other provinces including Kata'ib Hezbollah, Asaib Ahl al-Haq, and the
Badr Organization. These groups are thought to have primary loyalty to Iran's
Supreme Leader. There are other non-Ninawan, mostly Shi'a, PMF whose
primary loyalty is believed to be to religious leaders in Najaf and Karbala
provinces. Also, there are local PMF that recruited from sects and ethnic

Table 6.1   *Provincial characteristics*

	Population (millions) (2018)	Human Development Index (2018)	Destruction by Al Qaeda (AQ)/ISIS	Major employer	Secondary school net enrollment M/F (2016–17)
Iraq	38.1	0.689			56%/54%
KRG					
Dahuk	1.3	0.702[a]		Government	NA
Arbil	1.9	0.704[a]		Government	NA
Sulaymaniyah (with Halabja)	2.2	0.706[a]		Balanced	NA
North of Baghdad					
Ninawa	3.7	0.698	AQ/ISIS	Trade/Manufacture	NA
Kirkuk	1.6	0.708[a]	ISIS	Oil	42%/41%
Salah ad Din	1.6	0.695	ISIS	Agriculture	55%/49%
Diyala	1.6	0.698	ISIS	Government	56%/60%
Anbar	1.8	0.683	AQ/ISIS	Agriculture	31%/28%
Baghdad					
Baghdad	8.1	0.706*	AQ/ISIS	Trade/Manufacture	68%/68%
South of Baghdad					
Wasit	1.4	0.660		Balanced	48%/44%
Karbala	1.2	0.670		Trade/Manufacture	53%/56%
Babil	2.1		AQ/ISIS	Agriculture	60%/52%
Najaf	1.5	0.658		Trade/Manufacture	46%/56%
Qadisiyah	1.3	0.681		Agriculture	61%/58%
Maysan	1.1	0.638		Balanced	46%/38%
Muthanna	0.8	0.647		Balanced	46%/44%
Dhi Qar	2.1	0.686		Balanced	54%/49%
Basrah	2.9	0.668	AQ	Oil/ Trade/ Manufacturing	61%/59%

Note: *Province is ranked "High" on Human Development Index.
Source: UNDP (2020).

groups in Ninawa. These mostly non-Shi'a PMF groups presented themselves as defensive forces for Sunni Arabs, Turkmen, Christian, and Yezidi residents.

Despite the defeat of ISIS, which was the justification for Kurdish Regional Government (KRG) intervention in Ninawa, the Kurdish armed presence in Ninawa is still substantial. This includes not only the Peshmerga and several other official KRG groups, but also Kurdish forces not openly supported by the KRG. Several of these less formal forces are affiliated with ethnic

Ninawa-based self-defense forces. In addition, there are a number of private firms that provide armed security for businesses and individuals.

Finally, there are foreign military forces in Ninawa. These include units of Iran's Quds Force usually joined to Shi'a PMF, Turkish Armed Forces, and the American and European Combined Joint Task Force (Ala'Aldeen 2020, pp. 24–6).

What can be done to reduce the capability of various armed groups to disrupt Iraq? There was a July 31, 2019 deadline for all PMF to be integrated into the Iraqi Army. However, the deadline passed without integration. An option short of integration that has been often proposed but never completely executed is to give the Iraqi Army control of salary or other payments to the various PMF units. Many PMF, that were organized during the ISIS crisis, are still paid directly by the Government of Iraq (GoI). Obviously, this gives them independence from the regular Iraqi security forces. Having the Iraqi Army act as paymaster for the PMF could be expected to be particularly effective given the continuing economic difficulties of Iran. Prior to about 2016, the Iranian government was able to heavily subsidize PMF and other organizations in Iraq that supported Shi'a and/or Iranian interests. However, the effects of US organized economic sanctions and the collapse of oil prices has severely damaged the Iranian economy. As a result, Iranian-associated PMF have been informed that they must become self-financing: subsidies from Tehran will decrease.

A longer-term solution to the problem of many armed groups in Ninawas and the rest of Iraq would be to create a sense of security that would make non-GoI security groups unnecessary. Essential to this increased security is more professional, more even-handed, less corrupt, government security forces. Ongoing efforts to increase the professionalism of the Iraq security forces have been uneven, with starts and stops often as a result of uncertain finance. While there have been multiple efforts by international organizations as well as foreign states to train Iraqi police and military, there is one substantial lacuna: the training of professional military officers.

In addition to developing skills in leadership and combat tactics, as well as a wide knowledge of military equipment, what are the other desirable characteristics that training should accomplish? Of course, such training of military officers should develop patriotism and a desire to serve all of the people of Iraq, not just particular ethnic or religious sects. In addition, the officer training academy should be mentally and physically demanding in order to establish a reputation for quality both within the security services as well as in the wider civilian society.

Equally important, although less noted, is that the academy should provide an education that would allow an officer to earn a middle-class living if they decide to leave the military. As exemplified by Iraq's experience since 2003, if military officers lack economic opportunities outside of working in the security services, this weakens civilian control of the military. Without the alternative of earning a good living outside the government, it becomes almost impossible to cashier officers who are insubordinate, incompetent, or corrupt. The entire officer corps resists such discipline, since they see themselves as vulnerable to a similar dismissal. However, there will be less sympathy among the officer corps if an officer threatened with dismissal has favorable opportunities in the civilian economy.

One option would be to require that every officer attend a military academy that, in addition to their military-specific training, would require them to earn an engineering degree that meets international standards. The necessity of completing a rigorous engineering program would serve two purposes. It would not only ensure that officers develop the reputation of being smart and hard-working, but would also provide valuable skills if the officer decides to seek employment outside the security services. But the current lack of trust in the GoI goes beyond the public's distrust of the military leadership.

DISMANTLING THE BARRIERS BETWEEN LEADERSHIP AND PEOPLE[1]

There is a widespread belief in Iraq that elected officials as well as the leadership of the political parties, ministries, and state-owned enterprises constitute an elite whose primary focus is maintaining their high living standards and other privileges, rather than seeking to improve the welfare of the rest of the Iraqi people. Objection to this privileged elite is one of the major motivations of the large protests in Baghdad and other cities that began in October 2019, and continue as this book is written at the end of 2020.

The distrust is reciprocated. Many members of the political elite have expressed, through words and actions, their belief that the rest of the population is not to be trusted to make even low-level decisions. If the country's elite intend to lead Iraq during this complex combination of economic, political, and social shocks, they must first regain the trust of the people. This will require unprecedented actions that can be readily observed and judged by the Iraqi people.

Probably the single action that would do the most to restore transparency and trust would be to open an investigation into the persons who were responsible for the killing of an estimated 600 protestors since October 2019. Not just an investigation into who fired the guns, but also investigating those who gave the orders. Since confidence in the probity of the government leadership is currently low, this investigation may require the involvement of an international organization. The results should lead at a minimum to the naming of those responsible, although prosecution would make a stronger statement that the government supports justice for the victims.

Trust would also increase if the GoI made a serious effort to reduce the influence of organizations such as religious groups, political parties, and foreign states in the day-to-day operation of the Iraqi government. This influence often crosses the line into corruption, which was the topic of Chapter 4. However, the lack of public trust is exacerbated by the defiance of anti-corruption polices by the elite. As noted in Chapter 4, almost 7500 national and regional officials are required to submit financial disclosure reports to the Commission of Integrity. According to the Commission, compliance varies greatly among these elected and unelected officials (Commission of Integrity 2020). Legal action should be taken against those officials who fail to comply.

The GoI should accelerate the slow process of withdrawing most financial privileges from current and former presidents, vice-presidents, prime ministers, and members of the Council of Representatives. This would include the large allowances for security, transport, hospitality, and accommodation. While, as a result of the fiscal crisis, there have been high-level discussions proposing across-the-board cuts in salaries as well as suspending raises and promotions for all state employees, there has been little public discussion of curtailing even some of the more outrageous benefits captured by the elite. As an example, official residencies associated with various government posts are often kept by officials after they leave their posts, upon the payment of a symbolic minimal rent. Reducing or stopping these privileges would go a long way in gaining the much-needed trust of the people.

Finally, the Baghdad bureaucracy must reduce its dominance over the life of the country. It has successfully stalled efforts to achieve the true federalism called for by Articles 105 and 114 of the Constitution of 2005. This would require shifting many economic and social responsibilities from Baghdad to the provinces.

PROVINCIAL GOVERNANCE

There are substantial social, political, cultural, and economic differences among the peoples in Iraq's 18 provinces.[2] As discussed in Chapter 3 and illustrated by the data in Table 6.1, the provinces differ widely in ethnic composition, which parties dominate their electorates, the composition of their economies, levels of literacy and higher education, which language is commonly used, as well as their histories. However, whether under Saddam, the US-led occupation, or the post-2005 democracy, there has been a strong tendency to ignore these differences and impose centralized governance from Baghdad. This tendency towards "one size fits all" policies has not only led to policies that are inappropriate for one or more provinces, but also may have slowed the country's progress towards a more vibrant democracy.

The Baghdad bureaucracy dominance of the everyday life of Iraq must be broken for three reasons. First, there is too much political, economic, and cultural variance among the provinces for a ministry in Baghdad to craft a single policy that will effectively meet the needs of each province. These differences go beyond physical characteristics. As an example, although Iraqis generally exhibit low levels of trust in other people, the regional differences are substantial. In Northern Sunni states including Arbil, Sulaymaniyah, and Kirkuk, 17 percent of the population express general trust in other people. However, among Southern Shi'a states such as Babil, Qadisiyah, Najaf, Karbala, Wasit, Dhi Qar, and Basrah, the percentage of general trust is twice as great (Teti and Abbott 2016, Figure 27, p. 29). The trust gap is smaller between Sunnis and Shi'a in the central provinces of Iraq.

Second, many of the Baghdad ministries are dominated by a political party with a regional power base. It is widely believed that such ministries favor the region where their political support is the strongest. Finally, regardless of the number of kilometers, there is a great psychological distance between Iraqis who live and work outside Baghdad and the ministries that determine so much of their lives. Decisions made at the provincial or district level are more likely to reflect the true preferences of the people (Gunter et al. 2020, p. 6).

While there are provincial governors and councils, their activities and budgets are substantially controlled by the national government. They often see their roles as simply to execute policies decided in Baghdad. This lack of responsibilities of provincial and district governments has at least two adverse effects on progress in Iraq.

First, service as a provincial governor or council member provides little practical education in governance. There could be a natural progression where a person who has a successful tenure as an elected official at the provincial level will then seek to bring their knowledge of government operations to an elected position with the national government dealing with more complex political, budgetary, and economic issues. However, elected officials at the provincial level gain limited insight into the complexity of government operations as well as the necessary compromises and tradeoffs that are necessary in democratic politics. As a result, if these former provincial officials are elected to the Council of Representatives (CoR), they have little capability to effectively meet the requirements of the constitution to "enact federal laws", "monitor the performance of the executive authority", or draft a budget (GoI 2005a, Articles 61 and 62).

Second, if regions and individual provinces are allowed the flexibility to make their own laws, plan and execute their own budgets, and experiment with different methods of providing public and merit goods, it is likely that there will be successful creative responses that could then be adopted by other provinces or the central government. However, the idea that provinces might be laboratories for experimenting in governance has little support among the Baghdad elite. As an example, when the author recommended that one province be allowed to try an alternative system of business regulation to see whether it would lead to less corruption and more rapid economic development, the ministerial response was: "If this is a good idea then it should be the rule for all provinces. However, if you are not sure whether or not this is a good idea then it shouldn't be tried anywhere".

The CoR passed two significant laws on decentralization in 2008. Law No. 21 discusses governorates that are not incorporated into regions, while Law No. 36 deals with provincial, district, and subdistrict elections. Article 45 of Law No. 21 created the High Commission for Coordinating among Provinces (HCCP) which is chaired by the Prime Minister (Ala'Aldeen 2020, pp. 11–12). Despite these laws with high-level oversight, there has been little progress in decentralization since 2008. Part of the reason for the lack of progress was the ISIS conflict and the financial crisis brought about by the 2014 oil price collapse (see Figure 7.2 in Chapter 7). However, the primary reasons are institutional.

After almost 50 years of centralizing policy creation and execution in Baghdad, there is little capacity at the provincial or sub-provincial levels to either create realistic budgets or efficiently monitor budget execution. This lack of capacity is especially obvious in continuing failures to deliver essential services without large-scale waste. Corruption is as serious a problem at the provincial and sub-provincial levels as at the national level.

However, the greatest roadblock to decentralization is the unwillingness of the Baghdad bureaucracy to give up power, influence, or rents. In some cases, ministries have simply refused to carry out mandates to devolve activities and the accompanying finance to provinces, without the ministries being sanctioned or even criticized. At least such defiance has the virtue of visibility. More common is that ministries will publicly state their support for decentralization, but in reality make such decentralization almost impossible. There are two interrelated methods that are used to thwart decentralization. First, the ministries will earmark funds before they are transferred to local governments. The provinces are told in great detail exactly how the funds are to be spent, which effectively makes the provincial authorities into agents of the ministries. The second method used to delay decentralization is to require provincial authorities to run an obstacle course of bureaucratic processes before funds are released. This can include the necessity for "wet" signatures and "live" rubber stamps on documents that can require provincial officials to make multiple trips to Baghdad to sign or stamp documents, rather than the ministries accepting the use of electronic signatures and certifications.

Finally, despite the relatively unambiguous constitutional support for decentralization, when disputes are brought to the HCCP it generally sides with the Baghdad bureaucracy against the provinces.

Until senior Iraqi officials are willing to expend political capital to achieve decentralization, little progress will be made. In addition to such political resolution, successful decentralization requires that provinces have a dependable revenue stream and an unambiguous delineation of provincial responsibilities (Gunter et al. 2020, pp. 6–7). Revenues could be assured by the CoR finally passing the Oil Law dividing oil export revenues between the national, regional, and provincial governments. As will be discussed in the next chapter, this law has been pending a CoR vote since 2007.

Regional, provincial, and sub-provincial government responsibilities should include both the provision of certain services and goods as well as regulatory activities. Sub-national governments should be given responsibility for providing primary and secondary education, local and provincial police, public health services as well as the operation of public clinics and hospitals, and providing welfare especially for widows and orphans. In addition, with the exception of the regulation of the financial system, the regulation of businesses, health, and education should be executed by the provincial or local authorities, not by Baghdad.

One of the arguments made to delay any decentralization efforts is that there will be provincial mismanagement and corruption of whatever resources are transferred from the national government. Initially, this will be true. However, there is currently large-scale mismanagement and corruption in the Baghdad ministries. It seems unlikely that the regional or provincial governments could be less capable and honest than the national government. And, since the regional and provincial governments are physically and psychologically closer to the citizens, it is likely that errors and crimes will be more rapidly uncovered and rectified. Also, as discussed above, if local governments are given more responsibility, then they should begin to act as a training ground for elected officials that will be more capable if they decide to run for national office in the future.

Among provinces and regions, Kirkuk and the Kurdistan Regional Government (the KRG; also referred to as the Kurdish Region of Iraq, KRI) have a long, complex, often adversarial relationship with the national Iraq government.

KIRKUK AND THE KRG

The reaction to the post-1991 vacuum in national authority that has the greatest potential to become a permanent part of Iraq's political future is the semi-independent KRG. The KRG is still part of Iraq. However, since the US and the United Kingdom (UK) established a no-fly zone over the region in 1991, the KRG has maintained its own army – the Peshmerga – and its own international oil trade and investment policies. The KRG has diverged from the rest of Iraq in several ways. The KRG is generally considered to be less sectarian than provincial governments in the rest of Iraq. It is more business-friendly and generally considered to be less corrupt. Education opportunities are uneven but are generally considered to be better. There are a few excellent universities such as the non-profit private American University of Iraq, Sulaimani. The KRG is really the only federal part of Iraq with authority divided between the GoI in Baghdad and the KRG; in the rest of Iraq federalism is in its infancy, with almost all important decisions made in Baghdad.

Possibly as a result of its relative security, the provinces of the KRG have been more prosperous than the rest of Iraq. The Human Development Index (HDI) is a composite index of life expectancy, education, and per capita income intended to give a richer indication of the quality of life than per capita income alone. As can be seen in the second column of Table 6.1, the three provinces of the KRG along with Kirkuk and Baghdad are ranked at the level of high human development by the United Nations. These provinces have a quality of life roughly equivalent to that in Indonesia, Samoa, or South Africa. Iraq as a whole, as well as the other provinces, are ranked as having medium human development. Before the oil price collapse and the rise of ISIS in 2014, the economies of the KRG and Kirkuk were growing rapidly, especially as a result of foreign investment-driven construction.

During the 2014–17 war against ISIS, while the economy suffered, the KRG achieved several symbolic goals. The Peshmerga and other armed groups associated with the KRG were not only able to defend the territory of the three provinces from ISIS incursion, but also played a significant role in the defeat of ISIS in Kirkuk and Ninawa provinces. The occupation of Kirkuk with its massive oil fields by KRG forces greatly strengthened the KRG as a petroleum power and improved its negotiating position with respect to Baghdad. Symbolically, its military successes, combined with its role as a sanctuary for millions fleeing ISIS, improved the KRG's international standing. Foreign investors interested in Iraq continued to perceive the KRG as being a more secure and well-run alternative to the rest of the country. Even in downtown Baghdad, Western visitors and investors had to hire personal security guards and exercise great care in something as simple as going to a restaurant. On the other hand, in the KRG, the author, on impulse, hired a car and driver for a two-day trip from Arbil to Sulaymaniyah to Halabja and back.

On the negative side of the balance sheet, the KRG has pushed the government as the "employer of first resort" to an extreme, resulting in a large budget burden and massive government inefficiency. At the same time, there is an ongoing political dispute over whether the KRG President Barzani was in violation of the region's Constitution. This dispute resulted in several – possibly extra-legal – extensions of Barzani's term of office and the failure of the KRG Parliament to meet for almost two years. With respect to corruption – the special sin of Iraq – progress has been uneven. In the 2009 provincial elections, the anti-corruption Gorran Party was surprisingly successful (Barkey et al. 2011, p. 55). However, much of the manufacturing and international trade of the KRG is still apparently controlled by the leadership of the major KRG political parties.

In September 2017, Barzani took the bold step of holding a regional referendum on independence for the KRG. His true motivations are unknown, but there are probably several contributing factors. First, Barzani had called for such a vote for over a decade and may have thought that it was time. Second, with the success of the Peshmerga and other KRG allied forces in the then ongoing war against ISIS, Barzani may have believed that he would have strong international support for independence. Finally, there was a political motivation. A successful vote for independence could be expected to improve Barzani's position with respect to the Gorran Party and others who argued that he was increasingly authoritarian in his leadership of the KRG.

Over 93 percent of the voters approved the statement: "Do you want the Kurdistan Region and the Kurdistani areas outside the Region to become and independent state?" (Rudaw 2017). There were widespread celebrations throughout the KRG. Despite – or possibly because of – the overwhelming support for independence, the results of the vote were devastating to the KRG. The GoI stated that the vote was unconstitutional and GoI security forces closed roads between Mosul and Dahuk. Further, the GoI demanded that the KRG give up control of the international airports in Arbil and Sulaymaniyah to the GoI. When this control was not immediately ceded, the GoI stopped most international flights into the KRG. Instead of international businesspeople traveling directly to the KRG from Vienna or Istanbul without having to obtain an Iraqi visa, travelers had to first obtain an Iraqi visa and go to Baghdad or another Iraqi city before flying to the KRG. In addition, Kurdish-affiliated businesses throughout the other 15 provinces were subject to protests and, in some cases, violent attacks. However, psychologically, the most devastating response was probably Baghdad's military response.

The month after the independence vote, Iraqi security forces pushed the Peshmerga out of Kirkuk, ending KRG control of this oil-rich province. There were few casualties, as the Peshmerga generally retreated as the government forces advanced. Not only did the KRG lose control of the province, but the reputation of the Peshmerga as a tough organization was lessened.

The end game of the independence vote debacle was relatively quick. In October, the KRG government "froze" the results of the independence vote and sought immediate discussions with Baghdad. A few days later, President

Barzani resigned his office after 12 years. And in November, after Iraq's Supreme Court ruled that no Iraqi province was allowed to secede, the KRG government quickly accepted this ruling. In mid-March 2018, Baghdad allowed the international airports in the KRG to reopen, and roads were no longer blocked. The almost six-month closure of the airports to international traffic and road closures had substantial adverse impacts on businesses, educational entities, and humanitarian work in the KRG.

After the independence vote debacle, the KRG is generally considered to have three possible futures. Most pessimistically for a united Iraq, the KRG might seek, at some time in the future, to again seek complete independence by separating itself from the rest of Iraq in order to form an independent Kurdish state, possibly including parts of Syria, Turkey, and Iran that have large Kurdish majorities. Some believers see the creation of an independent Kurdish state as a correction of a historical error when the Kurds were left out in the early twentieth century great-power delineation of borders that created states for Arabs, Persians, and Turks (Stansfield and Ahmadzadeh 2008, pp. 129–30).

However, the separation of the KRG into an independent state is extremely unlikely for several practical reasons. As demonstrated after the September 2017 vote, any attempt to form such a state will face strong and probably violent opposition not only from the GoI, but also from Turkey and other neighboring countries concerned about possible insurgencies among their own Kurdish populations. In addition, one of the advantages of the landlocked KRG within Iraq is that it can play off the Iraqi and the Turkish governments against each other, especially with respect to oil production and exports. If an independent Kurdish state fails to obtain reliable access to pipelines in either Iraq or Turkey, its population will experience a substantial decline in living standards. Finally, over the last century, the Kurdish communities in Iraq, Turkey, Syria, and Iraq have gradually grown apart and now possess substantially different attitudes and loyalties. They may be "one people" in theory, but any attempt to create a Kurdish state will require the reconciliation of fundamentally different visions of what such a state represents.

Another, less threatening although still destabilizing, vision of the future of the KRG is that it will continue to be a sanctuary region for Kurds within Iraq with a loyal Peshmerga to offset any threat from the Arab majority. Protection for Kurds would not be limited to the KRG but it would also, despite its military reversals of 2017, include being seen as the symbolic protector of Kirkuk and other areas that have a large Kurdish population or are important to Kurdish history and culture. In fact, the KRG leadership has reserved the right to intervene outside the KRG if Kurdish interests are threatened, although this commitment is now less credible. This version of the future of the KRG will be fundamentally a continuation of the pre-independence situation.

But there is a more optimistic future. The leadership of the KRG have spoken about a regional government that will represent the interests of all of the residents of an Iraqi region: not just Kurds, but also Arabs whether Sunni or Shi'a, Chaldo-Assyrian Christians, and Turkmen (Stansfield and Ahmadzadeh 2008, pp. 123–5). In other words, in the KRG there would be less emphasis on the "K", Kurdistan, and more emphasis on the "R", Regional. Critics see such statements by the KRG leadership as cynical political ploys intended to disguise their real goal of independence. However, there have been attempts to incorporate non-Kurdish groups such as Chaldo-Assyrian Christians and Turkmen into positions of responsibility in the KRG government, while attempting to reduce loyalty to entities outside the KRG. There is a long tradition of Chaldo-Assyrian Christians joining the Peshmerga. Which vision of the future will actually occur depends in large part on whether or not there is a successful resolution of the three issues dividing the KRG and the GoI: oil, the Kirkuk referendum, and the 17 percent rule.

Kirkuk, although not legally in the KRG, is important for two reasons. First, Kirkuk province contains most of one of the largest "supergiant" oilfields in Iraq (and the world), with an estimated 8.6 billion barrels in reserves. This huge field extends into two KRG provinces, Arbil and Sulaymaniyah, but the primary pumping stations and other facilities are in Kirkuk province. In October 2019, exports from this oilfield were about 410 000 barrels per day (bpd), about 12 percent of Iraq's total exports (Ministry of Oil 2020). As can be seen in Chapter 7, Figure 7.1, the oil from the Kirkuk oilfield can be pumped north into Turkey before being piped west to the Mediterranean Port of Ceyhan. Previously, oil from the KRG was refined at the Baiji refinery, one of the largest in Iraq, before this refinery was wrecked by ISIS. Pending the restoration of this refinery, oil that is shipped south by pipelines can either be refined or exported through the Persian Gulf. Now that the KRG is cut off from Kirkuk, its ability to control oil exports will be severely constrained.

Second, Kirkuk province previously had a large non-Arab population, primarily Kurds. However, beginning in the 1980s, Saddam instituted a policy of ruthless Arabization of Kirkuk city, forcing Kurds to flee, and giving their jobs and homes to Arabs who received incentives to move into the city. Following the 2003 collapse of the Saddam regime, the process of ethnic change has reversed with the return of the Kurds and the departure – sometimes forced departure – of Arabs. The current Arab–Kurd population ratio is unknown, since the last reliable census of Kirkuk province was 25 years ago. However, both Arabs and Kurds claim to be a majority.

Article 140 of the Constitution was intended to resolve the status of Kirkuk province through a two-step process of a census followed by a referendum to determine residents' preferences for the future: affiliation with the KRG or continuation as an independent province. Over the 15 years since the approval of the Constitution, there has been much discussion and posturing by the parties involved but no census, and therefore no referendum. Initially, the KRG sought to delay the census pending the "return" of Kurds who claim that they were forced out by Saddam. More recently, the GoI appears to be delaying the census, believing that ambiguity over the Arab–Kurd population ratio is better than a possible KRG victory. It should be noted that Turkey opposes Kirkuk province joining the KRG, since it believes that it would increase the likelihood – despite the 2017 independence referendum debacle – of a further attempt at KRG independence (Barkey et al. 2011, pp. 22–23).

In addition to Kirkuk province, the GoI and the KRG continue to feud over contracting with foreign oil companies and the KRG's share of national budget expenditures. As discussed in Chapter 7, the GoI has yet to pass the National Oil Law written in 2007. This Oil Law was required by the 2005 Constitution and was intended to delineate national, regional, and provincial responsibilities for oil and gas exploration and production, especially where foreign oil companies are involved. While the passage of the National Oil Law was delayed, the KRG signed contracts directly with foreign oil companies. When oil produced by the KRG is exported, the payments are deposited in the GoI's Development Fund for Iraq account at the New York Federal Reserve or directly with the Ministry of Finance. The KRG then expects a portion of these funds to be transferred back to the regional government.

The GoI sees these KRG contracts with foreign oil companies as violating its authority over the exploitation of Iraq's oil reserves, but lacks a means to force the KRG to comply. If the GoI delays or restricts payments to the KRG for "illegal" oil exports, the KRG responds by reducing such exports, which not only cuts GoI export income but also decreases confidence in Iraq as a reliable trading partner. The GoI has also attempted to constrain the KRG by banning foreign oil companies that contract with the KRG from participating in oil production in the rest of the country. However, some major foreign oil companies, such as Exxon Mobil, have been willing to run the risks of signing contracts with the KRG. These oil companies appear to think either that the GoI is bluffing about excluding such companies from contracts in the south (especially where the foreign companies are already engaged in production activities) and/or that they can negotiate more favorable agreements with the KRG, and therefore would be willing to forego doing business in the south of the country.

The third and final quarrel between the GoI and the KRG concerns national budget expenditures. There is an understanding that the KRG should receive national budget expenditures based on its proportion of the nation's population. But since it has been decades since a complete national census was taken, the true population ratio is unknown. While the KRG authorities claim that their region accounts for 17 percent of the population, Baghdad claims that the proportion is closer to 13 percent (Raphaeli et al. 2020). As discussed above, holding a new national census is controversial, since it would probably also provide evidence on the relative Sunni–Shi'a populations. While the Shi'a believe that they account for about two-thirds of the population, the Sunnis think that the true populations of the two sects are closer to parity. It is expected that the census, which has already been delayed for 13 years, will continue to be rescheduled.

The difference between the KRG receiving 13 percent or 17 percent of GoI expenditures is critical due to the ongoing budget crisis. Since the oil price collapse that began in 2014, the KRG government has reduced pay to teachers, medical personnel, pensioners, and others dependent upon the government. What was expected to be a temporary loss of salary or pension for current or former government employees has now continued for years. To raise the necessary funds to continue even partial payments, the KRG has not only borrowed from private individuals but also delayed or reduced contractual payments to international oil companies. While records are incomplete, it is estimated that the KRG owes an estimated $26 billion (Raphaeli et al. 2020).

There have been several failed attempts to resolve the budget impasse. In December 2019, there was a reported agreement that Baghdad would transfer 12.7 percent, approximately $14 billion, of the federal budget to the KRG in exchange for 250 000 bpd of oil (Reuters 2020). Since then, Baghdad has claimed that the KRG failed to deliver the oil, while the KRG claims that Baghdad has shorted the promised payments. The situation is further complicated since neither Baghdad nor the KRG has approved a 2020 budget, nor does either entity use transparent accounting.

These disputes between the GoI and the KRG are unlikely to be resolved in the near future and are complicated by political issues. Post-2005 Iraqi prime ministers have depended on the support of both the KRG and Shi'a representatives to remain in power. This puts prime ministers on the horns of a dilemma: favor the KRG in the disputes and face weakened support from the southern Shi'a that are the usually the largest component of a coalition; but if an attempt is made to rein in the KRG then its representatives will withdraw their support and the government might fall. As a result, the most likely outcome is that the GoI and KRG will attempt to muddle through their budget issues without any clear-cut resolution. Recriminations and accusations will continue, accompanied by a series of short-term political and economic compromises. Of course, both the GoI and the KRG have to be aware of the ambitions of other nations with respect to this dispute.

EXTERNAL TENSIONS

Kenneth Pollack, of the Brookings Institution, half-humorously stated at a 2012 Middle East Institute discussion that Machiavelli's *Florentine Histories* (written in the sixteenth century) might be the best guide to understanding the complex relations between Iraq and its six neighbors. The Florentine reference was to the fact that each of the countries contains several official or private groups that directly or indirectly engage with foreign governments or organizations, often in opposition to the official policy of their governments. For example, with respect to Syria, one might think of at least four Iraqi foreign policies: the official policy of the GoI; the policy of the Shi'a groups that – although members of the current Iraqi government – are more supportive of the Syrian authorities; the policy of the Sunnis that favor the Syrian rebels; and, finally, the KRG who see the conflict in Syria as leading to closer KRG–Turkey relations.

This "Florentine" complexity originates from a combination of weak national governments and country borders that cut across ethnic and religious divisions. Iraq is primarily ethnically Arab – speakers of Arabic – and an estimated two-thirds of the population is Shi'a. To the east, Iran is mostly Shi'a but not Arab: the primary language is Persian. To the west, Saudi Arabia, Kuwait, Syria, and Jordan are Arab, but mostly Sunni. In Syria, the minority Shi'a still control the government despite a vicious civil war. Finally, Turkey is neither Arab nor Shi'a; its primary language is Turkish, while most of its Islamic population is Sunni. There is also a large population of ethnic Kurds that are mostly Sunni in Turkey. These Kurds are an estimated 18 percent of the Turkish population; in Iraq, Kurds are 17 percent of the population; in Syria, 8 percent of the population; and in Iran, 7 percent of the population.

In addition to the complexity of intersecting national, ethnic, and religious interests, countries in the region vary greatly in their degree of corruption, type of government, and proportion of youth. As can be seen in Table 6.2, Saudi Arabia and Jordan seem to have the healthiest political economies in the region, with reasonably honest governments as well as a good age distribution. At the other extreme is Syria, which has a corrupt and authoritarian government and a large youth bulge. Since 2012, Syria has suffered from large-scale government-inflicted violence. The other countries have combinations of internal strengths and weaknesses that, along with ethnic and religious divisions, influence their disputes and misunderstandings with Iraq and their other regional neighbors.

Among the many cross-border disputes, there are seven that are particularly troubling to the stability of Iraq. First, as discussed above, there are groups in

Table 6.2   *Living in an interesting neighborhood*

	Corruption 2018	Freedom status 2018	Population 2017 (Ages 0–14)
Iraq	168th	Not free	38 million (40%)
Iran	138th	Not free	81 million (24%)
Turkey	78th	Not free	81 million (25%)
Syria	178th	Not free	18 million (37%)
Saudi Arabia	58th	Not free	33 million (25%)
Jordan	58th	Partly free	10 million (36%)
Kuwait	78th	Partly free	4 million (21%)

Source: Corruption – Transparency International (2020); freedom status – Freedom House (2021); population – World Bank (2020c, Table 2.1).

the KRG that support an independent state uniting the Kurds of Iraq, Turkey, Iran, and Syria.

Second, both Iran and Turkey have made post-2003 cross-border military incursions into Iraq. Iran temporarily seized certain Iraqi oilfields on or near the border. Iran's incursion was possibly in response to Iranian domestic political pressures to demonstrate dominance towards Iraq, or to improve Iranian negotiating positions with respect to the exploitation of these fields. Turkish military forces have attacked towns in the KRG with artillery and ground forces. The Turkish government claims that these towns were harboring anti-Turkey Kurdish terrorist groups.

Third, as will be discussed in detail in Chapter 8, there have been dramatic decreases in the water flow of the Euphrates river as a result of the diversion of water for agriculture in both Turkey and Syria. In the absence of treaties guaranteeing a minimum flow, Iraq faces the potential of accelerated desertification and salinization of agricultural land in the western and southern parts of the country. Without a substantial increase in the efficiency of Iraqi water use, combined with treaty guarantees of future water flows from its neighbors, this will lead to both a reduction in Iraqi agriculture and further migration of poor farmers into urban areas.

Fourth, as a result of extremely confusing negotiations in the late nineteenth and early twentieth centuries among the UK, Russia, the ailing Ottoman

Empire, and Persia, Iraq was left with limited access to the Persian Gulf. To the east, the Shatt al-Arab waterway that connects Basrah with the Gulf is shared with Iran for much of its length, while traffic from the Gulf to Iraq's other major port in the west, Umm Qasr, must travel in the Khawr Abd Allah waterway that is shared with Kuwait. (For a more detailed discussion, see Chapter 8). As Iraqi export and import trade increases, it can be expected that both Kuwait and Iran will become less compromising with respect to the use of water routes that they share with Iraq. In fact, Kuwait has already announced plans to build a port on one of the major water routes from the Persian Gulf to Basrah, while Iran has made threats that it will close the Persian Gulf to oil tankers in response to increasing pressure from the West.

Fifth, major disputes related to the 1990–91 Iraqi invasion of Kuwait continue to fester. Kuwait currently receives 5 percent of Iraq's oil export revenues to compensate for the human and physical damages of the invasion. Iraqi authorities tend to see these reparations as temporary: when approximately $53 billion in estimated damages are paid, then the reparations will end. On the other hand, several prominent Kuwaiti leaders see the invasion losses as much larger. As a result, they expect the reparations payments to continue almost indefinitely. At the same time, Iraq and Kuwait are engaged in complex negotiations over the exploitation of oilfields that are located under the border.

Sixth, there is widespread cross-border smuggling between Iraq and all of its neighbors, but particularly with Iran and Syria. This smuggling is facilitated by tribal relationships, such as the close relations between Iraqi and Syrian branches of the same tribe (Visser 2008, p. 23). Smuggling undermines the GoI's management of the economy. For example, ongoing sanctions in Iran and political turmoil in Syria have greatly restricted trade and investment in both countries resulting in severe dollar shortages. As a result, groups are smuggling products across the borders into Iraq in order to obtain Iraqi dinars (ID) and then using intermediaries to exchange these dinars for dollars and smuggling these dollars back to Syria or Iran. As will be discussed in greater detail in Chapter 14, in order to prevent a destabilizing loss of its dollar reserves, the Central Bank of Iraq (CBI) has been forced to sharply devalue the dinar.

Finally, both Syria and Iran appear to be entering long-term periods of internal political instability. With respect to Syria, Iraq may seek to be a destabilizing force to weaken a Sunni neighbor that was a major source or transit station for foreign fighters as well as financial and logistical support for insurgencies in Iraq. With respect to Iran, the flow of instability is likely to go the other way. Surprisingly, Iran is also undergoing a period of unprecedented demographic collapse that will affect its long-term influence in Iraq and the region.

Demographic Divergence

Possibly related to the religious and ethnic differences between Iraq and its neighbors, there is evidence of a substantial divergence in fertility, the number of children that a woman is expected to bear. While almost all countries experience a decline in fertility rates in the initial stages of economic development, Iran is experiencing a collapse in fertility rates almost unprecedented in recent world history.

Iraq's fertility rate (number of children born to the average woman in her child-bearing years) declined from approximately 7.2 in 1970–75 to 3.7 in 2018. It should be noted that a 3.7 fertility rate is still well above the 2.1 rate needed for the population of a country to gradually stabilize with zero population growth. Also, Iraq currently has a higher fertility rate than any of its neighbors. During the same period, the fertility rate of women in Iran, Kuwait, and Turkey decreased to 2.1 (World Bank 2020c, Table 2.14). Iran is an extreme world outlier with respect to the rapid drop in fertility. While Turkey and Kuwait were on a gradual path to zero population growth, Iran experienced a demographic collapse. Only one generation ago, its fertility rate was 6.2. As a result, Iran's population is aging rapidly. According to the United Nations Population Division, the median age in Iran is 32 years, compared to only 21 years in Iraq.

Why are fertility rates falling in Iraq and among its Middle East neighbors? Possible explanations range from increased availability of contraceptives that allow women greater control over their fertility, to a hypothesis that such a sharp drop in fertility means that a people have lost faith in the future. If the future is hopeless, then why have children? The usual examples given to support this hypothesis of despair are the sharp drop in fertility (and life expectancy) in Russia after the collapse of the Soviet Union, and the depopulation of Italy in the later Roman Empire. In a controversial analysis, Goldman (2011, pp. 1–7) argues that fertility falls when women become educated and their religious fervor decreases. One of these causes is not sufficient to bring about a sharp fertility reduction.

The Goldman (2011) hypothesis appears consistent with both Iraq's and Iran's demographic transition to lower fertility. Female education has increased substantially over the last decades, while only 44 percent of Iraqi youth say that their religion is important to their identity (Arab Youth Survey 2020, p. 42); while in Iran, only about a quarter of the young regularly attend a mosque. Is this rapid fertility drop in Iran dangerous to Iraq?

While Iran's situation is almost unique in modern times, the precedent from ancient history is not favorable. Countries that are strong but expect to substantially weaken in the near future tend to be dangerous to their neighbors. There is a temptation to strike while they still have strong hopes of obtaining advantages that will prevent or at least delay their decline (Goldman 2011, p. xi). This may increase Iran's motivation to weaken Iraq's long-term political stability.

The complexity of Iraq's external challenges might be a reason for optimism. For example, if Iraq's only point of contention with Turkey was the divergence of water from the Euphrates and the Tigris rivers, it is a zero-sum game where the only way that one country can win is if the other country loses. However, Iraq and Turkey have a variety of other issues besides water. In addition to the Turkish desire that elements in the KRG be constrained in their independence movement, Turkey is a major oil and natural gas importer, and there is interest in obtaining natural gas from Iraqi fields for transshipment across Turkey to the European market (the Nabucco pipeline) (Barkey et al. 2011, pp. 57–8). Further in the future, rail transportation from Basrah to Rabiya (on the Syrian border) has the potential of becoming a major transit corridor for container shipments from the Persian Gulf to eastern Turkey. Similar webs of complex relationships exist between Iraq and each of its neighbors. These complex relationships could provide a framework for mutually beneficial compromises. But who will negotiate for Iraq? In view of the many complex internal and external challenges to the GoI, can another regime change be avoided?

WILL IRAQ'S DEMOCRACY ENDURE?

What is the likelihood that increased political instability in Iraq will lead to a reversion to an authoritarian state? Attempts to answer this question have taken two approaches. First, one can look at the probability of a democracy surviving under various economic conditions. Surprisingly, this research is fairly optimistic about the future of Iraqi democracy. Second, one can compare the capacity of the Iraqi government to the expected pressures imposed on it by internal and external shocks. A weak government may survive if it is never tested, and a strong government may fail if pressures are too great. This second approach is more pessimistic.

Economic Growth and Democracy

If "democracy is a regime in which government offices are filled by contested elections" (Przeworski et al. 2000, p. 19), Iraq should be classified as a democracy. Especially, the contested elections of 2010 and 2018 show that Iraq has broken the curse of Middle East democracy which often follows the pattern of: "one man, one vote, once". Some have argued that the meaning of democracy is broader than just contested elections, and in fact the Economist Intelligence Unit classifies Iraq as a "hybrid" regime containing both democratic and authoritarian elements, while Freedom House evaluates the country as "Not free". As shown in Table 6.2, Iraq has the same ranking as its neighbors, with the exceptions of Jordan and Kuwait which are considered to be "Partly free". Regardless of definitions, Iraq has made dramatic progress in both civil and political rights since Saddam was overthrown. But how vulnerable is Iraq to a reversion to an authoritarian state run by another strongman?

Iraq's per capita income favors the survivability of its democracy. Democracies with a per capita income of less than $1000 (at purchasing power parity, PPP) tend to survive an average of about eight years before being replaced by an authoritarian regime, while those with a per capita income between $3001 and $4000 (PPP) tend to survive an average of 36 years. There are no cases of a democracy with a per capita income of over $7000 (PPP) ever being replaced by an authoritarian regime (Przeworski et al. 2000, Table 2.3, p. 93). According to this study, a democratic nation with Iraq's current per capita (PPP) income of about $8900 would be expected to survive indefinitely.

The survival of democracies is also sensitive to economic growth. When democracies experience economic growth, their odds of being replaced by an authoritarian regime in any given year are only 1 in 66, but income declines reduce these odds of regime survival to 1 in 20: "most deaths of democracy are accompanied by some economic crisis" (Przeworski et al. 2000, pp. 109, 117). As discussed in Chapters 2 and 7, Iraq's economic growth is dominated by the price of oil, and this results in sharp year-to-year changes. As can be seen in Figure 7.2 in Chapter 7, the collapse in oil prices that began in 2014 was partly offset by an increase in oil export volumes. However, combined with rapid population growth, there was a sharp drop in national income per capita from about $15500 in 2013 to an estimated $8900 in 2020. Such a large extended economic decline put severe stress on Iraq's political system through its immediate impact on muhasasa.

As discussed earlier, *muhasasa* is the post-2003 system of dividing the ministries and their associated government jobs among the various ethnic groups in Iraq. After each election, the actual formation of a government is delayed while the parties affiliated with various ethnic groups haggle over the division of government posts and jobs. As a result, forming a new Iraqi government is difficult under normal conditions, but in a declining economy the difficulty increases substantially. Not only does the drop in government revenues mean that most ministries will have smaller budgets and therefore will be less able to hire more party members, but also the relative "value" of ministries will change. For example, in a growing economy, controlling infrastructure investment is a powerful position, since it provides opportunities for both increasing the number of jobs for supporters as well as corruption. But with the fall in revenues, non-oil-related infrastructure investment has sharply decreased, rendering its control almost valueless.

In mid-2020, it is hoped that the new Prime Minister, Mustafa Al-Kadhimi (hereafter referred to as Kadhimi) will be able to partially dismantle *muhasasa* by appointing technocrats without strong party or ethnic ties to key ministerial positions. While the public seems to be willing to give him an opportunity to fix the dysfunctional system of public finance in Iraq, Kadhimi is already facing strong opposition from various groups that are seeking to maintain their control over government expenditures and jobs despite the shrinking real economy. There is a danger that as soon as the budget constraint relaxes, as a result of either good management or good luck, these groups will withdraw their support from Kadhimi, hoping for a better deal under his successor.

Another characteristic that tends to be associated with a higher rate of regime change is income inequality. Research suggests that democracies are more stable in egalitarian societies, but the effect is not strong (Przeworski et al. 2000, p. 120). As discussed in Chapter 3, relative poverty is low in Iraq. The 10 percent of the population with the lowest incomes receive an estimated 4 percent of the national income. While far from equality, this proportion of national income received by the poorest Iraqis compares favorably with income distribution in most developing countries. Of course, substantial increases in per capita income are unlikely to have a stabilizing influence if most of the increase is diverted by corruption or wasted by internal conflict.

To the extent that the results of these studies of a large number of developed and developing countries are applicable to Iraq, the nation can expect its nascent democracy to survive as long as real economic growth greater than population growth rate can be restored, and this rise in average per capita income is not accompanied by a substantial deterioration in income shares of the poorest Iraqis. But statistical results across a large sample of countries can provide only limited comfort. What can be said about Iraq's specific situation?

Governmental Capacity and Pressures

A more complex method of examining Iraq's political stability is to compare the strength or capacity of the country's state institutions and civil society to the adverse pressures or forces faced by the country. The Fund for Peace has estimated such a comparison for 177 countries. Since the defeat of ISIS in 2017, Iraq has been steadily improving in these rankings. In fact, out of the 12 indicators monitored, since 2017, Iraq has improved in ten. It is only in the indicators of "Fractionalized Elites" and "Public Services" that Iraq has either stalled or deteriorated. The primary driver of this nearly across the board improvement in Iraq was the decrease in violence (Fund for Peace 2020). As shown in Figure 5.1, civilian deaths by violence have fallen to the lowest level since the 2003 invasion. Another indicator that has shown surprising improvement since 2017 is respect for "Human Rights and Rule of Law".

Iraq's improvement should not be overstated; the apparent progress has been from a very low starting point: Iraq is still only 17 steps from the bottom of the rankings, and the Fund of Peace still includes Iraq on its "Alert List" of fragile states. Political stability in Iraq is still threatened by many internal and external challenges. However, the capability of the Iraqi state to cope with these challenges appears to have improved. And if the economy can return to real growth accompanied by increasing employment opportunities, without a worsening of income distribution, then the internal pressure leading to serious political instability should diminish further. In terms of the Huntington (1968) hypothesis discussed in Chapter 5, Iraq must simultaneously accelerate economic development, facilitate mobility, and ensure that the political process is open to new participants. If Iraq succeeds in these very difficult tasks, the odds will further improve that it can avoid another authoritarian government; another Saddam.

NOTES

1. This section draws on Gunter et al. (2020, pp. 4–5).
2. Throughout Iraq's history, there have been multiple divisions of existing provinces into two or more new provinces, or mergers of existing provinces into a new larger province. Most recently, in 2014, the KRG voted to take part of Sulaymaniyah and create a new province of Halabja. If ratified by the federal government than this would mean that there would be 19 provinces. However, as of mid-2020, this ratification has not occurred so most studies, including this one, treat Iraq as currently divided into 18 provinces. Based on questionable historical speculations, Saddam considered Kuwait as Iraq's nineteenth province.

7. Oil and gas

A place called Ardericca [near Babylon] ... forty [furlongs] from the well which
yields produce of three different kinds. For from this well they get bitumen, salt, and
oil ...

(Herodotus 420 BC, in Godolphin 1942, Paragraph 119, p. 380)

Iraq possesses large reserves of low-cost crude oil that are so great as not only
to dramatically impact the nation's economy but also to shift the worldwide
balance of demand and supply of oil. It is a tremendous opportunity, but if
oil sales are to provide more than a temporary boost to Iraq's economy then
serious engineering, management, economic, political, and social problems
must be solved. Of these, the engineering problems might be the easiest.

EXPLORATION, PRODUCTION, AND EXPORT

Oil

Iraq is afloat on a sea of oil. Despite large-scale production beginning in 1927,
its proven reserves are still huge. Although estimates of proven reserves have
a degree of uncertainty (Cordesman and al-Rodhan 2006, p. 229), Iraq is
believed to have about 153 billion barrels. This quantity of reserves puts Iraq in
fourth place in the world behind Iran (estimated proven reserves of 156 billion
barrels), Saudi Arabia (267 billion barrels), and in first place, Venezuela (303
billion barrels) (GoI 2018, p. 135; IMF 2019b, Figure 1, p. 23). At current rates
of production, Iraq's oil will last for almost 50 years.

By 2015, only about 50 percent of the country had been explored for oil
or gas (Vogler 2017, p. 237). In addition, the existing estimates of oil and
gas reserves are outdated, based on obsolete technology. Ongoing attempts
to more accurately estimate Iraq's reserves may result in a 50 to 250 billion
barrels increase in the country's proven reserves. If the adjustment adds over
160 billion barrels to Iraq's proven reserves, Iraq will supplant Venezuela as
the country with the largest proven petroleum reserves. Along, with the United
States and Brazil, Iraq has the potential of being one of the major contributors
to oil production growth over the next decade (IEA 2019, p. 8).

Iraq's crude is relatively inexpensive to get out of the ground. It is not only
near the surface but also generally concentrated in large fields. Iraq has nine

"supergiant" (over 5 billion barrels) fields and 22 "giant" (between 1 and 5 billion barrels) fields. As a result, along with Saudi Arabia, Iraq has some of the lowest production costs in the world. With reasonable efficiency, Iraq can break-even exporting oil at a world price of only $12 per barrel.

Crude oils are compared and described by density and sulfur content, with the most desirable crudes having a low density (light) as well as low sulfur content (sweet). Light crudes are cheaper to distill into more valuable fuels, while excess sulfur must be removed to prevent pollution and equipment corrosion. As a result, the market pays a premium for light-sweet crudes. The standard for density is that of the American Petroleum Institute (API) which lists light crudes as °API 35 to 45, and average crudes as 25 to 35. With respect to sulfur, sweet crudes have less than 1 percent sulfur by weight (Hyne 2001, pp. 4–5).

Iraq's two major export crudes are both somewhat sour, but they are on the border between average and the desirable light density. Basrah light crude is listed at °API 34 while Kirkuk crude has a density of °API 36. Both have about 2 percent sulfur contamination.

This is important because many of the world's sources of light crudes such as those of Norway, the United Kingdom, and Algeria have already been substantially exhausted, while new discoveries such as those of Brazil and Canada tend to be very heavy and very sour. Iraq is therefore in an enviable position of not only having the potential of increasing its petroleum production and exports, but also having a relatively desirable crude to sell.

This desirable product is unevenly distributed across Iraq. In fact, one can describe much of Iraq's troubled past and possible future in two data-laden figures: the map of Iraq's oil and gas infrastructure in Figure 7.1, and the relationship among Iraq's production, exports, and world oil prices in Figure 7.2. These figures provide insights into not only the country's economy but also its politics. Both deserve close study.

Oil and gas reserves are primarily found in three basins. The Northern Zagros Fold Belt includes the supergiant Kirkuk field as well as smaller fields in the Kurdish Regional Government (KRG). In the middle and south of Iraq is the Mesopotamian Basin. So far, the largest fields have been found in the southern part of this basin near Basrah. Finally, in western Iraq is the Widyan Basin (Interior) located primarily in Anbar Province. This is the least-explored basin and is thought to contain mostly gas (IEA 2019, pp. 22–23).

As we can see in Figure 7.1, the most productive oilfields are primarily in two provinces: Kirkuk and Basrah. The greatest reserves include the supergiant Kirkuk oilfield (estimated 8.6 billion barrels in reserves) with Iraq's other supergiant oilfields in Basrah province which include the Rumaila field (over 17 billion barrels) and West Qumac (8.7 billion barrels). In fact, Basrah

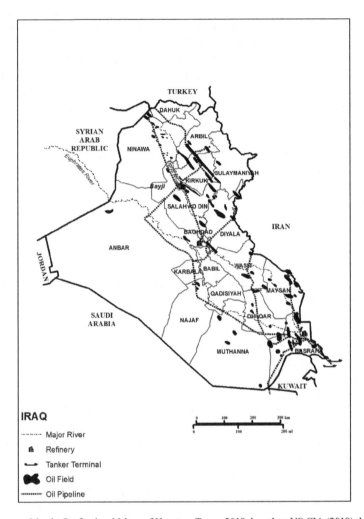

Source: Map by Professional Maps of Houston, Texas, 2019, based on US CIA (2019) data.

Figure 7.1   Oil and gas

and Kirkuk provinces account for almost 94 percent of the nation's petroleum production.

     While the developed oil reserves with currently producing wells are concentrated in two provinces, undeveloped reserves are much more widely distributed throughout Iraq. There are substantial undeveloped reserves in Maysan province (estimated 8.5 billion barrels), Baghdad province (6.5 billion barrels),

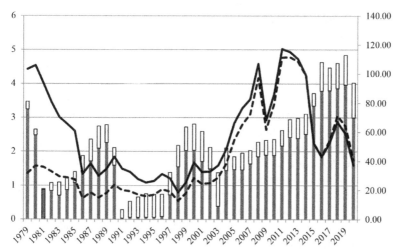

Note: Columns represent total annual oil production in millions of barrels of oil per day (mbpd, left axis). The black lower portion of each column shows exports while the white upper portion shows oil processed within Iraq. Lines represent the Brent Blend oil price ($ per barrel, right axis). The heavy solid line is the real, inflation-adjusted, price of oil while the dashed line is the nominal price.

Source: Oil production, exports, and nominal Brent Blend oil price: EIA (2020). Real oil price calculation by author.

Figure 7.2   Oil production, exports, and process

and Dhi Qar province (5.1 billion barrels). In fact, only six provinces have less than 100 million barrels each of undeveloped petroleum reserves. Even Anbar province with few oil reserves has substantial natural gas in the Akkas field (IEA 2019, p. 37).

The country's earnings from crude oil exports are dependent, of course, on both the world price of oil and the volume of oil exported. In Figure 7.2, the lines show the real (solid line) and nominal (dashed line) Brent Blend index of oil prices (Iraq earns about 90 percent of the Brent Blend price), and the columns show the total volume of oil either exported or processed domestically. Note that the solid line shows oil prices in 2015 dollars, adjusted for inflation. When both prices and export volumes are high, such as in 1979, 2008, and 2011, the nation's crude oil export earnings account for a very large proportion of GDP. For example, in 2008, with average crude exports of 1.82 million barrels per day (mbpd) and an average price of $91.50 per barrel (pb), Iraq earned almost $74 billion, equal to almost 86 percent of gross domestic product (GDP). However, in 2009, although export volumes increased to 1.88 mbpd, the drop in world oil prices to $55.60 pb resulted in a drop of oil earnings to $45.6 billion which contributed to a 25 percent fall in real GDP (IMF 2011a, p. 8 and Table 1, p. 16).

Unlike most oil exporting nations, Iraq will not reach its "peak oil" point for at least three decades. Therefore, one would expect that improvements of technology and increased infrastructure investment would have led to a steady increase in the volume of the nation's crude exports. But as Figure 7.2 illustrates, the volumes of Iraq's crude production and exports from 1979 through 2007 were extremely uneven, since they were determined by conflict, not technology or investment. Following the "golden year" of 1979, crude oil production and exports collapsed in the early 1980s as a result of the Iran–Iraq War, in the 1990s as a result of Saddam's invasion of Kuwait, and in the 2000s as a result of the invasion by the United States-led coalition and the accompanying insurgency. It is only with the reduction in al Qaeda violence in 2007 that Iraq's oil production and exports returned to "normal" times.

Since then, despite the ISIS insurgency that at its peak controlled about 40 percent of Iraq's territory, the volume of Iraq's oil production and exports increased sharply, with total oil production doubling between 2008 and 2019 when it reached 4.8 mbpd. Oil exports during the same period increased by 114 percent to 4.0 mbpd. However, despite this dramatic increase in the volume of oil exports, total export revenue increased by only 16 percent to an estimated $86 billion, as a result of the decrease in oil prices (Brent Blend adjusted for inflation) from $107.07 in 2008 to $59.21 in 2019.

In summary, over the last two decades Iraq has increased its export volume, even though, as is argued below, a rapid increase in oil production may not have been optimal. However, in the short run, Iraq's national income is driven by changes in world oil prices, not by increases in oil production and export volumes.

This truth is illustrated by the interruption of the trend of increasing oil production and exports in 2020. In response to the collapse in oil prices, driven in part by the worldwide Covid-19 recession, the Organization of the Petroleum Exporting Countries (OPEC) and a few other countries agreed to a sharp decrease in oil exports. Consistent with this decision, Iraq oil exports decreased by about 1 mbpd to an estimated 3 mbpd in 2020. Since domestic demand and storage were limited, the decrease in exports required a substantial decrease in total oil production from 4.84 mbpd in 2019 to an estimated 4 mbpd in 2020.

Oil Infrastructure

Extracting and processing oil for either export or domestic refining requires an extremely complex infrastructure. After the crude oil is extracted from the wells, it is pumped to a gas–oil separating plant to remove any associated gas, water, and salts from the crude. Much of the associated natural gas is wastefully burnt – flared off – since most Iraqi fields lack the infrastructure to process, store, or transport the gas. The water may be either the result of a natural water migration into an oilfield as field pressure drops, or it may result from water deliberately forced into a field to facilitate oil recovery. The salts must be removed to avoid excessive corrosion of pipes, pumps, and so on. The next step depends on whether the oil has measurable amounts of the very corrosive and poisonous gas hydrogen sulfide (H_2S). If the oil contains this dangerous gas with its rotten egg smell, then it must be sent to a stabilization plant to remove the H_2S before it can be delivered to a pipeline. Finally, the crude will be pumped to an Iraqi refinery, or to Turkey through the Kirkuk–Ceyhan pipeline, or to crude carriers at the oil terminals on the Persian Gulf (IEA 2019, pp. 26–30; Hyne 2001, pp. 10–11, 409–10).

The major links in oil production, refining, and exports can be seen in Figure 7.1. About 90 percent of Iraq's oil exports comes from the oilfields in the southeast of the country and is pumped aboard crude carriers in the ports south of Basrah City to be shipped through the Persian Gulf. The remaining 10 percent is exported through a pipeline from the supergiant oilfield in Kirkuk province to Turkey. There, crude from the KRG is piped west to the Mediterranean port of Ceyhan.

Also important is the strategic pipeline from Basrah to Kirkuk that allows crude from the southeast to be exported to Turkey, or crude from Kirkuk to be exported from Basrah. The other major pipelines to Syria and Kuwait are currently not in commission as a result of lack of maintenance, sabotage, and political disputes. While the process from oil production to refining or export is not as time-sensitive as that of electricity generation (see Chapter 12), storage is limited, capable of storing only about a week's production (GoI 2018, p. 135). As a result, any "break" in the process can lead to shutdowns up and down the supply chain. This supply chain fragility is illustrated with three recent examples.

First, storms in the Persian Gulf can substantially slow or stop the loading of crude carriers and result in rapidly filling onshore storage facilities. If the Basrah–Kirkuk and Kirkuk–Ceyhan pipelines are near capacity, it is necessary to cease pumping oil at the wells in the Basrah fields until the storms end and loading can resume.

Second, Iraqi refineries are inefficient. Refineries use heat and chemical processes to break crude into valuable fuels such as gasoline and diesel (gas oil). Iraqi refineries tend to produce substantially greater proportions of heavy fuel oil (HFO) than foreign refineries using the same crude feedstock. In 2018, HFO accounted for an estimated 50 percent of Iraqi refinery production, while higher-value products – gasoline, diesel, and kerosene – are about one-third. The global average is 10 percent HFO and 70 percent higher-value products

(IEA 2019, p. 30). Excessive production of HFO has led to the periodic shut-downs of refineries as they reach their HFO storage capacity.

Since it must be heated to flow, HFO is difficult to transport. In addition, since burning HFO tends to be very polluting, the International Maritime Organization beginning in 2020 has imposed limits on sulfur content that will reduce export demand for HFO. Using it domestically for electricity genera-tion tends to increase both maintenance costs and air pollution. Previously, Iraq tried to get rid of HFO by adding it to crude oil intended for export. Of course, this blending of HFO with export crude reduces the market price of the crude.

Finally, there are continuing difficulties coordinating the activities of the Ministries of Oil and Electricity. The Ministry of Electricity argues that inade-quate fuel quality and quantity reduce electrical generation, while the Ministry of Oil states that inadequate electrical quality (frequency drops) and quantity constrain oil production and refining.

Based on oil viscosity, the permeability of the reservoir, and pressure, only a fraction of the oil in a field can be recovered under normal conditions. This is referred to as the recovery factor (Hyne 2001, pp. 431–2). In Iraq, these conditions limit the quantity of oil that can actually be recovered from oilfields to 50 percent to 70 percent of the total reserves. In addition, there is an optimal range of oilfield exploitation that maximizes the recovery factor. If the rate of production of a field is above this range, it is likely that there will be a perma-nent decrease in the total quantity of oil that can be extracted without the use of aggressive techniques to increase oil flow (Vogler 2017, pp. 227–8).

Unfortunately for Iraq's ability to extract the maximum amount of oil, it has had to alternate periods of aggressive exploitation of its primary oilfields with periods of suboptimal production. The story is shown in Figure 7.2. Actual oil production peaked in 1979, before declining sharply during Saddam's war with Iran (1980–88). During this conflict, Iraq was forced to abandon production from oilfields in southeast Iraq, and many of these oilfields were not properly prepared for shutdown or the surface equipment was destroyed. At the same time, Iraq attempted to maximize short-term production from those oilfields that were more distant from the Iranian front. The destruc-tion sustained in the Desert Shield and Desert Storm operations of the early 1990s, combined with the lack of an export market before the United Nations Oil-for-Food Programme, and the post-2003 looting, sabotage, mismanage-ment, and corruption, have all kept production from some oilfields low while others were pumped at rates high enough to reduce the long-term recovery rates, especially at the older oilfields near the Iraq–Iran border (Cordesman and al-Rodhan 2006, pp. 225–7).

In addition, the oil and gas infrastructure in Iraq, from wellhead to sea ter-minal or cross-border pipeline, is in a bad state of repair as result of delayed maintenance, looting, and deliberate destruction by al Qaeda (2005–07) and ISIS (2014–17). Probably more costly than the destruction of the physical oil and gas infrastructure was the loss of human capital, as skilled managers and technicians in energy fields were threatened, fired, or driven to quit their jobs because they belonged to the wrong ethnic group or party, or refused to enthusiastically support corruption, or to free up jobs for supporters of the dominant group. As an example, for a period of time, al Qaeda made the decisions on who should be hired and promoted at the Bayji refinery. Some managers fled the country to avoid retaliation against themselves or their families (Vogler 2017, Chapter 16, pp. 184–205).

In the short term, the binding constraint on substantially increasing Iraq's oil exports is the limited capacity of oil pipelines, tanker moorings, and storage facilities in Basrah province. Over the last decade, the Government of Iraq (GoI) spent an estimated $8.4 billion project to build new storage facilities on the Faw peninsula, three new pipelines to the water's edge, and four floating terminals in the Gulf to speed the filling of large oil tankers (Hafidh 2011; Vogler 2017, Chapter 17, pp. 206–23).

One of the most serious remaining challenges is determining the true condition of the old and poorly maintained dual 48-inch pipelines from the Basrah shore to the Al Basra Oil Terminal (ABOT) in the Gulf. If these pipelines fail, Iraq loses a great deal of its export capacity. Experts believe that the steel in these pipes should have failed by now, and yet there is no evidence of leakage. Possibly, sludge in the pipes combined with the fact that the mud enclosing the pipes is under great pressure from the weight of the water above, has prevented any large leaks even after corrosion has weakened or penetrated the steel pipes. To prevent further damage, this pipeline is operating with reduced pressure and therefore less capacity (Vogler 2017, pp. 209–12, 217–18). The authorities dare not perform a physical inspection of the pipes' interiors because of the fear that the inspection equipment might rupture the corroded pipe.

As expected, attempts to rapidly expand the oil infrastructure over the last decade led to severe bottlenecks as increased imports of equipment and supplies overwhelmed existing port facilities. In addition, political groups, unions, tribes, and government officials have sought to profit from the urgent activities by demanding "cooperation payments" or bribes.

In the long term, the most serious challenge to expanding oil production is the shortage of clean low-salt water. Depending on the viscosity of the oil and the nature of the oil reservoir, it can require one to two barrels of water to produce each barrel of oil. Most of this water is injected into Iraqi oilfields in order to maintain pressure and therefore production levels. Depending on the oilfield, most of this water might be recovered during oil production and reused. However, there remains a large net demand for water to maintain – much less increase – oil production. This water has to be filtered and deoxygenated before it is used, to reduce pipe and pump corrosion. If water from the Persian Gulf is to be used, it must also be desalinated. The cost of filtering, de-oxygenating, or desalination depends on how pure the water was originally. Therefore, oilfield operators prefer to use clean – potable – water to reduce their costs. A related issue is the disposal of contaminated water. Water used in oil recovery is often severely contaminated and will pollute river or ground water if it is dumped.

As discussed in Chapter 8, official water prices in Iraq are low or zero and, as a result, water is distributed by political fiat rather than by a market. At the same time, existing environmental laws are poorly written and rarely enforced. The national oil companies have taken advantage of this situation by diverting clean water for use in the oilfields, and casually disposing of contaminated water. In view of the growing shortage of clean water for family use and agriculture, increased water demand for oil production will be opposed by households, agriculture, and other industries.

In order to meet the need for water, the GoI began the Common Seawater Supply Project (CSSP) in 2011. This project was intended to eventually provide 7.5 mbpd of desalinated seawater to be used in the southern oil fields. Originally, the CSSP was to be completed in three years, but contract disputes, security issues, corruption, and bureaucratic delays mean that it is still in construction. Optimistically, the first phase for 5 mbpd of desalinated seawater will be completed in 2022, eight years later than originally planned (IEA 2019, pp. 24–5).

The long and convoluted effort to create the CSSP exemplifies a serious barrier to future development of Iraq's oil industry, as well as to the non-oil infrastructure investment discussed in Chapter 12. When faced with decisions, bureaucrats face unbalanced incentives when dealing with the international oil companies. If a government official makes a quick decision, it is unlikely that the official will directly benefit. However, if an official makes any decision concerning a financial commitment with a foreign entity, they are extremely vulnerable to having their decision second-guessed and possibly overturned. In addition, having made a decision that may be considered favorable to a foreign entity, they run the high risk of being accused of corruption or mismanage-ment. In view of these possibilities, bureaucrats tend to avoid making even relatively low-level decisions in a timely manner. Instead, such decisions are sent up layer by layer in the organization for the minister's personal approval. The opportunity costs of these delays to the GoI and/or foreign partners are ignored (Vogler 2017, pp. 218–19).

Natural Gas

In addition to oil, Iraq also has huge reserves of natural gas. Natural gas is mostly methane and, unless treated, is odorless and invisible. About 12 percent of Iraq's gas (5.6 trillion cubic feet or tcf) is found in four major fields while the remaining 88 percent (42.2 tcf) is associated gas. Associated gas is mixed with crude oil, and when the oil is pumped to the surface the associated gas evaporates. Since this vapor is explosive, it must be either captured or flared off – that is, wastefully burned. As expected, the proportion of flared gas from strictly gas fields is relatively low; most of the flared gas is associated with oil production. Depending on the field, an estimated 40–60 percent of all associated gas is flared, with much of the remainder re-injected into oil wells to maintain the pressure needed to force the oil to the surface (IEA 2019, Figure 13, p. 37). The wasted flared gas is enough to fuel about 4.5 gigawatts of electricity generation, which could power 3 million homes (IEA 2019, Box 2, p. 36). There are only two options to reduce the wasteful flaring of natural gas.

The first option is to take the gas to the user. Because of its characteristics, there are only two reasonably inexpensive methods of transporting gas: by pipeline or by ship. Pipelines must be dedicated to gas transportation; one cannot use a fuel or crude oil pipeline without extensive modification. Prior to the 1990–91 Gulf War, Iraq exported natural gas to Kuwait through a 170 km (105 mile) long pipeline to Ahmadi, Kuwait. However, this gas pipeline is currently shut down due to damage from conflict, sabotage, and looting. In addition, renewed gas exports would also require the passage of a petroleum law. There has also been discussion of an Arab Gas Pipeline project that would transport gas from the Akkas field (a 2.1 tcf field in Anbar province) to Syria. However, because of the ongoing conflict in Syria, this project is stalled. Exporting gas by ship requires special port equipment as well as specialized ships, since to liquefy gas requires a temperature of −162°C (−259°F).

The alternative option is to take the user to the gas, that is, to locate large users of natural gas near the fields. Since 40–60 percent of associated gas is being flared off, the shadow price of energy from gas is almost zero. This would provide a strong comparative advantage for the production of products such as electricity, aluminum, or cement that are energy-intensive. To take cement as an example, its manufacture requires the heating of raw materials (limestone, clay, additives) to about 980°C (1800°F). In Western Europe, energy accounts for almost one-third of the total cost of cement production. Unfortunately, there have been few successful attempts to locate energy using factories near to where gas is flared. In Iraq, I saw a cement plant using refined fuel as an energy source while, across the road, gas was wastefully flared off.

Ironically, because of the lack of the necessary infrastructure to process and transport natural gas from the oil fields to end users in Iraq, the country simultaneously flares off natural gas while importing large quantities of gas from Iran. For example, a power plant in Mansuriyah near the Iranian border was supposed to receive gas from a nearby oilfield. However, as a result of construction delays, the power plant runs on imported Iranian gas, while gas from the nearby field is flared (IEA 2019, p. 34). In addition to the waste involved, paying for the imported Iranian gas is a major budget expense.

Of course, the flaring of natural gas along with methane vented during oil production has adverse environmental effects. Currently, Iraq has the second-worst volume of carbon dioxide (CO_2) emissions – after Venezuela – of the major oil producers. In addition to the impact on public health, there is an increasingly serious reputational effect. Iraq is increasingly viewed as seriously damaging the world environment. However, if Iraq could sharply reduce flaring and methane emissions using existing technology, then the country has the potential of the lowest emissions in the world (IEA 2019, Box 2, p. 36).

OIL LAW CONTROVERSIES

Production, transportation, domestic refining, and export of petroleum products are complicated by the absence of governing laws and regulations for the oil and gas industries. In the continuing controversy over Iraq's proposed Oil Law, much attention has been paid to constitutional ambiguities over both the right to sign oil contracts with foreign companies and the divisions of oil export revenues between national and provincial governments. However, the debate over the oil clauses of the Constitution is more important than simply resolving textual uncertainties. To the provincial, regional, and national governments, decisions about control of production, contracts with foreign companies, and the division of oil revenues are existential: critical and irrevocable determinants of the future welfare and political power of the negotiating parties. Provinces or regions that "win" the negotiations can look forward to increasing prosperity for the foreseeable future. "Losers" will increasingly fall behind. It has become a battle that is too important for any party to even consider substantial compromise.

Any paraphrase of the relevant portions of the 2005 Constitution (GoI 2005a) would add to the distortion caused by translation. The key articles are in Section Four, Powers of the Federal Government: Articles 111, 112, and 115:

> Article 111: Oil and gas are owned by all the people of Iraq in all the regions and governorates.
> Article 112: First: The federal government, with the producing governorates and regional governments, shall undertake the management of oil and gas extracted from present fields, provided that it distributes its revenues in a fair manner in proportion to the population distribution in all parts of the country, specifying an allotment for

a specified period for the damaged regions which were unjustly deprived of them by the former regime, and the regions that were damaged afterwards in a way that ensures balanced development in different areas of the country, and this shall be regulated by a law.

Second: The federal government, with the producing regional and governorate governments, shall together formulate the necessary strategic policies to develop the oil and gas wealth in a way that achieves the highest benefit to the Iraqi people using the most advanced techniques of the market principles and encouraging investment. Article 115: All the powers not stipulated in the exclusive powers of the federal government belong to the authorities of the regions and governorates that are not organized in a region. With regard to other powers shared between the federal government and the regional government, priority shall be given to the law of the regions and governorates not organized in a region in case of dispute.

With respect to Article 111, "governorate" means the same as "province", while regions reflect the constitutional right for two or more governorates/provinces to join together as a governing body (see GoI 2005a, Section Five, Articles 117 and 119). As of mid-2020, only three of Iraq's 18 provinces have agreed to form a region. This is, of course, the KRG in northern Iraq. The importance of Article 111 is in the statement that it refers to "all of the regions and governorates", not just those that are currently producing oil or gas or are expected to produce them in the future. This is in contrast to Sections 112 and 115 that explicitly (Article 112) or implicitly (Article 115) refer to producing regions and governorates.

Among the ambiguities of Articles 111, 112, and 115 are the following. Both the producing regions/provinces and the federal government are assigned responsibility for the "management" of existing fields. But the incentives of sub-national political bodies differ greatly from those of the national government with respect to sequencing of production, expense reimbursement, and contracting. Sub-national authorities generally want the production from their areas to be first in line for development, they want generous definitions of allowable costs that would be subtracted from export earnings before the remainder would be transferred to the national treasury, and they want the authority to negotiate directly with foreign companies. The national government wants a national oil and gas development strategy, a narrow definition of allowable costs, and national negotiation of all contracts with foreign entities.

Another conflict exists between oil producing regions/provinces and the other provinces. As noted above, the provinces of Basrah and Kirkuk account for almost 94 percent of the nation's petroleum production. But Article 111 states that: "Oil and gas are owned by all the people of Iraq". Producing provinces want the bulk of domestic and foreign investment – and the accompanying employment – to be concentrated near existing fields, arguing that such a concentration will lead to the most rapid increase in exports and national income. These provinces also argue that efficiency would be increased if the production of supporting infrastructure, including everything from cement and pipe manufacturing to training for oil workers, was located near the producing fields. Non-producing provinces want a substantial portion of investment to be in new fields to more evenly distribute the gains from petroleum. For the same reason, they want the production of supporting infrastructure to be more widely distributed.

One interpretation is that the management of fields that are currently in production will be shared between producing regions or provinces and the national authorities. On the other hand, new fields will be the sole responsibility of the national authorities. As expected, this has led to arguments concerning fields that were previously productive but were closed because of conflict or the former oil embargo. If these fields are returned to production, are they considered new fields or not? Also, there are disagreements over whether fields that have been professionally evaluated including the drilling of test wells, but have not begun commercial production, should be considered as currently producing or new fields.

The final, and probably most serious, dispute concerns the use of the export earnings. While there was some discussion during the writing of the Constitution about establishing a sovereign wealth fund (SWF) – providing annual payments directly to all eligible Iraqis from oil export earnings – this was never really a serious option. Instead, the Constitution in Article 112 states three priorities for the distribution of oil revenues. These are: (1) revenues will be distributed on the basis of population; (2) with a temporary allotment for regions that were severely damaged during Saddam's regime; and (3) another temporary allotment for regions that were damaged during the 2005–07 al Qaeda insurgency. The intent is to balance development across the nation.

Immediately after the passage of the Constitution, in the absence of any reliable estimates of reconstruction costs or even accurate population figures for each province, the GoI responded to these priorities by budgeting $1 billion to be allocated to provinces on the basis of estimated population – with a 17 percent allotment to the Kurdish region – and an additional $1 billion for reconstruction in the provinces that were perceived to have suffered the most under Saddam and during the al Qaeda insurgency. As expected, this decision left all of the provinces complaining that either their population or their reconstruction needs (or both) had been severely underestimated.

In order to resolve (or possibly perpetuate) these conflicts and ambiguities, four major pieces of petroleum legislation have been under consideration by the GoI since 2007. These are: (1) the basic oil and gas framework law; (2) the revenue-sharing law; (3) the law reorganizing the Iraqi Ministry of Oil; and (4) the law reconstituting the Iraqi National Oil Company (INOC). (For a detailed discussion of these laws and their extremely complex interactions and implications, see the excellent analysis by Zedalis 2009, especially Chapter 3).

As of mid-2020, only the law re-establishing the INOC appears to be close to approval by the Council of Representatives (CoR) (GoI 2020a). The original INOC existed from 1966 until 1987 when it was merged with the Ministry of Oil (MoO). The new INOC, if approved by the CoR, will control nine existing companies involved with oil exploration, drilling, production, transportation, and the State Organization for Marketing of Oil (SOMO) (Reuters 2018). With the transfer of these companies to the INOC, the MoO will focus on strategic planning and regulatory issues.

Pending the final approval of the basic oil and gas framework law, both the KRG and the national government have signed contracts with foreign companies for petroleum exploration and production. Periodically, the GoI has publicly questioned the legality of contracts signed by the KRG. It is widely expected that the final passage of the basic oil and gas framework law will put an end to these controversies. However, a more realistic view is that the national, regional, and provincial governments can be expected to continue to seek advantage by advocating conflicting interpretations of the basic law that will have to be gradually resolved, possibly by judicial interpretation but more likely by political compromise. While the political sector continues these debates, Iraq continues to legally and illegally produce, refine, and sell refined oil products in domestic and foreign markets.

DOMESTIC REFINING AND FUEL SALES

Once Iraq's oil has been pumped from the ground and, if necessary, excess sulfur removed, then there are three possible destinations. As Figure 7.2 shows, much of the country's officially estimated oil production is exported; the remaining crude oil is transported by pipeline or truck either to be burned for electrical generation or to one of the domestic refineries to be processed into more valuable fuels. Refinery capacity is currently insufficient to meet Iraq's fuel requirements. In part this is a function of inefficiency in the refineries themselves as well as in the supply and product chains associated with each refinery. As an example of the latter, the author toured a modern refinery near Sulaymaniyah and was surprised to discover that both the crude oil inputs and refined fuel outputs were transported not by pipelines but much less efficiently by tanker trucks.

The refinery situation worsened as a result of deliberate destruction by ISIS between 2014 and 2017, followed by looting when ISIS was defeated. Prior to 2013, the country's 14 refineries had a capacity of 945 000 barrels per day (kbpd), which was more than sufficient for domestic needs and actually allowed a small volume of fuel exports. However, by late 2017, all seven refineries located in the north of the country were damaged or destroyed. As a result, refinery capacity was cut in half to 465 kbpd (GoI 2018, p. 136). As a result, oil-rich Iraq imports \$2–\$2.5 billion of fuel, mostly from Iran (IEA 2019, p. 29).

The biggest setback occurred with the 2014 destruction of the Bayji Refinery in Salah Ad Din on the north–south strategic pipeline. This refinery, formerly the largest in Iraq with a nameplate capacity of 310 kbpd, one-third of national capacity, was devastated in 2014 by ISIS and then looted again by militia forces in 2015 after ISIS had been driven out (Vogler 2017, pp. 276–7). It was only in 2018 that the refinery resumed partial operation. Currently, Iraq can still only refine about 60 percent of its pre-ISIS volume.

What is missed in the official estimates of the production of crude oil and refined fuels is the large-scale divergence of crude and fuels into the black or underground market. This has made it almost impossible to reconcile physical flows of crude and processed oil with the resulting financial transactions. The divergence of oil into the black market may occur at many different points between the wellhead and the ultimate consumer. Crude oil is smuggled over the Iranian and Turkish borders by truck, as is refined fuel from Iraqi refineries. In addition, some of this refined fuel appears in the domestic black market: directly from the refineries, or diverted during transportation, or out the "back door" of official fuel retailers. There have been cases where GoI fuel refined at Iraqi refineries is smuggled into Iran and then sold back to the GoI by Iranian middlemen.

The booming business in black market crude and fuel is driven by country-wide fuel shortages and facilitated by a general lack of crude and fuel monitoring and auditing. While the dramatic rise in official fuel prices between 2005 and 2007 severely reduced the profits possible from black market diversion, the political unpopularity of these increases has resulted in an unwillingness of the GoI to allow official prices to rise enough to set amounts demanded of gasoline and diesel (gas oil) equal to amounts supplied in urban markets. This provides profit opportunities to anyone who can redirect fuel from official sources; and consumers are willing to pay a high price for black market fuel when official sources run out.

Diverting crude or fuel is simplified by the lack of reliable basic metering. Pipeline meters are either missing or miscalibrated, while cargoes of fuel trucks are often either estimated or roughly calculated by dipping a marked stick into the tank. There are common reports of the drivers or others selling fuel and then diluting the remainder to maintain the expected volume. Progress is being made, with all exports through the Basrah–Kirkuk and the Kirkuk–Ceyhan pipelines finally being metered. However, completing the metering of domestic fuel usage is expected to take several more years. Of course, the supply of crude and fuel to the black market is facilitated by the widespread corruption.

In view of the very inefficient public sector refining, transportation, and sale of fuel, should Iraq engage in any of these activities? A 2008 study estimated that while the Iraqi Treasury earned – net of costs – approximately $37 per barrel of crude exports, it received only about $2 for each barrel of crude that was refined and either sold to state-owned enterprises (SOEs) or sold by official dealers to customers. If these estimates are still even roughly accurate then the Treasury and the national economy would be better off simply exporting all crude and importing all the fuel that the country needs.

There are two counterarguments. First, the cost estimates were made when Iraq was still recovering from large-scale insurgent attacks on fuel refining and transportation. With a restoration of relative peace, it is expected that security-related costs will decline, and refinery efficiency will increase. Second, there are no technical reasons why Iraq cannot dramatically increase its efficiency in refining, transporting, and selling fuel. Like much of the oil industry, refineries suffered from foregone maintenance and severe parts shortages as a result of the insurgencies of 2005–07 and 2014–17. If refineries are repaired and necessary upgrades, especially increasing desulphurization capacity, are made then there should be not only an increase in refinery throughput but also a decrease in the production of low-value HFO.

THE OPTIMAL RATE OF EXPLOITATION

The challenge of Iraq's oil future is to exploit its tremendous reserves so as to maximize the long-term benefit to its people. Dealing with this challenge raises theoretical, technical, policy, and legal issues that have divided Iraq's polity.

In late 2010, the GoI announced an extremely aggressive medium-term production plan for oil. National oil production, which was an estimated 2.63 mbpd in 2011, was to increase to 12.20 mbpd by 2017; a 360 percent increase in six years. After further review, and possibly in response to International Monetary Fund (IMF) criticism, in late 2011 the GoI revised the 2017 target for oil production down to 10.0 mbpd. The IMF criticized this revised plan as overly optimistic and estimated that Iraq might be able to achieve 5.35 mbpd by the end of 2017 (IMF 2011a, Box 1, p. 6). Of course, as can be seen in Figure 7.2, all of these estimates turned out to be very wrong. Total oil production in 2019 before the OPEC-mandated production reduction was only 4.8 mbpd.

It has been argued that Iraq's plans for a large increase in oil production were defeated by two unpredictable occurrences. First, the decline in oil prices that began in 2012 and accelerated in 2014 resulted in a substantial drop in government revenue from oil exports. As will be discussed in Chapter 14, government investment is the "shock absorber" of the Iraqi economy.

The decline in government revenue led to a sharp decrease in the necessary infrastructure investment needed to expand oil production. Second, the war against ISIS from 2014 to 2017 had both direct and indirect effects. Directly, ISIS destruction of the pipelines, the Bayji refinery, and several small oil fields reduced both the production of crude oil and refinery throughput. Indirectly, increased GoI spending to fight ISIS further reduced the funds available for oil and gas-related investments.

However, even if the necessary funds were available, there are two arguments that Iraq should not adopt an aggressive strategy to rapidly increase oil production and exports. First, technical constraints may make it difficult to achieve GoI oil production goals at a reasonable cost. These constraints have been the subject of extensive studies. But the second argument, that a slower rate of crude oil production is desirable even if a more rapid rate is technically possible, has received much less attention. These two arguments will be referred to as the technical constraints and market constraints arguments.

Technical Constraints

A fundamental truth of project management is that of the three desirable characteristics of a project – good, fast, and cheap – you can only choose two. If Iraq seeks to rapidly increase its oil exports, its cost per barrel will rise sharply for several reasons.

First, domestic producers of the necessary equipment and supplies for expanding oil and gas production are either non-existent or are very inefficient. Almost all of this domestic capability is in SOEs that, as discussed in Chapter 10, will require substantial improvements in managerial incentives for any substantial efficiency gains to occur. Second, attempts to import necessary equipment and supplies are already running into bottlenecks. The cargo throughput of the Basrah ports is still restricted by damage from conflict, severe mismanagement, and labor disruptions. As a result, Iraq's port of Umm Qasr is periodically inundated with imports of oil industry equipment, and getting needed supplies off ships and onto the roads has become a difficult challenge. There are options for speeding up delivery of needed materials but, consistent with the fundamental truth, such work-around methods tend to either be more expensive or result in lowered quality. For example, some oil companies are attempting to avoid port congestion by transporting equipment overland, by arranging for air transport, or by looking the other way while their agents bribe government officials to speed up processing.

Third, aside from the difficulty of obtaining the necessary equipment and supplies either from domestic producers or as imports, technological constraints must be faced. It takes time to efficiently build pipelines, oil–gas separation plants, or even roads. Iraq combines a high unemployment rate with a severe shortage of skilled labor. There are large numbers of unskilled workers, but electricians, plumbers, and other types of skilled workers are in short supply. Iraq must either train the needed skilled workers or hire expatriate electricians, plumbers, and so on.

Finally, and most importantly, there is a shortage of necessary managerial human capital. In particular, Iraq has very few individuals who are currently capable of efficiently managing billion-dollar projects. And the few individuals qualified to manage smaller projects are severely overtasked. To efficiently manage a billion-dollar project requires not only a first-class education but also 12–20 years of "on the job" experience with regularly increasing responsibilities and quality mentoring on complex projects. During the 30 years that Saddam's Iraq was substantially cut off from the world, the quality of higher education, particularly in the engineering and management fields, deteriorated. The socialist economy of the Saddam era rewarded managers more for their political capability than for their ability to achieve technological or market efficiencies. If a rapid expansion of crude oil production and exports is attempted without possessing quality management to oversee the process, then the country can expect huge waste, substantial delays, and severe corruption.

Any attempts to overcome the shortage of domestic big-project managers by a "temporary" reliance on expatriate managers will not only be expensive but also raises other issues. Who will manage the managers? There is a principal–agent problem, since the motivations of the GoI and expatriate managers are different. It is difficult or impossible to draft large project management contracts for expatriate managers and ensure their compliance with these contracts without substantial technical and management capability. With a few exceptions, these capabilities are lacking at the upper levels of the GoI. Retaining foreign consultants to write and monitor agreements with expatriate managers, or breaking the project into easier-to-understand pieces, are options but these just push back the question by one more level. Who has the skills to contract and monitor the consultants who have been hired to contract and monitor the expatriate managers? Or who knows enough about the large project to know how to break it down into manageable pieces?

In view of the technical constraints, how rapid an expansion of the country's crude oil production is possible? A reasonable estimate is that Iraq could produce 6.0–6.5 mbpd by 2035 (IEA 2019, p. 7; GoI 2018, p. 137). However, even this conservative plan will require substantial expansion of the oil industry's infrastructure. Most critical is completion of the CSSP and similar projects to provide the necessary water to pressurize existing oilfields. In addition, there is need for expensive maintenance of the Al Basra Oil Terminal, pipelines, and a great expansion of oil storage facilities. While this conservative scenario would still result in almost a 50 percent increase of 2035 production compared to that of 2019, it will be painful for the Iraqi people to accept the necessity of a reduced rate of increase of crude oil production. The expectation of an oil-driven increase in the living standard of the average Iraqi has both fueled regional disputes and provided a reason for optimism for the average Iraqi. It will be difficult for elected politicians to try to convince the Iraqi public that a slower growth rate is optimal for technical reasons, not to mention the impact of a rapid expansion of oil exports on the world oil market.

Market Constraints

There are two general forms of oil production agreements. In a production sharing agreement (PSA) an oil company agrees to bear the exploration and production costs in return for receiving a fixed percentage of production. After a contracted period of time, usually five years, the oil company transfers ownership of the field back to the government or mineral rights owner. Under a technical services agreement (TSA), an oil company performs exploration and production for a flat fee. And the fee is only paid if production takes place. As an example of a TSA, the supergiant Rumaila oilfield produced about 1 mbpd in 2009. At the first oil auction, British Petroleum agreed to increase production to about 2.9 mbpd in return for a per-barrel fee of $2 and GoI reimbursement of certain expenses (Vogler 2017, Chapter 18, pp. 224–43). Of these two forms of oil production agreements, international oil companies (IOCs) prefer the lower risk of a PSA (Downey 2009).

Outside of the KRG, the GoI has used TSAs, beginning with the first bid rounds in 2009. One of the motivations was political: officials could argue that they were not selling the country's natural resources to IOCs but rather simply hiring them to perform certain services. One of the dangers to Iraq of relying on TSAs is that they are procyclical. Assume that the flat fee is $2 pb. In mid-2014, when the price of oil was $100 pb, the fee was only 2 percent, but in April 2020 when the price of oil fell for a short period of time to $20 pb, the fee was equivalent to 10 percent. In other words, while the cost of a PSA rises when oil prices are high because the barrels of oil promised to the IOC are more valuable, the cost falls during a period of low oil prices. But with a TSA, the cost stays the same regardless of the price of oil. During a period of low oil prices and therefore lower government revenue such as began in 2014, Iraq struggles to pays the fees.

Near the end of 2020, the GoI proposed an oil pricing initiative intended to partially relieve the government's desperate revenue shortage. The GoI would guarantee 130 000 bpd for five years to the buyer, most likely one of the IOCs. Unlike existing oil sales contracts, there would be no restrictions on the initial buyer selling the crude oil anywhere in the world. In return, the IOC would be required to pay for the first year's shipment in advance, approximately $2 billion at end-2020 oil prices. While a new initiative for the GoI, this form of advance payment contract has been used by Chad, the Republic of Congo, and even the KRG (Al Ansary and Ajrash 2020; Blas and Hurst 2020). However, any oil exports under this initiative will be constrained by the GoI's attempts to fulfill its responsibilities to OPEC.

Although Iraq was a founding member of OPEC in 1960, its production quota was suspended in the 1990s. Reinstating a production quota for Iraq was expected to be controversial, since by the end of 2017, Iraq was the second-largest producer in OPEC. Although Iraq declined to participate in earlier OPEC-organized production cuts, it agreed to reduce production by about 1 mbpd in 2020. However, Iraq found it difficult to meet its OPEC commitments. In May 2020, Iraq production decreased by only about one-third of the promised amount, 311 kbpd. In June, the reduction was also too small, and oil production actually increased in July (Lawler 2020; Smith et al. 2020).

One reason for Iraq's failure to reduce production in the initial stage of the OPEC-organized cuts is its difficult relationship with the international oil companies. The IOCs resisted requests from the GoI to reduce production to meet the OPEC quota, because under the existing TSA agreements the IOCs were contractually required to maintain or increase production. Also, due to the financial crisis that began in 2014, the GoI was often in arrears to the IOCs. In 2015, these arrears reached $3.8 billion (IMF 2019a, Table 4, p. 30). This increases the unwillingness of the IOCs to reduce production and therefore reduce their fee income. Therefore, any production cuts have to come from the few smaller oilfields still operated by Iraqi firms.

Even if Iraq is unable or unwilling to follow its OPEC agreement, then the GoI will still have to decide on the optimal rate of expansion of its oil production and exports. In theory, there are three options: leave the oil in the ground as an investment; pump it now and use the funds to establish a sovereign wealth fund; or pump it now and use it as an income stream.

The choice of whether to leave the country's petroleum in the ground or pump and export it depends not only on the current price of crude oil or natural gas, but also on the expected risk-adjusted prices of crude and gas in the future as well as the expected risk-adjusted return on alternative investments. For example, if the inflation-adjusted price of oil is expected to rise from $40 pb to an inflation-adjusted $80 pb in the next decade, then this is equivalent to a 7.2 percent annual real return. If the country's next-best investment option – say, a diversified portfolio of European, Japanese, and American stocks – is expected to have an annual real return of only 5 percent over the next decade, then Iraq's best investment strategy is to leave the crude in the ground. Not pumping the oil has a higher expected return. On the other hand, if the next-best investment pays a real return of 10 percent then Iraq should pump and export now, since it has the option of investing crude export earnings in a higher-earning alternative.

The GoI could sell its oil reserves in the ground – a version of PSA – but this is extremely unlikely. This unwillingness to allow foreigners substantial control over Iraqi oil and gas is a political necessity in Baghdad. However, while the exact terms are rarely made public, it appears that some of the oilfield development contracts signed by the KRG with foreign oil companies contain PSA clauses. This discrepancy is one of the major causes of the disputes between the GoI and the KRG over oil exports.

Therefore, if Iraq must finance its needed investment through production and sale of oil and gas, then the critical question is about choosing the optimal rate of exploitation of its huge oil reserves. Iraq accounted for 16 percent of OPEC's total production in 2019, second behind Saudi Arabia's 33 percent (OPEC 2020, Table 5-9). Since most of the other large OPEC and non-OPEC producers, with the exception of the United States and Saudi Arabia, are already producing at near maximum capacity, Iraq is one of the few countries capable of substantial increases in crude exports over the next decade. Therefore, because of its huge reserves of reasonably good-quality oil and the potential of substantial increases in exports over the next decade, Iraq must confront the fact that its production decisions will substantially impact the world price of oil.

In the short run, the demand for crude oil is inelastic. In fact, a study of 21 countries showed a very inelastic demand for oil, with an average short-run elasticity of only 0.05 compared to a long-run elasticity of 0.23 (Cooper 2003, Table 4). Therefore, an increase in total crude oil exports, everything else unchanged, will lead to a decline in the total revenues of oil exporting countries. For example, if the long-run elasticity of 0.23 is accurate then a 10 percent increase in total world oil supply will lead to more than a 40 percent drop in oil prices.

The supply of crude oil is also inelastic. A large proportion of the costs of getting crude oil out of the ground and to a market are fixed. Examples are the drilling, oil–gas separation plants, pipelines, and associated pumps. Oil companies and oil exporting nations will require the expectation of high enough prices to provide a profit after accounting for all costs before they begin investing in increased production. But, once production has begun, lower prices – even dramatically lower prices – will not lead to a substantial reduction in oil production. The fixed costs are sunk and cannot be recovered. As long as the price of oil covers the variable costs of oil production, countries will continue to produce oil.

As a result of the oil market having both inelastic demand and supply, the marginal revenue – additional revenue of exporting one more barrel – will be substantially less than the world price of oil. As a result, a rapid expansion in Iraqi oil exports will not only drive up the cost of producing each barrel but also reduce the net earnings from oil exports. Even if total oil export revenues increase, a rapid expansion of oil exports may injure the production of other export goods and services by reinforcing Iraq's already bad case of the "Dutch disease". The term "Dutch disease" was first used by *The Economist* business magazine to explain the decline of the manufacturing sector in the Netherlands following the exploitation of a large natural gas field. It occurs when a country has a single dominant export product that leads to a large current account surplus and currency appreciation. This appreciation reduces the competitiveness of the country's exports while encouraging increased imports.

One of the surprising policy successes of the GoI since 2003 had been the fact that the Iraqi dinar (ID) not only avoided losing its value but also actually appreciated by over 17 percent. The causes and impacts of this phenomenon are discussed at greater length in Chapter 14, but the role of crude oil exports is clear. In addition to receiving large amounts of foreign aid, the uneven rise in oil prices from 2003 through 2013 combined with a gradual rise in crude exports (see Figure 7.2) resulted in an unintended current account surplus. It was unintended because the GoI had planned large-scale investment projects, especially in the areas of essential services provision, and these investments would have required a large increase in imports, pushing the current account into deficit. But because of limited bureaucratic capacity to execute these investment projects, the GoI was unable to accomplish them and therefore there was no sharp rise in imports. The resulting current account surplus led to a rise of Iraq's international reserves to about $68 billion at the end of 2019 (World Bank 2020b, pp. 8–9), and an increased confidence that the dinar would maintain its value or possibly increase.

But the appreciation of the dinar has had an adverse impact on non-oil employment in Iraq. Iraq's exports, from carpets to dates, became 17 percent more expensive and, as a result, have failed to regain even a fraction of the pre-invasion share of foreign markets. At the same time, imports have flooded the Iraqi market. While in Iraq, I once attempted to find some "Made in Iraq" consumer goods to use for a display for visiting VIPs. However, Baghdad shops were filled with imported appliances, tools, canned goods, and household items. Few items were made in Iraq.

In the short term, Iraq's problem with the Dutch Disease was solved by the 23 percent dinar devaluation of late December 2020. Admittedly this devaluation was motivated more by a large budget deficit than by concern over the lack of non-oil exports. (See Chapter 14 for a more detailed discussion). But this solution is temporary.

An option for reducing the adverse long-term impact of the Dutch Disease while preserving wealth for future generations of Iraqis is a sovereign wealth fund, as discussed earlier (IMF 2011a, p. 14). Some of the earnings from petroleum exports would be deposited in an Iraq SWF. The SWF would then invest in a diversified portfolio of financial assets from other countries.

Since these earnings would not be converted into Iraqi dinars, they would not cause currency appreciation. At some future time, the SWF could begin to make periodic payments to Iraqis. Establishing a SWF was one of the eight objectives for the financial and monetary sector in the National Development Plan: 2010–2014 (GoI 2010, p. 53). However, little progress has been made in making this fund operational.

Opposition to the establishment of an Iraqi SWF is based on two related arguments. First, it is argued that the development needs of Iraqis are so great that the GoI should rapidly expand development expenditures to both reduce poverty and alleviate political instability. Second, the high level of cynicism among Iraqis concerning the honesty of their leadership leads to the widespread fear that the political elite would loot a sovereign fund.

CONCLUSION

From 2007 until 2013, the combination of increased oil export volumes and high international oil prices provided Iraq with half a decade to build a new economy. This time and money could have been used to lay the foundation for a diversified economy with reduced dependence on oil exports and gradually replace the GoI as the "employer of first resort". Unfortunately, driven more by political incentives and constraints then by economic efficiency, the GoI sought a rapid expansion of government employment and compensation, and spent massive amounts on oil infrastructure investment and essential services subsidies.

Unfortunately, as will be discussed in Chapter 15, Iraq faces at least a decade of an oil prices of $60 or less per barrel and slow export volume growth. This will result in an extended period of low oil export revenues. Diversification of the Iraq economy away from its excessive dependency on oil is critical. Options for diversification are discussed in Chapter 8 on agriculture, Chapter 10 on state-owned enterprise reform, and Chapter 11 on entrepreneurship. But substantial diversification will require Iraq to overcome a long-standing dilemma. When oil prices are high then there are funds to support diversification, but there is little motivation to diversify. And when oil prices are low then there is a strong motivation to diversify, but there are no funds to support the effort.

8. Agriculture and the environment

> The Tigris and Euphrates rise without warning; are always abrupt; carry five times the sediment of the Nile, have their annual flood in March, April, and May, too late for the winter and too early for the summer crops; traverse a country where the temperature rises to 120 degrees [49 Celsius] in summer and falls to 20 degrees [−7 Celsius] in winter ... In spite of the many drawbacks, the ancient Babylonians made of the Euphrates delta a country so rich that Alexander the Great conceived the project of making Babylon the capital of the world.
>
> (Lord Salter 1955, p. 40, quoting Sir William Willcocks)

Throughout history, Iraq, along with Turkey, was one of the few countries in the Middle East with sufficient water for large-scale agriculture (Ahmad 2002, p. 170; Savello 2009a, Table 2, p. 2). In fact, for most of its history, Iraq was an agricultural cornucopia. However, from approximately 1980 until roughly 2010, Iraqi agriculture stalled. During this period, there was little progress in adopting new techniques, seeds, and so on, and delayed maintenance of the irrigation infrastructure. As a result, Iraq became a large net importer of agricultural products.

As can be seen in Figure 8.1, Iraq has three different climates. The southwest region is a relatively flat, very arid desert. Mean annual rainfall is only 100–170 millimeters (4–7 inches). At the other extreme, the northeast of the country is mountainous with a Mediterranean type climate. In a typical year, annual precipitation ranges from 760 to 1000 millimeters (30–40 inches), sufficient for rain-fed agriculture. In between is a band of semi-arid climate that stretches from the Syrian and Turkish borders in the northwest to the Persian Gulf in the southeast. Two of Iraq's largest cities – Baghdad and Basrah – and most of the country's population are in this semi-arid region. In central Iraq, typical temperatures range from 40°C (100°F) in July and August to 17°C (64°F) in the winter, although highs of 48°C (120°F) and lows below freezing have been known. The climate of central Iraq is similar to that of the southwestern United States of America (USA), with hot summers, cool winters, and agreeable spring and autumn months.

Iraq's total land area is approximately 44 million hectares (440 000 km^2 or 176 million donums) (donums being the most common Iraqi measurement of area). Only about 16 percent (7 million hectares or 28 million donums) is arable land. The rest is too dry, salt-poisoned, or otherwise unusable. However, only about half of this arable land (4 million hectares or 15 million donums)

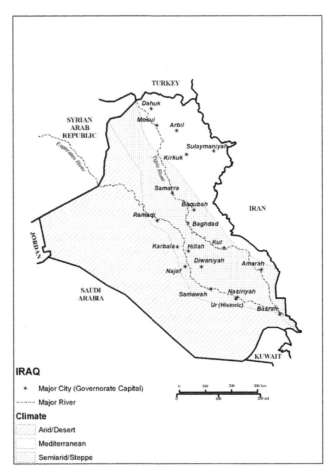

Source: Professional Maps of Houston, Texas, 2012.

Figure 8.1   *Climate map*

has actually been cultivated in the recent past (GoI 2018, p. 131). Much of this land is used for meadows and pastures rather than for crops. With respect to crops, grain production accounts for most of the cultivated area, while vegetables, fruits, oil seeds, tubers such as potatoes, legumes, industrial crops such as cotton, and forage crops have a higher value.

The scale of farming ranges widely in Iraq, from extremely large wheat farms to family farms of less than 1 donum (0.24 hectares or 0.62 acres). The most common farms are small in size – about 15–45 donums (4–12 hectares or

10–30 acres). They typically grow a small crop of grains (wheat, barley, rice, and some corn) along with vegetables (tomatoes, cucumbers, melons, egg-plant, squash, onions, and potatoes). Often there will be a small flock of sheep (20–30), or a few cattle. Production per hectare is substantially below that of other Middle Eastern countries due to the use of outdated farming techniques, poor-quality seed, and old equipment.

In part, this inefficiency is driven by rules on inheritance. For example, a landowner only has the right to transfer one-third of his property to one relative in order to ensure that other inheritors are not deprived of their rights under Islamic law (USAID 2005, p. 6). Such rules lead to a fragmentation of agricultural lands, resulting in farms too small for modern agricultural techniques (GoI 2018, p. 131).

Agriculture's share of gross domestic product (GDP) varies significantly from year to year depending on world oil prices. When oil prices are high then oil's share of Iraq's GDP grows, and consequently agriculture's share falls. But to provide a sense of scale, prior to the ISIS insurgency, agriculture accounted for between 8.5 and 10.0 percent of GDP.

Although agricultural employment has fallen by almost half since the late 1970s, it is still the largest source of employment in Iraq after the government. In 2017, agricultural jobs accounted for 19 percent of all employment. In comparison, trade accounts for only 14 percent of total employment, while construction provides another 12 percent. Employment in farming is particularly important in four provinces: Salah ad Din where agriculture accounts for 40 percent of all employment, Babil at 39 percent, Anbar at 33 percent, and Qadisiyah at 28 percent (FAO 2020c; COSIT 2008, Table 5-20A, p. 325).

Cash income received from agricultural labor is below the national average of all industries. For example, about one-quarter of households that report agriculture, hunting, forestry, or fishing as their major source of income received less than 400 000 Iraqi dinams (ID) per month (about $340 per month). The average rural household has about eight members, of which half are 15 years or older, which is considered adulthood for employment purposes; household income averages for these low-income farmers was about 100 000 ID (or $85) per worker per month.

Agricultural employment differs by gender. Among men, agriculture accounts for only 13 percent of all employment, which puts it behind trade at 19 percent and construction at 14 percent. Particularly on smaller farms, males of farming families will seek other employment in nearby semi-urban or urban areas, while helping out on the family farm during labor-intensive periods such as planting and harvesting. Since 1990, the number of males engaged in farming has decreased by over 40 percent, while the number of females has remained almost unchanged. In fact, agriculture accounts for one-quarter of female labor force activity, more than any other industry (FAO 2020a; COSIT

Table 8.1   *Iraq agricultural production (metric tons)*

	2014	2018	Change (%)
Wheat	5 055 000	3 211 000	-26
Barley	1 278 000	571 000	-54
Rice	403 000	275 000	-32
Maize	289 000	209 000	-26
Potatoes	402 000	295 000	-27
Tomatoes	771 000	266 000	-65
Onions	75 000	15 000	-80
Dates	662 000	615 000	-7

Source: GoI (2018, p. 133).

2008, Table 5-20A, pp. 324–5). The importance of agricultural employment for females is, in part, a function of cultural restrictions on women working outside the home. Male workers earn a larger percentage of all agricultural income than women. In many cases, women receive no pay for working on family farms.

The widespread destruction by ISIS, water shortages, increased salinization (salt poisoning of soil), ambiguous land ownership, government policies towards crop prices, and the tendency towards more destructive droughts are the major challenges facing Iraq farmers. If these enormous challenges can be met, Iraqi agribusiness has the potential not only of becoming a significant source of employment, but also of helping to reduce Iraq's dependency on imported food.

AGRICULTURE PRODUCTION

Iraq's overall agricultural production contracted sharply from 2014 through 2018 as a result of drought and the ISIS-related destruction in the north and west of the country. As can be seen in Table 8.1, primarily rain-fed wheat and barley production decreased by 26 percent and 54 percent, respectively. Rice production which is mostly an irrigated crop in the southeast decreased by a third. Tomato production collapsed by two-thirds while the recovery of date production faltered.

While productivity has gradually increased in Iraq, it remains substantially below that of other nations. For example, cereal yield in Iraq was approximately 2800 kilograms per hectare (kgph), compared to an average of about 4600 kgph in the Middle East and North Africa (MENA) countries and 3200 kgph in countries at a similar level of income (USAID 2020). As expected, the collapse in domestic agriculture production led to a 280 percent increase in food imports to $9.7 billion.

In addition to adverse effects of drought and the ISIS-related destruction, soil fertility continues to decline as a result of salinization, mismanagement, and overuse. Multiple revisions in land ownership since 1958 have created great uncertainty and disruption that has encouraged a short-term perspective among cultivators. Bureaucratic interference from the Ministry of Agriculture and six other ministries that impact farming waxes and wanes unpredictably as the price of oil changes (Owen and Pamuk 1999, p. 169). Inadequate maintenance and low levels of new investment have reduced both the volume and, more importantly, the predictability of irrigation flows. Finally, the 2006–08 appreciation of the Iraqi dinar of about 17 percent, combined with (since 2003) open markets for imported agricultural products, have substantially increased import competition for domestic production. It will be interesting to see to what degree the 23 percent depreciation of the dinar in December 2020 will lead to higher prices for agricultural products and an eventual increase in production.

Grain production has a millennia-long history in Iraq and has dominated Iraq's agricultural production based on land area. To a great extent, the dominance of grain production reflects a long-standing Iraqi government policy of encouraging grain self-sufficiency. Much of the concern with grain self-sufficiency is a function of Iraq's history since the beginning of the Iraq–Iran War in September 1980. This war severely constrained food and other imports through Iraq's Persian Gulf ports, especially in 1984 when both sides began to target tankers and merchant ships. The ceasefire in 1988 was followed in 1990 by Iraq's invasion of Kuwait that resulted in an embargo and almost a decade of sanctions. Despite the corrupt United Nations Oil-for-Food Programme, there were severe food shortages of basic foods. As a result, the Government of Iraq (GoI) engaged in a complex and extensive program to encourage grain production.

Both the production and the marketing of cereals are heavily influenced by the national government. The Ministry of Agriculture provides heavily subsidized inputs including fertilizer, irrigation water, and electricity. The subsidy of the water supply is particularly important since about 60 percent of wheat and 100 percent of rice fields are irrigated (World Bank 2006b, p. 55). The harvested wheat and rice are purchased by the national government at prices that rarely equal those of imported products. Of course, this distorts farmers' incentives (Ridha 2020, p. 13).

Grains continue to have the highest import value. In 2017, wheat imports were valued at almost $2.6 billion (FAO 2020a), while grain exports were *de minimus*.

Fruit and vegetable production receive much lower levels of GoI subsidies and enjoy less stringent regulation than grains. Most of the annual production is immediately transported and sold in nearby urban areas. Only a relatively small proportion of fruits and vegetables are processed.

Dates have historical importance in Iraqi as well as in Arabic and Islamic cultures. While tomatoes are a relatively recent import to Iraq, dating from the eighteenth century, date production in Iraq/Mesopotamia goes back to the fourth millennia BC. Dates are mentioned multiple times in the Qur'an and, in fact, the peak season for date consumption among Muslims is during the month of Ramadan when most of the entire Muslim community around the world – over 1.5 billion people – celebrate the end of each day's Ramadan fast with dates. Dates are also important to other religions and cultures; date consumption is high during the Christian Christmas and the Hindu festival of lights, Diwali.

Dates were formerly Iraq's single most valuable agricultural export. In fact, in 1980, Iraq produced more dates than any other country. However, decades of conflict, sanctions, and mismanagement have substantially reduced both the volume and the quality of Iraqi dates. From an estimated 30 million date palms prior to the Iran–Iraq war (1980–88), the number of date palms decreased to about 10 million (Benhaida 2018). As a result, by 2018, Iraq ranked sixth in the world with respect to date exports. Iraq's approximately 600 000 tonnes of exports were about a third of the number one producer, Egypt (FAO 2020d).

It takes 7–10 years for a newly planted date orchard to produce quality fruit. It is a labor-intensive crop, since producing quality dates requires cultivators on ladders or on cranes to manually fertilize the female palms. There are a variety of pests that infest date orchards and consume or damage the fruit. These pests are generally controlled by either aerial or ground spraying. The profitability of date orchards is enhanced by the practice of growing complementary citrus crops amongst the date palms. In view of the long period before new orchards bear quality fruit, and the severe competition from other date exporting nations, the future of Iraqi date production is mixed. Rehabilitating existing or abandoned orchards will probably be profitable; planting new orchards will probably not be.

ISIS

ISIS, the Islamic State of Iraq and Syria, was also known as the Islamic State of Iraq and the Levant, the Islamic State, and Daesh. Beginning in 2014, ISIS conquered almost 40 percent of Iraq with between 25 and 35 percent of the country's population. Areas entirely occupied by ISIS were the northern and western provinces of Anbar, Diyala, Ninawa, and Salah Ad Din. There was also substantial ISIS activity and associated destruction in Babil, Baghdad, and

Kirkuk. Before its defeat at the end of 2017, ISIS was not only responsible for mass atrocities and widespread urban destruction but also devastated agriculture in the occupied areas. In the short term, ISIS reduced Iraq's agricultural production by an estimated 40 percent (USAID 2019, p. 5). And the long-term adverse effects are worse.

The immediate effect of ISIS occupation was the widespread confiscation of crops both in the fields and in silos. In some cases, payment was offered but at prices substantially below the pre-ISIS market price or even production costs. Violence and threats of violence were used to keep farmers producing, although ISIS confiscated agricultural land and withheld resources and equipment in order to intimidate. Urban dwellers under suspicion were denied food; farmers who were less than enthusiastic about the new ISIS order were denied water, seed, and fertilizer. In other cases, anything of value possessed by the farmers was looted. In Ninawa, it was estimated that 90 percent of the irrigation infrastructure was looted or destroyed (USAID 2019, p. 17). In Anbar, population displacement, casual executions, and looting led to a significant reduction in agricultural output (World Bank 2018b, p. 47). The other occupied provinces suffered similar destruction.

As ISIS was gradually driven out of occupied farming communities, it engaged in an orgy of destruction. Irrigation pumps were wrecked, tractors were stolen or destroyed, greenhouses were torn down, silos were blown up, livestock was stolen or slaughtered and left to rot, century-old orchards were cut down, improvised explosive devices (IEDs) were planted in fields, land ownership records were destroyed, and, worst of all, farming families were forced to flee the land.

Post-ISIS, the collapse of government revenues has retarded the rebuilding of agricultural-related infrastructure, especially with respect to power generation, flood control, and irrigation. In the areas occupied by ISIS, three out of seven dams and 56 out of 67 flood control barrages were destroyed, as well as 38 out of 47 pumping stations (World Bank 2018b, Table 22, p. 53). Even if Iraq did not face the fiscal constraints discussed in Chapter 14, reconstruction on this scale will take a decade or more.

Ironically, the Covid-19 epidemic is having a favorable short-term impact on Iraqi agriculture. As discussed in Chapter 3, Iraq has increased its efforts to close its border with Iran in order to reduce the likelihood of infection crossing the border. At the same time, the GoI has banned imports of over two dozen crops from the Kurdish Regional Government (KRG). This ban is based on the belief that much of this production are not of KRG origin but rather transshipped products from Iran and Turkey. Collecting tariffs, limiting smuggling from Iran, and controlling transshipments through the KRG has led to higher prices and lower quantities for imported agricultural products. This has brought large quantities of Iraqi-grown fruits and vegetables into the major urban markets for the first time since 2003 (Saadoun 2020). Of course, higher food prices that favor rural farmers are injuring urban consumers. And the long-term success of Iraqi agriculture will depend on the supply of water.

WATER

Water Demand

The demand for water is still dominated by agriculture, although the degree of this dominance is decreasing. The Tigris, Euphrates, and their tributaries provide an estimated 44 billion cubic meters per year (m³/yr) (GoI 2018, p. 132). In view of growing water demand for agriculture (about 75 percent of total water demand), municipal use (10 percent), and industrial use (15 percent), the Tigris and Euphrates are expected to run dry in southern Iraq by 2035.

Demographic changes are driving the increased municipal demand for water. Not only does the population continue its rapid growth – the population of Iraq has almost doubled in the last three decades – but also this population is increasingly urban. A rise in urbanization tends to increase the volume of water usage. According to one study, in a nomadic society, water use per person per day is an estimated 10–30 liters. In a village, per person water use averages 60–80 liters per day; while in a modern urban area, water use per day might reach 400–800 liters per person (Agnew and Anderson 1992, pp. 276–7).

The increase in industrial water withdrawal has been even more dramatic. Over a period of 15 years, industrial use increased 33 times. Much of this industrial water use is wasted, in two ways. First, since water usage fees are extremely low and often not collected, industrial water users have little incentive to avoid using excess amounts of water. Second, environmental standards are rarely enforced in Iraq and it is common for industrial users to dump water contaminated by factory use back into the nearest river. This decreases the quality of river water for downstream users.

In addition to the continuing demand for water for agriculture, industry, and urban uses, there is the complicated issue of the restoration of the large Mesopotamian Marshes located in southeast Iraq between the cities of Nasiriyah and Basrah. These three marshes covered an estimated 20 000 km² (7800 square miles) and were the home of a large human population as well as many rare birds and mammals. Efforts to drain the marshes in order to create farmland and facilitate oil exploration began in the 1950s, but relatively little progress was made until the 1990s. However, following the First Persian Gulf War 1990–91, Saddam diverted the Euphrates and the Tigris rivers so as to rapidly drain the marshes in order to punish the marsh people for supporting opposition to his regime. By 2003, an estimated 90 percent of the marshes had been drained causing great environmental damage and a forcible relocation of many of the marsh people.

Following the 2003 invasion by the US-led coalition, water has been diverted back into the marshes and some parts have begun to recover. However, a return to their original size will require substantial water diversion – possibly 10 billion m³/yr – for at least another decade. Even when the marshes are restored to approximately their original size, the necessity of keeping the marshes a vibrant living area will require a substantial annual flow of water from the Euphrates and the Tigris rivers. It is not clear whether the water supply will be sufficient.

Water Supply

Mesopotamia is from the Greek for "land between two rivers". For over seven millennia, almost all of Iraq's water has come from the Euphrates and Tigris rivers. The Euphrates (in the western part of Iraq) accounts for about one-quarter of the nation's usable water, while the Tigris accounts for the remainder. Groundwater is a source of less than 1 percent of Iraq's water. Both rivers originate in the mountains of Turkish Anatolia before flowing through Turkey, Syria, and Iraq. The Euphrates receives 100 percent of its flow from Turkey and Syria while the Tigris receives about two-thirds of its flow from Turkey and Iran. The northeast of Iraq annually receives 30–40 inches of rain, much of which drains into the Tigris. On the other hand, the Euphrates River flows through arid and semi-arid regions of Iraq and therefore receives little rain. The two rivers merge north of Basrah to form the Shatt al-Arab waterway that flows for about 150 km (roughly 90 miles) to the Persian Gulf (World Bank 2006b, p. 12).

The flows of the Euphrates and Tigris rivers are subject to substantial variances in timing, volume, and quality. The annual peak flow of both rivers occurs in late winter and early spring following the snowmelt and rainy season in Turkey. As a result, almost 50 percent of the flow into the Euphrates and the Tigris occurs in just two months: April and May. This flow is too late for winter crops in Iraq and too early for spring crops. In addition, there is a large year-to-year variance in the volume of water of approximately plus or minus 8 billion m³/yr (plus or minus 12 percent of average annual water resources) (World Bank 2006b, p. 14; Savello 2009b, Table 1, p. 2). The timing and wide variance in annual flows in both rivers have led to the construction of dams and large water storage facilities (barrages), not only to capture some of the April and May floods and release the water during the main growing seasons, but also to store some of the water from years of larger-than-average flows until they are needed. As discussed above, three out of seven dams and 56 out of the 67 barrages in Ninawa, Anbar, and Salah as Din were destroyed by ISIS (World Bank 2018c, p. 53).

Aside from the annual variations, Iraq expects a large reduction in future water volume of the Euphrates and Tigris rivers as a result of dam construction and irrigation projects in Turkey, Syria, and Iran. Dams constructed as part of Turkey's Southeastern Anatolia Irrigation Project (the Turkish acronym is GAP) have already resulted in a significant reduction in the volume of the Euphrates. The ongoing GAP calls for the eventual construction of 22 dams and 19 hydroelectric plants on the two rivers (Yilmaz 2003, p. 81).

The completion of the GAP with its expanded irrigation in Turkey, and similar smaller-scale projects in Syria, will lead to a further decrease in river flow, especially in the Euphrates. If the flow in the Euphrates deceases to less than 10 billion m^3/yr, it is likely that the Euphrates will run dry before it joins with the Tigris north of Basrah (World Bank 2006b, Figure A2.2, p. 51). Not only will this reduce water availability for irrigation and municipal consumption, but also there will probably be an adverse impact on electricity generation at the Haditha Dam, the only major Iraqi dam on the Euphrates.

The problem is not as severe for the Tigris river. Although Turkey and Iran contribute almost two-thirds of Tigris water, their planned irrigation projects over the next decade will reduce flow by only about 4 billion m^3/yr or roughly 10 percent. Since the Tigris currently has about 50 percent more water than is required for agriculture, municipal, and industrial use, this reduction will not cause a severe water shortage over the next decade like that of the Euphrates. However, by the mid-2030s, water in both rivers will be insufficient for Iraq's expected agricultural and other needs (Yilmaz 2003, pp. 86–8).

Will the diversion of water by upstream nations for irrigation and power generation lead to conflict? Many commentators have pointed to the competition among Turkey, Syria, Iran, and Iraq for the waters of the Euphrates and Tigris as a sign of impending war.

The counterargument is that, as discussed in Chapter 6, Turkey, Syria, and Iraq have a variety of other issues of mutual concern, such as the Kurdish question. During negotiations, upstream nations have been willing to make concessions on water flow in order to achieve other goals.

There is currently no agreement on the use of the Euphrates or Tigris rivers like those that govern water relations between Egypt and the Sudan (1959 Agreement) or the 1994 Agreement between Israel and Jordan (Richards and Waterbury 2008, p. 174). However, there are at least two "rules" or norms that Turkey, Syria, and Iraq have roughly followed in the past. In July 1987, Turkey and Syria agreed to a Protocol of Economic Cooperation that committed Turkey to maintaining a flow in the Euphrates of at least 15.7 billion m^3/yr at the Turkish–Syrian border (Dolatyar and Gray, 2000). During most months, the flow has greatly exceeded this level, and during the filling of the Ataturk

Dam reservoir in Turkey that temporarily reduced the flow below the promised level, Turkey provided advance notice to downstream nations and increased the flow before and after to bring the multi-month average to promised levels.

The second norm is a result of 1990 negotiations under the Secretariat of the Arab League. It was agreed that Syria could keep up to 42 percent of the Euphrates water received at the Turkish–Syrian border, and allow the remaining 58 percent to flow into Iraq. Pessimistically, if Turkey releases the minimum and Syria utilizes the maximum, then Iraq will receive 58 percent of 15.7 billion m^3/yr water in the Euphrates, or about 9.0 billion m^3/yr. At this minimum, the Euphrates will run dry south of Baghdad.

The second argument against the thesis that disputes over the Euphrates river will lead to conflict is that all three nations are extremely inefficient in their use of water. Irrigation canals open to the sun allow evaporation, cracked canals permit leakage, and flood irrigation not only wastes water but also contributes to salt poisoning of the soil. As a result, it is estimated that about half of water reserved for agriculture is wasted. In addition, since water is priced substantially below its marginal cost, there is little incentive to conserve water not only in agriculture but also in municipal and industrial uses. Potentially, if water were used more efficiently this would more than offset the expected reduction in the Euphrates and the Tigris flows over the next two decades.

Like many downstream nations in the world, Iraq avoids emphasizing possible efficiency gains in order to avoid weakening its negotiating position. It is thought that if a downstream nation is moving towards increasingly efficient use of water, then this will be taken as a sign to upstream nations that it is unnecessary for them to make concessions to maintain existing flows. However, if Iraq can be assured that upstream nations will not abrogate existing norms, such as the 15.7 billion m^3/yr and 42 percent/58 percent norms discussed above, then the incentives to improve water use efficiency will be clear.

Water rights are officially controlled by the GoI (Unruh 2018, pp. 265–7). Once the water of the Euphrates and Tigris rivers crosses the borders into Iraq, it flows into a complex government-controlled irrigation system. Four major dams and two reservoirs are used to control the rate of flow and, in the case of the Haditha Dam, to create electricity. There are over 50 large-scale gates and regulators that direct the flow of water to over 200 major irrigation and drainage pump stations with over 1000 pumps. These stations direct the water into an estimated 127 000 km (76 000 miles) of primary and secondary irrigation canals. As a result of over three decades of neglect as well as the ISIS-related destruction, the efficiency of this complex system is low. The pumps are generally worn out and obsolete, resulting in extensive downtime for maintenance or repairs. Most of the irrigation canals lose large amounts of water in leaks that not only reduce irrigation flow but also increase salinization by raising the water table.

Because of the geology of northwest Iraq, drainage is generally difficult. Drainage water, contaminated with salts and other agricultural waste, requires its own system to prevent it mixing with river water. This is a more severe problem in the Euphrates watershed and, in response, the GoI constructed the East Euphrates Drain to capture drainage water and move it south of Baghdad where it flows into the Main Outlet Drain (MOD). The drain system is almost a third Iraqi river, that sometimes has a greater flow than the Euphrates where the two flows cross north of Basrah city. The MOD has its own discharge into the Persian Gulf.

The reduction in river flow combined with leaks and evaporation in the irrigation system results in severe water shortages in many parts of irrigated Iraq. Water feuds are not uncommon as farmers associated with various tribes or clans attempt to increase the flow to their fields by reducing the water available to others. In many rural areas, tribes or clans threaten or bribe government officials to divert water flows. Violence is not uncommon. Water officials and their families were murdered because of water disputes near Abu Ghraib in the first half of 2010 (IRIN 2010c).

In addition to the expected reduction in the volume of the two rivers, both rivers are experiencing a fall in water quality as a result of increased irrigation runoff in Turkey, Syria, and Iran. Since irrigation runoff has increased amounts of dissolved salts, the already difficult challenge of reducing salt poisoning of the soil – salinization – in Iraq will only get worse.

SALINIZATION

The most serious challenge facing Iraqi agriculture is salinization, with about 70–75 percent of the nation's irrigated land suffering from various degrees of salt poisoning, and an estimated 161 000 donums (39 000 hectares) of agricultural land lost annually from production (World Bank 2017, pp. 96–7; GoI 2007, p. 36; GoI 2010, p. 63). These salts are natural elements in the soil, and all water, even rainwater, contains salts. However, arid regions are especially vulnerable to an excessive buildup of salts near the ground surface and that reduces soil fertility.

Salinity is measured by the way in which salt changes electrical conductivity. Sea water, for example, is about 30 dS/M. Unfortunately, the water in the Euphrates and Tigris rivers is already moderately saline, 4–8 dS/M. At this level of salinity, the yield of many crops is reduced since most fruits and vegetables require salinity of less than 4 dS/M (Christen and Saliem 2013, Table 2, p. 11).

The major effect of salt poisoning is that it makes it more difficult for plants to absorb water. Plants differ widely in their tolerance of excessive salts; for example, rice is more salt tolerant than other grains. However, even moderate salinization stunts the growth and reduces the productivity of most crops (Abrol et al. 1988, Section 3.1). The evidence of high levels of salt poisoning is obvious even to a casual observer; from a helicopter, one observes plant discoloration, barren spots, and white salt crusts.

Irrigation, unless carefully managed, can accelerate the salinization process. As irrigation water is absorbed by plants or evaporates, most of the salts are left in the soil. Ideally, there would be sufficient additional water to dissolve these salts and adequate drainage to carry this salt-laden water away. This process is most successful when the moisture content of the soil is low and the groundwater table is deep (Abrol et al. 1988, Section 3.2.1). Therefore, the process can fail in three ways. First, excessive evaporation can result in salts in the soil increasing faster than they can be leached away. Second, there may not be sufficient water to carry the dissolved salts away. Finally, even if there is sufficient water, poor drainage will lead to a re-contamination of the soil. If the salt-laden water table pools within 2–3 meters of the surface, the salts will return to the surface through capillary action.

Unfortunately, much of the farmland in Iraq has poor drainage and heavy levels of salt-contaminated soil (Evans et al. 2013, p. 9). Excessive irrigation only exacerbates salt poisoning. As excess irrigation water evaporates or seeps into the ground, it not only contributes additional salts to the soil but also raises the water table in areas of poor drainage. If the irrigation water was heavy in salts to begin with, the rate of contamination is more rapid. Fields that are further above the water table not only require pump irrigation but also tend to have better drainage. As a result, such fields tend to have less salt poisoning than flow-irrigated land (Mahdi 2000, p. 118). The salinization problem is not new. There are regions of Iraq that in ancient times supported a large agricultural population but are now barren due to salt poisoning (Evans et al. 2013, p. 9; Agnew and Anderson 1992, pp. 158–9).

Once salt contamination has occurred, the most common means of restoring soil fertility is by leaching. This requires flooding contaminated fields and then removing the salt-laden water through artificial or natural drainage. The amount of relatively salt-free water required is large. A useful rule of thumb is that it requires enough water to flood a field 1 meter deep in order to remove 80 percent of the salts in the top meter of soil. A more sophisticated estimate would depend on the initial amount of salt contamination and the type of soil. A study of leaching in three areas of Iraq shows that the higher amount of clay in the soil means that more water will be required (Abrol et al. 1988, Section 3.2.1.i).

To reverse the salinization of Iraqi agricultural land will be difficult. Leaching out salts requires large quantities of low-salinity water. In the absence of such quantities of water, Iraq should adopt triage.

First, severely contaminated land, say dS/M greater than 16, should be abandoned as farmland. Second, moderately and highly contaminated land, dS/M between 8 and 16, should be shifted to those crops, mostly forage, that are more salt tolerant. Finally, desalinization efforts should be concentrated on restoring moderately saline land, dS/M less than 8. Salt poisoning of this land could be reduced by a combination of improved drainage to lower the water table, and reduced irrigation to reduce the quantity of salts that must be dealt with. One possibility would be to replace the current dependency on flood irrigation with drip irrigation where pipes take water directly to the plants. In other countries, the shift to drip irrigation has reduced water expenditure by 80 percent.

Of course, the burden of mitigating salinization will fall unevenly. Despite low yields and the likelihood of further deterioration, many small-scale farmers can be expected to continue traditional flood irrigation since they lack both the capital and the incentives to adopt less wasteful methods. And preserving the fertility of soil requires a long-term perspective that Iraqi government policy towards land ownership discourages.

LAND OWNERSHIP

Iraqi agriculture policy over the last 70 years reflects a tug of war between the statist DNA of the Arab Ba'athist Socialist Party and the country's need for domestic food production. Prior to 2003, when government oil revenues were high, the GoI pushed cultivators into collective farms or state farms and heavily regulated any remaining private farmers. Not only did these government policies weaken the influence of large landowners, but also there was a widespread belief, at least in the 1950s, that government-guided agriculture would be more productive than private cultivation. Of course, the latter belief turned out to be false, as developing countries around the world have discovered. Government control of farming tends to lead to reduced quantity and quality.

This truth was periodically recognized in Iraq whenever government revenues fell or access to foreign food markets was reduced. At such times, the government would create a window for a more private sector, entrepreneurial approach to farming. As a result, cultivators experienced an irregular cycle of policy initiatives and reversals in land rights, subsidies, water and equipment availability, and market access. Policy uncertainty has added to the severe impact that the years of conflict have had on Iraqi agriculture by delaying improvements in farm productivity and motivating a large rural-to-urban migration. With respect to land ownership, this policy tug-of-war has resulted in a transition from dominance by large landowners to dominance by government bureaucrats.

Current land ownership in Iraq defies easy categorization. Ownership is a bundle of rights. Ownership of farmland includes: the right to decide on what crops will be cultivated; the means of cultivation; the right to benefit from the sale of the farm's production; the right to benefit from any improvements in the farm; the right to use the land as collateral for loans; and the right to sell or otherwise alienate the land.

While in some countries the same individual or entity will have all of these rights, in Iraq the rights tend to be controlled by a variety of persons and agencies. Cultivators, tribal and other traditional groups, absentee landlords, and a variety of government agencies may all influence or control one and more of Iraqi land ownership rights. Only some of these land ownership rights are recorded in writing; many are matters of tradition or political influence. Many of these rights overlap, and when they conflict there is often no accepted adjudication process. It should also be noted that procedures in the KRG differ from those in the rest of Iraq.

Keeping in mind the complexities of agricultural land ownership in Iraq, and that a large portion of this land is not currently under cultivation, there are four general types:

1. Private ownership. About 10 million donums (2.5 million hectares) or 32 percent of all non-KRG agricultural land is privately owned. Iraqi law governing privately owned land reflects somewhat contradictory Ottoman, British, and Islamic influences. Subject to legal limits on the maximum size of individual land holdings, privately owned land can be sold or otherwise alienated.

2. Free distribution. Beginning in 1958 and continuing through 1987, there were a series of land reform laws and modifications whose varied impacts are discussed below. Under these laws, the GoI confiscated large private land holdings and then, usually after a substantial delay, redistributed the land to individuals or groups of cultivators. While such free distribution land reached around 12 million donums at its peak, it is currently believed to be about 5.9 million donums (1.5 million hectares) or about 19 percent of non-KRG agricultural land. Cultivators can generally pass this land on to their heirs but otherwise it escheats to the GoI.

3. Rented. The Ministry of Agriculture rents to farmers around 15.3 million donums (3.8 million hectares), equivalent to roughly 48 percent of all non-KRG agricultural land. The cultivator has the right to use the land (usufruct) but the farmer is unlikely to be reimbursed for any improvements if they leave the land.

4. Other. Religious endowments, *waqfs*, control 0.2 million donums, which is about 1 percent of all non-KRG agricultural land (COSIT 2012, Table 3/24).

Iraq has experienced a series of land reform initiatives. In general, these initiatives have had two substantial effects. There have been significant changes
in the distribution of land ownership and the compensation of cultivators.
Also, the series of often contradictory land reform initiatives over the last
seven decades have created uncertainty over the future of land ownership
rights in Iraq. Cultivators and others respond to this uncertainty by adopting
a short-term outlook for agricultural investment. The stated motivation for the
first substantial Iraqi land reform in modern times, the Agrarian Reform Law
Number 30 of 1958, was the severe concentration of Iraqi agricultural holdings. Almost two-thirds of holdings were small and accounted for less than 3
percent of the country's total agricultural land. At the other extreme, less than
3 percent of the holdings were very large and accounted for about two-thirds
of Iraq's agricultural land. In fact, some of the land holdings were huge, with
about 0.1 percent of all holdings accounting for 28 percent of Iraq's total
agricultural land. There were less than 300 of these very large holdings, with
an average size of 26 000 donums (exactly 6500 hectares, or approximately
16 100 acres).

At the other end of the scale, there were very small farms with an average
size of 1.4 donums (one-third of a hectare or 0.9 acres). Even with intensive
agriculture, it was almost impossible to make a living on such small farms.
These small farmers carried the burden of short and precarious leases, high
rents, and high debts (Issawi 1982, p. 138). Most of the farmers of this scale
were forced to supplement their farm earnings with other sources of income.
Over three-quarters of the rural population were landless.

There was also a strong political motivation for land reform. Large rural
landowners, often tribal sheikhs, had been supported by the British and, later,
by the Hashemite Monarchy (1921–58) as a counterweight to more nationalist
urban elites. Therefore, urban elites saw land reform as a means to weaken
their political opposition (Richards and Waterbury 2008, pp. 158–9).

After rebellious Iraqi army officers overthrew the monarchy in 1958, they
took advantage of outrage over the wide disparity in land ownership to pass
the far-reaching 1958 Land Reform Law. Promulgated less than three months
after the revolution, this law set an upper limit of 1000 donums (250 hectares
or 620 acres) for irrigated land, and 2000 donums for un-irrigated land. The
landowners who owned more than the legally allowed amounts of land chose
which land to keep, and the rest of the land was turned over to the authorities
for later distribution to landless cultivators. Compensation was to be paid
for the expropriated land. Government-sponsored co-operatives were to be
established to provide services to the newly independent cultivators, especially
seed, fertilizer, access to equipment, and marketing services. Because of the
wide variance discussed above, the redistribution of land affected relatively

few landowners – less than 3000 – but it did affect a great deal of land: almost two-thirds of all agricultural land.

The intention of the 1958 Law was to provide land to all cultivators. The authorities were to distribute 30–60 donums of irrigated land or 60–120 donums of un-irrigated land to each cultivator. With reasonably fertile soil, sufficient water, and equipment, this was sufficient land for cultivators to make a living even if they were growing grain.

Severe opposition to this law meant that the initial wave of land reforms was not imposed on the Kurdish areas or in Maysan province on the Iranian border. Even where the law was imposed, progress was very slow. After five years, only 6 percent of the confiscated land had actually been distributed to 29 000 cultivators, and only 25 co-operatives had been formed, so new small landowners lacked financial and technical support (Owen and Pamuk 1999, pp. 164–5). Since Iraqi agriculture still reflects the effects of the 1958 Law, it would be valuable to discuss five of the unexpected consequences.

First, the division between irrigated, un-irrigated, and non-farmland was often difficult to establish. If annual rains are early and heavy, some fields that are usually irrigated might be rain-fed. On the other hand, during a drought, irrigated areas will increase. Since large landowners were generally educated, politically influential, and had access to financial assets, they were often able to get favorable rulings on the nature of land in order to maximize the amount that they were allowed to retain.

Second, the large landowners tended to choose their 1000 or 2000 donums portions so that they controlled access to water and transportation. As a result, cultivators on the remaining land often had difficulty in obtaining access to water and markets on a timely basis. As a result, many recipients of land were unable to support their families by farming and were forced to abandon the land and seek a better living in urban areas. The total amount of land under cultivation declined.

Third, the 1958 Law was silent on tractors and other equipment. As a result, many large landowners took all or the best equipment to work their remaining holdings. The 1958 Law called for the establishment of farmers' co-operatives to provide equipment to small cultivators, but these co-operatives generally failed to meet cultivators' requirements.

Fourth, on large land holdings, fields were previously allowed to remain fallow every other year in order to restore their fertility. Small cultivators could not afford to do without earnings during a fallow year, and the resulting constant cropping led to a rapid loss in fertility or required increasing amounts of costly chemical fertilizer.

Fifth, the government was unable to rapidly distribute the expropriated land to cultivators. As a result, the government became the largest landowner. Also, the government-sponsored co-operatives lacked both the knowledge and the equipment to support the small cultivators that had received redistributed land. Over time, cultivator support for these co-operatives decreased and they became part of the government's efforts to control farmers (Mahdi 2000, Chapter 7; Owen and Pamuk 1999, pp. 169–71; Rivlin 2009, pp. 139–40).

Attempts to deal with these unintended consequences of the 1958 Land Reform Law as well as the impact of the Iran–Iraq War, invasion of Kuwait, sanctions, and the 2003 invasion by US-led coalition forces led to a complicated series of often reversed, often contradictory agricultural policies. (For a detailed discussion of the twists and turns of land reform, two works by Kamil Mahdi stand out: *State and Agriculture in Iraq* from 2000, and *Iraq's Economic Predicament* from 2002). The most substantial impact of the extreme uncertainty concerning land ownership rights has been to create a short-term perspective among farmers.

In addition, the ISIS occupation of 40 percent of Iraq, including some of the most fertile land in the north and west of the country, was a disaster for farmers. As discussed above, there was widespread destruction of irrigation canals and pumps, silos, and other agricultural equipment. In addition, there was increased land ownership uncertainty. Many farmers were forced to "sell" their land to ISIS individuals or groups. Others fled the occupied areas, and their land was taken over by others. Finally, land records were lost or deliberately destroyed as ISIS sought to ensure that its defeat would lead to chaos. As a result, in Ninawa and the other provinces formerly occupied by ISIS there is great uncertainty concerning who is the legal land owner. Those currently in possession are often afraid that a judicial decision or political gambit will transfer their land to someone else. Of course, this discourages farmers making long-term investments to increase agricultural production.

PUBLIC DISTRIBUTION SYSTEM

Due in part to the devastations caused by ISIS and the associated increase in internally displaced persons, as well as the Covid-19-related disruptions to the food supply, an estimated 4.1 million persons require humanitarian assistance, including 920 000 Iraqis who are food-insecure (USAID 2020). Currently, the majority of the food-insecure are located in provinces previously occupied by ISIS: Ninawa, Anbar, Salah ad Din, Diyala, and Kirkuk (FAO 2020c). To reduce incidence of hunger, the Public Distribution System (PDS) provides a "basket" of basic foods to every Iraqi family.

The PDS, which is probably the most popular government program in the country, had its origin in the mid-1990s. The international sanctions imposed on Iraq after its 1990 invasion of Kuwait led to severe food shortages. To maintain political support, the GoI began to distribute certain foods. This program expanded after the 1995 establishment of the corrupt United Nations

(UN) Oil-for-Food Programme, which allowed Saddam's Iraq to export oil in exchange for imported food. Under the PDS, every Iraqi is supposed to receive a monthly "basket" of 11 basic food items, of which the most critical are wheat flour (9 kg), rice (3 kg), sugar (2 kg), and vegetable oil (1 liter) (World Food Programme 2015, Table 1). The wheat flour component alone amounts to 435 000 tons per month (Ridha 2020, p. 12).

Most of the PDS food is imported and then delivered to over 400 warehouses. Then thousands of local grain merchants obtain the food from the warehouses and distribute it once a month to each family's representative. It is a paper-driven process, and it is difficult or impossible for families to receive their PDS allotment outside of their hometown, which caused a severe food shortage among persons displaced by ISIS. In addition, there have been reports that some families were denied their PDS baskets because they were of the "wrong" ethnic or religious group (Woertz 2017, p. 515). The PDS, although managed by the Ministry of Trade, is actually executed by a combination of private firms and government organizations. On the government side, there are an estimated 27 organizations including ministries, state-owned enterprises (SOEs), and other government units which are engaged in this complex process, with one regional, 18 provincial, and about 120 district-level authorities (Fathallah 2020). As expected, the incredible complexity of the PDS system results in inefficiencies, opportunities for corruption, and much higher costs.

In 2019, the cost of the PDS program was an estimated $1.4 billion (Fathallah 2020). While the food in PDS baskets is supposed to be free, there are various official fees, and many Iraqis report paying unexplained excess fees.

As a result of corruption on the part of the Ministry of Trade and the entities that operate the PDS, some of the better-quality food imports intended for the PDS are diverted into the Iraqi or foreign markets, and either lower-quality items were substituted, or the items were not provided that month. One survey showed that over 80 percent of the baskets were missing at least one item or had less quantity than promised. The Ministry of Trade is considered to be one of the most corrupt ministries in Iraq.

It is estimated that the PDS provides 70 percent of the calories consumed by households whose incomes are the bottom 40 percent (Krishnan et al. 2019, p. 92). Calculating the cost of each basket is a difficult question. Published data on the PDS by the Ministry of Trade is incomplete and is generally considered unreliable. And these estimates exclude the salaries, pension costs, and so on, of the Ministry of Trade and other government employees. Even if employee compensation were included, this would still be an underestimate since it fails to include subsidies received by these ministries and PDS-related SOEs for electricity, water, and fuel. One estimate is that the GoI spends about $6.30 for every $1 worth of food provided by the PDS (Ridha 2020, p. 13).

Iraqi households purchase many food items that are not available in the food baskets, especially fresh fruit and meat. As a result, despite receiving heavily subsidized food, most Iraqis still spend a significant portion of their household income on food. While there is a wide divergence across provinces and income levels, the average Iraqi household spent almost 43 percent of its total expenditures on foodstuffs and non-alcoholic beverages (Krishnan et al. 2019, Figure 1, p. 94). The next-highest categories of expenditures were dwellings (including water, gas, electricity, and fuel) and transportation.

What is Wrong with the PDS?

The PDS, as currently structured, is an expensive program that distorts incentives for both domestic food production and consumption. However, modifications of the PDS run the risk of increasing the food vulnerability of the poor and will be resisted by both the public and the bureaucracy.

The impact of the PDS on GoI budget expenditures can be viewed from two contradictory perspectives. First, the PDS represents a transfer of oil export earnings to (until the recent exclusion of higher-income government workers) almost every household in Iraq. It is possible to defend this transfer as favoring more diversified economic development. One might think of the monthly PDS basket as being the Iraqi equivalent of the annual check that each Alaskan resident receives from that state's oil production. A second perspective on the impact of the PDS on the GoI budget is that the PDS funds could be better used for other government expenditures.

The PDS is an inefficient welfare program in that the cost per basket is high compared to the market value of the food provided, and the baskets go to many high-income individuals. The GoI with the assistance of the UN World Food Programme, has begun a multi-year program to decrease the cost of each basket while excluding high-income individuals as recipients. But even if these cost reductions are achieved and the number of recipients reduced, the adverse impact of the PDS on both consumption and agricultural production will continue.

Although there are exceptional years, most of the cereal distributed as part of the PDS is imported by the Ministry of Trade, which has resulted in lower domestic food prices (World Bank 2017, p. 95). There are regulations that forbid the export of grains. As a result, Iraqi grain farmers sometimes find it difficult to sell their production in Iraqi markets at a price that covers their costs of production. In order to prevent the collapse of domestic grain production, the Ministry of Agriculture not only provides water, fertilizer, and seed free or at heavily subsidized prices, but also purchases the output at non-market prices. With the GoI determining input and output prices for grain producers, these farmers are really government employees, one step removed.

The PDS also distorts consumers' incentives. Since the items in the basket are available at a fraction of market prices, these increase the relative prices of items such as fresh fruit and meat. As expected, this has led the poor to consume a diet deficient in these relatively expensive items (World Bank 2006a, p. 52).

Of course, the PDS also represents a substantial increase in family income, in that families do not have to purchase PDS items, which frees up income for other purchases. Such transfers almost double the effective household income of poor households. If, instead of a PDS basket, the head of a family received its cash equivalent of approximately 36 900 ID ($31) per person per month, then they might prefer to purchase non-food items such as medical care, education, and so on. While recipients have the option of trading or selling items from their basket, the local markets for these items are so inefficient that there is a strong incentive for Iraqi families to consume the basket items rather than sell or trade them. Therefore, the PDS probably increases the proportion of income that is spent on food.

AGRICULTURAL POLICY

Government's Role in Agriculture

The recent history of agricultural policy among developing countries is riddled with market and government failures. Programs introduced with great enthusiasm are quietly abandoned years or decades later after their unexpected consequences become too serious to ignore (Stiglitz 1987, pp. 400–403). Economic development theory identifies three possible Iraqi government roles in agriculture that have alternated over the last six decades (Timmer 2005, pp. 394–6). First, there is benign neglect. This approach is usually accompanied by the belief that industry is the key driver of economic development, and trust that open market economies will outperform economies closed to international trade and finance. This policy of benign neglect roughly describes the situation in Iraq through the 1950s.

The second possible role for agriculture is as part of an integrated development plan. The government develops, coordinates, and executes a detailed plan that incorporates all sectors of the economy including agriculture. Agriculture product and input prices are managed along with quantitative restrictions, especially with respect to foreign trade and sometimes with respect to regional trade within a country. To the extent that the Arab Ba'athist Socialist Party had an explicit agricultural policy, it was one of state control primarily to achieve food security (GoI 2010, Section 5.1, p. 62). However, as was discussed above, this policy was subject to many contradictions and reversals based on the state of government finances and access to imported food. It gradually became apparent that government planners lacked both the necessary information to efficiently manage agriculture as well as effective tools to force farmers to act precisely according to the plan.

The third possible role is an awkward combination of the first two. Having recognized that direct government intervention to correct market failures in agriculture frequently leads to a further deterioration, while the free-market approach miscarries due to "missing" markets especially with respect to risk contingencies, a combined approach is attempted. The government reserves the right to intervene but limits its intervention to influencing key agricultural prices rather than direct quantitative intervention. At this high level of abstraction, Iraq might be viewed as gradually moving from the direct quantitative intervention of the second approach to the "market policy" third approach. However, if the National Development Plan: 2018–2022 is an accurate guide to GoI agricultural policy, then the government reserves the right to continue to intervene in agricultural markets, especially in the areas of irrigation infrastructure (GoI 2018, pp. 133–4). Progress will be uneven not only because the third approach is very demanding with respect to data and analysis, but also because there is a large sector of the bureaucracy that sees the abandonment of the direct intervention approach as reducing their authority.

As in previous chapters, there is a strong temptation to provide a long list of difficulties along with possible methods of ameliorating them. However, in view of the limited capability of the GoI for making timely structural changes, if might be more useful to focus on a few initiatives that aim at relaxing the more serious binding constraints.

Water Price Reform

The increased water usage in Turkey, Syria, and to a lesser extent Iran, combined with the increased Iraqi household and industrial use and widespread irrigation waste, are leading to a growing water shortage. The sheer complexity of these water-related challenges facing Iraq would probably defeat any attempt to resolve them by better planning and coordination at the ministerial level. Even disregarding the continuing problem of corruption at the Ministries of Trade, Agriculture, and Water, there is a shortage of personnel with the appropriate skills set to accurately estimate the costs and benefits of water usage throughout Iraq or to manage this complex system even if the cost–benefit analysis is completed in a timely manner.

The alternative of establishing a formal market price for water would be extremely controversial. Of course, water is already bought and sold informally under several guises. In urban areas, families often buy potable water from tanker trucks rather than risk the possibility of disease from the low-quality free water available from the public system (World Bank 2017,

p. 98). Entities including tribes that control the distribution of irrigation water expect financial, physical, or political support in return. Government officials are widely believed to be open to bribes to divert water. Finally, access to cheap (or free) irrigation water has become capitalized into the value of land. In fact, cultivators can be expected to resist creating a positive water price or raising existing water prices since this will lead to a drop in the rental value – a partial expropriation – of their land (Richards and Waterbury 2008, p. 173). In addition, at least one ministerial spokesperson stated publicly, without collaborating details, that selling water violates the Qur'an.

However, a formal pricing of water would provide the proper incentives for not only the efficient division of the scarce resource among alternative uses but also the efficient use by end users. Water charges should reflect not only the volume of water but its quality and timeliness as well.

One option is to divide the provision of water into a merit good amount that should be provided to everyone at low or zero price, with any demand above this merit good amount priced at least at its marginal cost. For example, every Iraqi would receive their minimum water right of, say, 125 m^3/yr but any use of water above this amount must be purchased (Richards and Waterbury 2008, pp. 165–6). Since the primary determinates of water supply – the dams, barrages, major irrigation canals, and urban water systems – will remain under public control, it will be necessary for the authorities to set an official price for all water above the minimum water right. The most efficient price would equal the long-run marginal cost of water: the cost of supplying an additional liter of water. Among Middle Eastern countries, only Syria charges such a price (Richards and Waterbury 2008, p. 178, endnote 21). However, as is discussed in Chapter 12, in view of the large-scale corruption, Iraq may be better off charging its average cost – including capital costs – than the theoretically more appealing marginal cost.

A positive price for water can be expected to lead to better service. For example, the cost of connecting an additional residence to piped water and wastewater systems ranges from an estimated 600 000 ID (about $500) to 3 500 000 ID ($3000) (World Bank 2017, p. 97). Currently, with the expectation that the consumer will pay almost nothing, this cost must be from the national budget. However, in view of the collapse in oil prices combined with the expense of dealing with the Covid-19 infection, ministerial budgets are lean. And, almost without exception, each ministry, including the Ministry of Water Resources, concentrates on paying salaries, with little left for other expenditures (Müller and Castelier 2018). However, if suppliers were able to collect the cost of water from consumers then not only would the funds be available to pay for connections, but also the government would have a stronger incentive to actually provide the connections. Currently, every new connection produces expenses but no revenues.

If water is sold, it is likely that Iraqi agriculture will become the residual claimant. Surveys show that household demand is more inelastic than that of industry, which is more inelastic than that of agriculture (Richards and Waterbury 2008, p. 167). Facing a substantially higher cost of water will have both an income and a price effect on farmers. Paying for water will not only reduce farmers' net income but also reduce the value of their land (by eliminating the land's capitalized water value). If this income effect is not offset, one can expect more farmers to abandon agriculture and migrate to urban areas. One method of dealing with this income effect would be to grant water rights to farmers and allow them to either use the associated water or sell the rights to others.

With a positive water price, strong incentives would be created to reduce water waste, possibly by covering canals to reduce evaporation, repairing leaks, utilizing sprinkler or drip irrigation, and selecting crops that are less water-intensive. Among grains, rice is extremely water-intensive and would not be as popular a crop in Iraq in the absence of large water subsidies. In view of the estimates that over half of current irrigation water is wasted, increases in efficiency should allow more land to be brought under cultivation.

However, fighting water scarcity through improved irrigation efficiency is not a cure-all. Unintended consequences may include a reduction in the quality of water, especially if there is an accelerated expansion of urban and industrial water use (Richards and Waterbury 2008, p. 170). In addition, while more efficient water use can be expected to reduce salt poisoning of the soil by lowering the water table, it will make it more expensive to flood fields in order to leach out existing salts.

The restoration and maintenance of the dams, barrages, and irrigation canals will be a major burden on both the budget and the administrative competence of the appropriate ministries. Since this subject is covered in Chapter 12, only a single example will be mentioned here. Due to poor site selection and inappropriate construction, the major dam on the Tigris northwest of Mosul has been in a process of slow-motion collapse since before it was completed. Daily reinforcement of the dam's foundations has so far prevented disaster, but a permanent repair may require temporarily draining the huge lake behind the dam, with implications for irrigation and other downstream users.

Land Reform

The series of contradictory land reform initiatives since 1958 have resulted in a great deal of uncertainty about agricultural land ownership. This uncertainty has encouraged a focus on short-term gains since cultivators and others (who uncomfortably share land ownership rights in Iraq) lack the confidence that current legal relationships will not be overturned. GoI proposals to further reform the legal environment of the agricultural sector (see GoI 2010, p. 75) further exacerbate this uncertainty.

The most-needed land reform in Iraq is a long-term commitment to cease land reform. This would gradually lead to a longer-term perspective on the part of the cultivators and others, which would increase their willingness to invest in the land. Since the GoI lacks credibility, any announcement that no further land reform initiatives are planned will probably have little effect on incentives. However, if the central government can avoid making any land reform proposals for an extended period of time, then gradually farmer behavior will change.

PDS

The PDS situation is complex. It provides essential nutrition to many poor families in Iraq; without the PDS, the rate of malnutrition would be much greater. However, as currently structured, it combines strong incentives for corruption with disincentives for domestic agriculture. Two overlapping recommendations have received attention. One option would be to exclude high-income families from receiving the PDS baskets. In addition to reducing the cost of the PDS subsidy, this would force higher-income families to increase their purchases in the local food markets, providing more opportunity for local producers.

In fact, the GoI is currently engaged in an effort to reduce the expense of the PDS program by reducing its coverage. Ministerial employees who earn more than 1.5 million ID per month (about $1300) are being excluded from the PDS distribution lists. About 70 000 government employees exceed this limit, which, since their family members would be excluded as well should lead to an eventual reduction of 300 000 to 500 000 persons on the PDS rolls. In 2010, the GoI began the difficult process of attempting to identify high earners in the private sector so their families can also be excluded from receiving PDS baskets (IRIN 2010b). By restricting this pruning effort to high-income families only, the GoI is seeking to reduce the burden of the PDS on the national budget and provide a significant increase in domestic food demand while avoiding any increase in malnutrition.

A second option would be to gradually replace the PDS baskets with their cash equivalent (GoI 2007, p. 17). For example, the head of the family would have the option each month of accepting the PDS basket or the value in dinars of the basket. If the choice, quality, and prices of food in the local market were acceptable, many families would choose the cash equivalent. This would not only increase demand for domestic agricultural products but also make corruption in the PDS system more difficult. It is easier for local grain distributors to pass a counterfeit food product (for example, substitute lower-quality cooking oil) then to pass counterfeit money. As was discussed above, it will be difficult to determine the cost of replacing each PDS item in the local markets. Offering the head of each family the choice of taking either the food basket or the estimated cash equivalent would provide a mechanism to ensure that the cash offered was sufficient. If most heads of families chose to take the food basket then this will make it clear to the government that the cash alternative is insufficient.

However, there may be an adverse shift in family spending. In traditional Iraqi families, food is a woman's responsibility while money is handled by the man. If instead of a PDS basket going to the woman to feed her family, there was a cash payment to the man of the family, it is possible that he will spend part of these funds on non-food consumption.

Agribusiness

While farming in Iraq is unlikely to provide substantial employment growth, there is potential for such growth from agribusiness involving the processing, packaging, transporting, and selling of agricultural products. A large amount of Iraq's food imports are items such as frozen chickens and canned tomatoes that could potentially be produced domestically. However, it is important to distinguish between the potential for grain and non-grain agribusiness development.

Because of the complex web of subsidies, non-market prices, and detailed regulations, it is extremely difficult to determine the true cost – the market cost – of wheat and rice production in Iraq. Complicating any analysis is the fact that average productivity of cereal growers in Iraq is poor even by Middle Eastern standards. It can be argued that if the web of regulations, subsidies, and price controls were rationalized then this would lead to a substantial increase in the productivity of cereal production.

However, the administered price for Iraqi wheat and rice purchased for the PDS is substantially greater than the import price; sometimes twice as great. Therefore, it is unlikely, after all costs are considered and adjusting for quality, that even with a great increase in productivity that Iraq will be able to produce cereals for the domestic market cheaper than imported wheat (from, say, Argentina, Australia, Brazil, Canada, and the USA) or imported rice (from Thailand and the USA).

Even if the policy-imposed inefficiencies are disregarded, the barriers to efficient grain production are still high. Water is scarce and expected to become scarcer in the future. This is especially true for rice, which uses about three times more water per hectare than vegetables and six times more water than wheat. In addition, Iraq lacks the large farms with advanced machinery that exemplify high-quality, low-cost, large-scale wheat production. Finally, from 2005 to 2020, Iraq suffered from a severe case of the "Dutch disease" as appreciation of the Iraqi dinar resulted in lower prices for imported products. While the December 2020 appreciation of the dinar is expected to eventually improve the international competitiveness of Iraqi agriculture, this process will take time.

One strategy for dealing with this long-term lack of competitiveness in Iraqi grain production is to accept the necessity of continuing large-scale subsidies in order to preserve domestic production, rural life, and employment. Of course, this is a political decision. But there are at least two economic issues that should be considered. First, as is discussed throughout this book, Iraq faces many expensive claims on its national budget. Unless there is a sharp increase in petroleum export volumes combined with higher world oil prices – a contradictory combination of events – the government and people of Iraq will have to choose. Funds spent on subsidizing grain production will not be available for other needs. Second, the subsidies and regulations necessary to preserve large Iraqi grain production have severely distorted the incentives facing cultivators.

While the trend is adverse, Iraq still possesses a generous water supply – compared to other countries in the region – and a large skilled agricultural labor force. Therefore, Iraq has the potential of becoming a large food source in the Middle East. But rather than focus on grain production for the domestic market, it is probably more beneficial for the country to follow its comparative advantage in labor-intensive fruits, vegetables, and non-cattle meat such as goats and chickens. This will provide for domestic consumption and also, if the challenges of processing and transportation can be solved, for substantial exports, especially since Iraq is surrounded by relatively prosperous nations that, with the exception of Turkey, are major food importers.

Iraq's crude oil exports supplemented by agribusiness exports based on fruit, vegetables, and meat will allow the importation of grain and grain products at lower opportunity costs than if these items were produced domestically. In one sense, Iraq is in an enviable situation with respect to international commodity trade because the relative price of its major export – crude oil – has remained relatively stable compared to that of its major import, wheat. This stability is partially a function of the fact that the international trade in both oil and wheat is denominated in US dollars. Therefore, changes in the value of the dollar tend to have offsetting effects on the international prices of oil and wheat.

Farmers' own-price elasticity is low; for example, a rise in grain prices will not lead to a significant increase in grain production. But, at the same time, farmers' cross-price elasticity of supply is high; for example, a fall in grain prices relative to fruit prices will lead to a significant shift from grain production to fruit production. In other words, farmers tend to be highly responsive to price signals among crops (Richards and Waterbury 2008, pp. 161, 163).

Therefore, in order to motivate cultivators to shift from grain production to products where Iraq has comparative advantage will require changing relative prices. Changing the relative prices of grain and other crops can be done either by reducing or eliminating the subsidies for grain production; or by imposing or increasing subsidies for non-grain agricultural production.

The first option should result in greater efficiency over time. The National Development Strategy 2007–2010 (GoI 2007, p. 36) noted that initial steps had been made towards enabling markets in agricultural inputs, but progress has been uneven. Unfortunately, the GoI is also using the second option. In early 2010, the GoI imposed an import ban on tomatoes, with the expected results. This reflects in part the inertia of statist economies that find it much more acceptable to impose a new distortion in an economy then to remove an existing distortion. The import ban resulted in a rise in tomato prices, and farmers seem to be working towards increasing future production. However, the rise in prices reduced tomato consumption among low-income families and led to an increase in corruption as a result of increased smuggling.

Of course, increasing reliance on market prices in Iraq's food economy will not guarantee that the country will experience higher quality, lower prices, and increased agribusiness employment. But without a rationalization of the prices of agriculture inputs and products, it is unlikely that substantial progress towards national agricultural goals will be met. Agricultural price reform is not a sufficient condition for economic growth, but it is necessary.

9.   Financial intermediation

In the eighteenth century BCE: Grain and other valuables in Uruk and other Sumerian cities were stored in temples. Some argue that these temples, reposing on the safest shores of the Tigris and Euphrates, became the first institutions to issue inscribed tablets, or tokens, as "receipts" for valuable deposits or trade purposes. It has been said that those secure and trusted institutions of safekeeping in ancient Mesopotamia were in fact the precursors of modern depository "banks".

(Edwin Black 2004, p. 17).

Along with corruption and a hostile regulatory environment (see Chapters 4 and 11), Iraq's dysfunctional banking system completes a trifecta of bad policy that both impedes improvement in the average Iraqi's quality of life and severely complicates the Government of Iraq's (GoI) attempts to diversify the economy.

Efficient financial intermediation serves several important purposes. It facilitates the conversion of private savings into much-needed investment. Also, by providing bill paying and other services, financial intermediaries reduce the transaction costs of business. Finally, it aids the monetary authorities in their difficult balancing act of providing sufficient liquidity to maximize real activity in the economy without fueling inflation. Providing the appropriate regulatory framework to allow financial intermediaries to achieve these ends is difficult, and especially so in countries such as Iraq that are making the transition from socialism to a more market-oriented economy.

As a measure of the long path ahead in achieving reasonable efficiency in financial intermediation, it should be noted that the contribution of the banking and insurance sector to Iraq's private sector is very small even by Middle East and North Africa (MENA) country standards. In 2018, it was estimated that Iraqi banks' credit extended to the private sector was less than 9 percent of gross domestic product (GDP). The average for MENA countries was about 55 percent, while for all countries at roughly the same level of economic development the average is 118 percent (World Bank 2020c, Table 5.5).

As a result of its socialist past, financial intermediation in Iraq is dominated by a moribund banking system. The country is severely "underbanked". In 2018, there were only 4.5 bank branches and 4.0 automated teller machines (ATMs) per 100 000 Iraqis. As a matter of scale, the average MENA country had 13.8 branches and 36.4 ATMs per 100 000 (World Bank 2020c, Table 5.5). Only 23 percent of Iraqi adults have any type of bank account. When asked why they do not have an account, 70 percent of Iraqis responded that they did not have sufficient funds to make it worthwhile. Other reasons given are the expense of an account, lack of trust in bank security, and distance to the nearest bank branch (Global Findex 2017). Only about 10 percent of all small and medium-sized enterprises (SMEs) in Iraq have a bank account.

Checking accounts, credit cards, and simple electronic transfers are generally unavailable; as a result, most transactions are in cash, with all of the associated security problems and inefficiencies. Self-employed Iraqis report that 81 percent of their earnings are in cash, while salaried employees are paid in cash 74 percent of the time. Even 40 percent of government payments are in cash (Global Findex 2017).

Mortgage loans are rare. For a private business to borrow is a complex and drawn-out process. Even if a loan is eventually approved, private businesses can borrow only small amounts at a high interest rate, and only after pledging substantial collateral. As a result, only 4 percent of Iraqis had an outstanding bank loan in 2017 (Global Findex 2017).

Even the use of banks as a safe place to keep funds is limited, since Iraq lacks core banking. With core banking, a person or business can deposit funds in one branch of a bank and withdraw these funds from another branch of the same bank. Without core banking, withdrawals can only be made from the same branch as the deposit.

Although it is technologically advanced, the Iraq Stock Exchange (ISX) has a relatively low market capitalization, only about 103 listed firms, very light daily trading, and almost all trading is in private bank and telecom shares (Rabee Research 2020a, p. 4). Future expansion of the ISX is dependent more on a substantial reduction in the GoI's regulatory hostility towards private business, rather than any technical fixes of the exchange's operations.

Like banking, the nation's insurance business and pension funds are dominated by state-owned entities, and these entities focus on providing insurance and pension services to other state-owned entities. Private sector insurance and pension services are in their infancy.

IRAQ'S UNBALANCED BANKING INDUSTRY

Public Banks

In 1965, the GoI nationalized all Iraqi private banks and all branches of foreign commercial banks. The government then gave Rafidain Bank, which had been established as a private bank in 1941, a monopoly of all banking in the country. In addition to providing banking services, Rafidain also acts as an executive fiscal agent for the GoI, responsible for making government disbursements, transferring tax receipts, borrowing on the government's behalf, and making policy loans to state-owned enterprises (SOEs). Due to its fiscal responsibilities, Rafidain administratively fell under the Ministry of Finance rather than the Central Bank of Iraq (CBI). Over the last two decades, Rafidain's net worth has deteriorated.

Public banks were forced to make loans to politically favored individuals and institutions and then roll over these loans in lieu of repayment. As a result, Rafidain's "bad loan" portfolio boomed during the economic turmoil of the 1980–88 Iraq–Iran War. In addition, Saddam and his family used the banks' deposits as personal sources of ready cash. Also, in a problem that continues until today, there was an absence of reliable timely audited accounting information. In 1988, the GoI attempted to deal with Rafidain's devastated balance sheet by establishing a new bank, Rasheed Bank, and transferring the bulk of Rafidain's bad loans to the new institution. Rasheed Bank is currently the third-largest bank in Iraq, after Rafidain and the Trade Bank of Iraq, with the second-largest number of branches (Central Bank of Iraq 2019, Tables 59 and 60, pp. 117–18).

At the end of 2018, there were 71 banks in Iraq: seven state banks, 46 private banks, and 18 branches of foreign banks. Despite the relative deterioration of the Rafidain and Rasheed banks, they still overshadow their private sector competition. In addition to Rafidain and Rasheed, there are five other state-owned banks that specialize in various sectors: Trade, Agricultural Cooperative, Real Estate, Industrial, and the Islamic State Bank. The Trade Bank, established in 2004, is the second-largest bank in Iraq by assets and focuses on supporting foreign trade and investment. It was created in response to Saddam-era lawsuits in several nations that made the assets of the existing financial intermediaries vulnerable to attachment. The Trade Bank is one of the few state-owned banks that not only has audited accounts but also publishes them online. Since it was originally intended to be a temporary fix pending a resolution of the foreign lawsuit problem, the Trade Bank of Iraq was not integrated into the existing financial regulatory system. As a result, the Trade Bank, until 2011, was perceived to have a degree of independence exceeding that of other state-owned or even private banks. However, in that year, the GoI replaced the Chairman of the Trade Bank with a former Rafidain Bank manager in an effort to reduce its independence (Gutman 2011, p. 3).

In 2018, Rafidain and Rasheed, along with one large and four smaller state-owned banks, possessed estimated total assets of about 94.9 trillion Iraqi dinars (ID), about $80.3 billion. This is equal to roughly 77 percent of all bank assets in Iraq (Central Bank of Iraq 2019, Table 59, p. 117). The state-owned banks have a nationwide system of branches that include most cities and large towns but generally exclude smaller towns and villages. With possibly one exception, these public banks are believed to be insolvent and grossly mismanaged. They generally suffer from large-scale overemployment and are extremely bureaucratic.

Private Banks

There were also 46 private banks in operation at the end of 2018, including 22 Islamic banks. These private banks account for 19 percent of total banking assets. While private banks had an estimated 389 branches, their branches are generally only located in major cities and provincial capitals. There are many large towns without any private bank branches. While it is difficult to obtain timely audited accounting statements, it is thought that about half of the private banks are insolvent. Most of the private banks' management personnel are former employees of Rafidain Bank, and the quality of management varies widely. The remaining 4 percent of total bank assets are held by the 18 banks in Iraq that are branches of foreign banks. Two of these foreign bank branches are Islamic banks (Central Bank of Iraq 2019, Table 59, p. 117).

There is a wide divergence among private banks in age, asset size, number of branches, organizational structure, and business strategy. The oldest private bank and the second-largest private bank by total assets, the Bank of Baghdad, was established in 1992; but almost two-thirds of the nation's banks are less than two decades old, established after the 2003 overthrow of Saddam's regime. Private banks range widely in size. There are three banks with more than 1 trillion ID ($846 million) in assets: the Mansour Bank for Investment (the largest private bank), the Bank of Baghdad, and the Kurdistan International Bank for Investment and Development. At the other extreme, there are three banks with less than 100 billion ID ($85 million) each in assets. There were formerly many smaller banks but these merged with their larger competitors after the CBI announced that all private banks must meet minimum capital standards by mid-2013.

Most private banks have their headquarters in Baghdad, although 11 have their headquarters in the Kurdish Regional Government (KRG). Individual private banks have relatively few branches. The Bank of Baghdad has the most branches – 30 – but 12 of these branches are in Baghdad. None of the private banks has branches in all 18 provinces of Iraq.

There are 22 private Islamic banks that are dedicated to Sharia-compliant operations. These Islamic banks have 119 branches and total assets of 8.2 trillion ID ($6.9 billion) which is equivalent to about 7 percent of the nation's total bank assets.

There are no Islamic banking regulations, nor even a consensus among Islamic legal authorities on the exact necessary characteristics of Islamic financial institutions. However, such institutions generally seek to avoid *riba* (usually translated as interest), *gharar* (risk or uncertainty), and *maysir* (gambling). In addition, Islamic financial institutions seek to avoid supporting activities that are *haram* (forbidden activities or products such as alcohol consumption, pork, or pornography). These restrictions are interpreted within the Sharia framework that good Muslims should practice brotherhood, ensure the fair remuneration of labor, and provide alms to the poor, *zakat*. Islamic banks have a Sharia committee to ensure that the banks' activities are compliant.

Restrictions on interest and some other financial activities should not be interpreted to mean a generalized religious hostility to capitalism. Visser (2009, p. 48) quotes Mahmud Ahmad on the differences between capitalist, socialist, and Islamic economics: "Capitalism accepts both profit and interest, socialism rejects both, and Islam accepts the profit motive but rejects interest". As discussed in Chapter 11, far from being hostile to business, many historically prominent Islamic religious leaders were successful businessmen.

The prohibition of *riba* or interest is the most prominent characteristic of Islamic finance. But interpreting this prohibition is complex, and permitted interest-related activities can vary not only among Islamic countries but also among different institutions in the same country. (For detailed discussions of the various Islamic views of economic activity, see Kuran 2004, Chapter 5, pp. 103–20; Visser 2009, Chapter 3, pp. 25–51).

For example, in Iraq, some Islamic banks permit interest to be paid on deposits, since this benefits an individual depositor rather than an institution. Restrictions on interest on loans are more common. But even for loan transactions, some institutions distinguish between consumption and production loans. These institutions may allow interest on consumption loans providing the loan terms are clear to the borrower, the loan only carries simple – not compound – interest, and the rates are not unacceptably high. With respect to loans for productive purposes, almost all Islamic financial institutions in Iraq seek to avoid the prohibition on *riba* by structuring transactions so that the lender shares the risk of loss from each loan transaction. One of the most widely used Sharia compliant means of structuring such a transaction is *murabaha*. An example of a *murabaha* transaction would be where a manufacturer wants to purchase raw material for his plant. Rather than a bank simply making a loan that would have to be repaid regardless of whether the manufacturer made a profit, an Islamic bank might purchase the raw material and provide it to the manufacturer in return for a portion of the profits. If the manufacturer's initiative fails, the bank also suffers a loss.

Islamic banking is a relatively recent innovation in Iraq; only two institutions predate the fall of Saddam. So far, the general public's interest in doing business exclusively with Islamic banks appears to be limited to some residents of the Shi'a sacred cities as well as conservative Sunni in Anbar province. Surveys show that most Iraqis consider Islamic banking to be one of several options; they will borrow from wherever they can obtain the best terms.

Table 9.1   *Bank assets, 2018*

Loans to	Central government	Public institutions	Private sector	Total
By state banks	15 595 bn	2 675 bn	12 877 bn	31 148 bn (81%)
By private banks	0 bn	0 bn	7 339 bn	7 119 bn (19%)
Total	15 595 bn	2 675 bn	20 216 bn	37 953 bn (100%)

Source: CBI (2019, Table 9, p. 29).

At the end of 2018, 16 private foreign banks – including two Islamic banks with partial foreign ownership – had licenses to operate in Iraq. Foreign ownership ranged from 49 percent to 85 percent, with three banks having majority foreign ownership. Banks with foreign ownership have 44 branches and total assets of 4.5 trillion ID ($3.8 billion) which is equivalent to about 4 percent of the nation's total bank assets. Aside from security issues – much improved since the defeat of ISIS in 2017 – the binding constraint to a more rapid expansion of foreign banking in Iraq is the difficulty in hiring skilled local personnel, especially for management positions.

Research has shown that when foreign ownership of a bank exceeds roughly 70 percent, there are measurable increases in banking efficiency. Of course, this increase in efficiency tends to promote economic development. However, in some former socialist countries the expansion of foreign-owned banks has tended to increase banking concentration as less-efficient domestically owned banks lose market share (Haiss et al. 2005, p. 4).

BANKING ASSETS AND LIABILITIES

With respect to state-owned banks, Tables 9.1 and 9.2 reveal a straightforward pattern of bank assets and liabilities: state-owned banks obtain deposits from both government and non-government entities, and then lend these funds to government entities. Almost 74 percent of the liabilities of these banks are government or other public institutions; while almost 59 percent of the assets are deposits at the CBI, loans to government agencies, and loans to SOEs. According to the official statistics, excess bank reserves reached 16.3 trillion ID at the end of 2018, about $13.8 billion (CBI 2019, Table 15, pp. 41–2). Due to the high interest rate paid by the CBI on reserves, all of the state-owned banks are profitable; and, of course, due to their large holdings of reserves, they are very liquid.

These excess reserves are equal to almost 81 percent of the amount of credit provided to the non-government economy. Despite their inefficiencies,

Table 9.2   *Bank liabilities, 2018*

Deposits by	Central government	Public institutions	Private sector	Total
To state banks	21 800 bn	27 069 bn	17 226 bn	66 095 bn (86%)
To private banks	472 bn	188 bn	10 138 bn	10 799 bn (14%)
Total	22 272 bn	27 257 bn	27 364 bn	76 894 bn (100%)

Source: CBI (2019, Table 8, p. 27).

state-owned banks receive a large inflow of deposits for which the banks pay zero or low interest. Not only do government entities and state-owned enterprises deal almost exclusively with state-owned banks, but also Rafidain Bank and Rasheed Bank receive large amounts of deposits from private firms and individuals. The primary motivation for these deposits is the lack of other secure savings instruments and the almost universal belief that deposits at state-owned banks, unlike their private sector competitors, are guaranteed by the GoI.

Interest rates are not market determined in Iraq. Following a 2004 agreement between the CBI and Ministry of Finance, the CBI announced that, pending the development of a liquid secondary market in short-term GoI debt, the CBI will announce a Policy Rate that will not only signal its target for interbank overnight transactions but also provide a benchmark rate from which the overnight deposit as well as the rates at which the CBI will lend to banks will be determined (CBI 2004, p. 6). This policy rate was initially 6 percent, rose to 20 percent for 12 months in 2007–08 as part of the anti-inflation campaign, before declining back to 6 percent in April 2010, where it remained until March 2016 when it was reduced to 4 percent (CBI 2020a).

The patterns of assets and liabilities of Iraq's private sector banks are very different from those of state-owned banks. As can be seen in Table 9.1, private banks' assets were dominated by loans to the private sector. However, because the total assets of state-owned banks are over four times larger, their total credit to the non-governmental economy is actually larger than that of private banks: 12.9 trillion ID ($10.9 billion) compared to 7.3 trillion ID ($6.2 billion). It should also be noted that private banks have a lower proportion of their total assets on deposit with the CBI.

On the liabilities side, Table 9.2 shows three sharp differences between state-owned and private banks. First, private banks report zero deposits from government agencies, since such agencies are legally forbidden to make deposits at private banks. Second, private banks have zero exposure to foreign entities. This is caused in part by legal limits on providing services to foreign entities. For example, the state-owned Trade Bank of Iraq has a monopoly on issuing letters of credit to finance trade, although it may delegate low-value

Table 9.3   *Bank assets by sector*

	Total	State banks	Private banks
Total	37 180 (100%)	29 849 (80%)	7 330 (20%)
Society services	14 509 (39%)	13 520	989
Construction	7 920 (21%)	6 672	1 248
Hotel/restaurants	5 656 (15%)	1 638	4 018
Transport	2 776 (7%)	2 543	233
Agriculture	2 129 (6%)	2 047	82
Manufacturing	1 855 (5%)	1 348	506
Water/gas/electricity	1 462 (4%)	1 379	82
Finance/insurance	842 (2%)	684	158
External world	32 (~0%)	18	14
Mining	0.8 (~0%)	~0	0.8

letter of credit authority to private banks. Finally, private banks have a greater percentage of bank capital, reserves, and loan-loss provisions than state-owned banks: 30 percent compared to 5 percent. Although, in the absence of quality auditing, these estimates of bank capital, reserves, and loan-loss provisions should be taken with more than a few grains of salt.

Another distinction between the assets of the state-owned and private banks is the breakdown of their loans. As can be seen in Table 9.3, three-quarters of the loans of state-owned banks were accounted for by three borrowers: social services that is, government (45 percent of all loans), construction (22 percent), and transport (9 percent). In contrast, two types of borrowers accounted for almost three-quarters of private bank loans: these were loans to hotels and restaurants (55 percent) and construction (17 percent). While private banks were more willing to make small commercial loans, most of their loans are to larger entities. In fact, less than 4 percent of SMEs in Iraq's formal economy have an outstanding bank loan (Global Findex 2017). It is believed that a large proportion of the assets of state-owned banks reported as loans to non-government entities are actually loans to either bank employees or politically connected persons. These loans facilitate corruption since, in many cases, neither the borrowers nor the bank actually expect the loans to be paid back.

The asset distributions of conventional, Islamic, and foreign-owned banks are fairly similar, with the possible exception of the fact that Islamic banks do not make loans to government agencies while foreign-owned banks lend more to such agencies than to non-government entities. This asset pattern of foreign-owned banks in Iraq differs from that in most post-socialist states, where foreign banks tend to lend more to the private sector (Haiss et al. 2005, p. 5). There is a similar story on the liability side. The distribution of liabilities is fairly similar among the three types of private banks, and very different from that of state-owned banks.

The loan process at both state-owned and private banks tends to be long drawn-out and bureaucratic, in part because a large percentage of all loans must be approved by the bank's board of directors. There is a widespread belief that to obtain a bank loan requires connections or a bribe. For all of these reasons, state-owned or private banks rarely lend to smaller firms in the formal economy, and entities of any size in the informal economy are excluded entirely.

Over the last decade and a half, private banks derived substantial profit by purchasing US dollars from the Central Bank and reselling these dollars to the private sector. During normal times, the spread on such transactions is 1–2 percentage points. However, when economic difficulties lead to a dollar short-age, such as in the second half of 2016 and the first half of 2020, the spread can widen to a very profitable almost risk-free return of as much as 10 percent (Tabaqchali 2020b; IMF 2019b, p. 34).

CHALLENGES FACING IRAQI BANKING

Need for Recapitalization

With the possible exception of the Trade Bank, the state-owned banks suffer from opaque accounting systems. The banks generally either fail to release the standard banking data in a timely manner or use entirely unrealistic assumptions that severely distort the bank's situation. The most extreme cases of unrealistic data involve the foreign exchange transactions of Rafidain and Rasheed banks. Acting as agents of the GoI, these banks borrowed large amounts denominated in foreign currencies and transferred the funds to the Ministry of Finance. However, as a result of the sharp depreciation of the dinar in the decade prior to 2003, the foreign liabilities of these banks – in terms of Iraqi dinars – exploded. Despite a detailed plan developed by the GoI in 2006, the necessary recapitalization of these banks by the GoI has not begun, nor have the discrepancies among state-owned banks' assets and liabilities been completely reconciled.

A 2008 study estimated that despite portfolios containing almost 60 percent of non-performing loans, the recapitalization requirements of Rafidain and Rasheed banks, although large, were lower than expected since both banks' loan portfolios were quite small. To bring both banks into compliance with Basel II (an international standard of safe levels of bank capital) would require 15.7 trillion ID ($13.4 billion) (IMF 2008b, Box 3, p. 16). This was equivalent to 18 percent of 2008 total budget expenditures or 14 percent of 2008 GDP. However, repeated initiatives to recapitalize these banks have failed (IMF

2019b, p. 35). The difficulty is not only the large expenditures required, but also a shortage of political will. It will be interesting to see the effect of the December 2020 dinar depreciation on state bank accounts. Based on fragmentary data, this depreciation may have substantially reduced the recapitalization requirements.

Lack of Deposit Insurance

Although non-government entities and persons account for almost two-thirds of private banks' total liabilities, further growth of such accounts is constrained by two factors. There is no program of deposit guarantees in Iraq, so many depositors choose to keep their funds in state-owned institutions with their implicit government guarantee. Also, the Ministry of Finance (MoF) has decided that checks drawn on private banks would not be accepted for tax payments. This decision was based on the difficulty that the MoF had cashing the checks of two small private banks. As a result, all private bank checks are excluded for an undetermined length of time from being used for tax and other payments to the GoI, which further reduces the attractiveness of having a deposit account at a private bank.

Lending to Private Entities

Private firms, especially SMEs, generally lack access to bank finance. Loans to individuals or non-SOE firms are generally only made for short-term trade financing. Generally, collateral – at a multiple of the value of the loan – is required. There is little cash flow lending. As a result, a survey of Iraqi businesses revealed that only 2.7 percent of firms were able to obtain bank loans to finance investment projects. Even excluding high-income countries, the average for the other MENA states was 23.7 percent (IMF 2019a, p. 38). Instead of bank loans, SMEs tend to rely on informal sources of finance such as from family or friends, the source of an estimated 52 percent of SME finance, and credit from suppliers, 33 percent (Global Findex 2017).

The dependency on informal finance has several implications for Iraqi SMEs. As in most trading cultures, larger merchants often give credit to smaller ones not as a distinct transaction but as part of a complex (especially to outsiders) business relationship. The fact that such credit transactions are part of more extensive relationships may explain why some borrowers prefer more expensive credit from their business partners than cheaper credit banks, microfinance institutions (MFIs), or even government-subsidized sources. However, financing from informal sources cannot fulfill all credit requirements because they tend to be small-scale; require intense family, social, or business connections; and are often expensive. There is also a concern that informal finance may involve criminal or insurgent groups. Difficulty of obtaining loans from any source was cited by almost 50 percent of Iraqi private businesses as negatively affecting growth of their companies. As a related note, issuing equity accounted for almost none of private firm financing.

The Iraqi Company for Bank Guarantees (ICBG) was established by 11 Iraqi private banks in August 2007. The ICBG is intended to reduce the risk to banks of making loans to micro, small, and medium-sized enterprises (MSMEs). The ICBG now provides partial guarantees of both principal and interest for loans to MSMEs by member banks. Member banks can obtain a guarantee for up to 75 percent of the loan principal, in return for a 2 percent loan guarantee fee. The allowable loans are extensive, with loans to manufacturing, services, tourism, trade, and agriculture entities allowed. The major exclusion is loans to purchase other financial instruments. Loans are permitted for either cash flow or collateral lending. Maturity is limited to one year for working capital and five years for loans for fixed assets (USAID 2007a, pp. 21–2).

Mobile/E-Banking

Among the complications of dealing with Iraqi banks is that not all banks have adopted "core banking". Core banking focuses on facilitating banking transactions by small businesses, such as being able to make a deposit in one branch of a bank and withdraw the funds in another branch of the same bank with little delay. In Iraq, some banks have a substantial delay before deposits are recognized, and any withdrawals must be from the same bank branch where the deposit was made. These delays and limited core banking help to support the dominance of cash in private and business transactions. In addition, the lack of core banking means that each bank branch keeps its own record of deposits and loans. This makes it difficult to audit an entire bank in a timely manner.

Not only are loans to private businesses rare, but it is also very difficult for private businesses to arrange electronic fund transfers to pay their suppliers or employees, to establish a checking account, or to obtain a credit card. Most banks do not offer ATM or credit card services, nor can they arrange for international payments. These inadequacies of Iraqi "brick and mortar" banks have created a competitive opportunity for mobile banking.

Mobile or e-banking refers to payment services performed from a mobile device. In Iraq, an increase in mobile banking can be expected to reduce the cost of individuals or businesses making payments and transferring funds. In addition, by providing an electronic trail that is lacking with cash payments, mobile banking can be expected to complicate corruption. Finally, mobile banking has the potential of rapidly reducing the proportion of the unbanked, without the expense of constructing more brick and mortar branch offices.

The potential for mobile or e-banking in Iraq is great (IMF 2019b, p. 37). Cellphone ownership is ubiquitous in Iraq, with about half of the phones "smart", that would allow mobile banking. However, only 4 percent of Iraqis have a mobile money account (Global Findex 2017). This is a substantially lower percentage than in countries with a comparative level of economic development. The binding constraint on an expansion of mobile banking are regulations that favor physical banks, especially if those banks are state-owned.

MICROFINANCE INSTITUTIONS

Since 2003, 12 microfinance institutions (MFIs) have been established in Iraq, specializing in small loans to individuals and entities that lack access to state-owned or private banks either because of lack of collateral or because small loans tend to be unprofitable because of higher costs. However, since 2010, the number of Iraqi MFIs has decreased sharply, for two reasons. First, some of the MFIs received their initial capital directly or indirectly from the US government as a way of supporting counterinsurgency operations. MFI loans were used as a means of reducing support for the insurgency by reducing unemployment among young men. (For an extended discussion, see Gunter 2009b). Following the withdrawal of US and other forces in 2011, these MFIs had difficulty replacing this capital. Second, the deterioration of security during the ISIS insurgency in 2014–17 led to severe loan losses, destruction of financial infrastructure, and the loss of key MFI personnel (Chehade et al. 2016). Going forward, it is expected that questionable physical security will cause Iraqi MFIs to experience increased operating expenses and lower loan quality as repayments are delayed.

As a result of these two shocks, there were only three MFIs still involved in large-scale lending in 2017. These MFIs had 63 500 loans outstanding compared to 75 200 loans in 2010, while the total 2017 loan volume was an estimated 128.1 billion ID ($108.4 million), about the same as in 2010. MFI loans tend to be outliers compared to those of other MENA country MFIs. Possibly reflecting their original counterinsurgency motivation, MFIs tend to make larger loans, with an average loan size of $1708 compared to the $560 average in the rest of the MENA countries. Also, Iraqi MFIs make fewer loans to women: 27 percent compared to almost two-thirds for MENA countries. Finally, Iraqi MFI loans are dominated by individual loans, rather than the solidarity group or Islamic loans that are more prevalent in the rest of the MENA countries (Microfinance Information Exchange 2018, pp. 25–29).

Iraqi MFIs also face two unique challenges. First, the Ministry of Labor and Social Affairs, the Ministry of Industry and Minerals, and the Ministry of Agriculture are making loans to the same types of borrowers as MFIs. The Ministry of Labor and Social Affairs was established to assist workers seeking employment and to protect Iraq's most vulnerable citizens through a network of social services. In October 2006, its responsibilities were expanded to create jobs by making low (or zero) interest loans (USAID 2007a, p. 2) on a very large scale. In fact, the first budget for this purpose was 600 billion ID ($510 million). This one-year budgeted amount is equal to roughly five times the total loans of all MFIs in that year. In 2015, the CBI launched an initiative to finance SME projects. Between 2015 and 2017, about 1100 projects received 1 trillion ID ($840 million) in loans, ranging from 5 million ID ($4200) to 50 million ID ($42 000) (IMF 2019b, p. 43). These loans have injured MFIs and private banks in several ways.

One concern is that the CBI and government bureaucracies such as the Ministry of Labor and Social Affairs, the Ministry of Industry and Minerals, and the Ministry of Agriculture will be more concerned with the volume – not the quality – of loans. Officials will see their primary responsibility in lending out the full amount as soon as possible, rather than ensuring that the borrowers will be able to pay it back. Also, interest rates on these loans are substantially below those charged by either private banks or MFIs. A further concern is that after a period of time the elected officials of the GoI will come under a great deal of political pressure to declare a debt holiday, suspending or canceling loan repayments to government entitles.

It has been argued that the need for loans to micro and small enterprises is so great that there is plenty of room for both ministerial lending as well as MFIs. However, the ministerial programs further weaken the culture of credit in Iraq. Individuals and businesses that have heard about low and zero interest government loans that allow repayment schedules to be continuously extended are confused when MFI officials explain that not only will interest be charged, but also it is expected that loans will be repaid according to the contracted terms. Some potential borrowers are suspicious – and maybe rightfully so – that they are being stuck with less advantageous terms on their loans because they lack the proper connections.

Second, the long-term sustainability of MFI depends upon being able to charge a high enough interest rate to cover both the cost of funds as well as the costs of operation. Since they were established, Iraqi MFIs obtained their loan capital mostly through increasingly scarce grants supplemented by retained earnings. In many MENA countries, MFIs also obtain loan capital by borrowing. However, if Iraqi MFIs seek to expand their loan activities by borrowing then they run the risk of reduced competitiveness. For example, in 2017, the yield on gross loan portfolio (interest and fees) was 26.5 percent (Microfinance Information Exchange 2018, p. 29). This is almost twice the rate charged by state-owned and private banks, and at least five times greater than the loan interest and fees charged on loans from the Ministry of Labor and Social Affairs, the Ministry of Industry and Minerals, and the Ministry of Agriculture . The limited access to loans from banks or MFIs would not be as severe a challenge to the private sector if it had access to equity finance. However, such finance is in its infancy in Iraq.

IRAQ STOCK EXCHANGE

The Iraq Stock Exchange (ISX) is relatively new, and its trading is dominated by private bank and telecom shares. While the structure of the ISX is well designed for facilitating equity finance of Iraqi businesses, the regulatory hostility towards private businesses in Iraq – discussed at length in Chapter 11 – severely limits the usefulness of the ISX. In the absence of a radical change in private sector business regulations, attempts to further improve the operation of the ISX can be expected to have limited impact. Government securities with a maturity of one year are traded on a separate system operated by the CBI. Only banks currently have access to this system. There is little trading in government bonds with a maturity of greater than one year, since the GoI has issued few such bonds. This partially explains the absence of a market in private sector bonds, since there is no information on the "risk-free" return on government bonds to allow realistic pricing of private bonds (World Bank 2011a, p. 54).

Founded in 2004, the ISX has a monopoly on securities trading in Iraq. In June 2020, there were 104 listed companies. The total market capitalization of listed companies was approximately 13.2 trillion ID ($11.2 billion), or 6.5 percent of GDP. Compared to the other stock exchanges in MENA countries, this ratio of market capitalization to GDP is extraordinarily low.

Relatively few shares are traded. As can be seen in Table 9.4, the shares of only 49 out of the 104 listed companies traded on even a single day during June 2020. One telecom corporation – Asiacell Communications – and four banks accounted for over 80 percent of total trading volume (Rabee Research 2020a, p. 3).

As expected, market capitalization on the ISX has mirrored the country's security situation. Market capitalization of the ISX reached a pre-ISIS high of 12.1 trillion ID ($10.2 billion) in June 2013. However, after the ISIS invasion, capitalization fell by almost 50 percent to 6.6 trillion ID ($5.6 billion) in May 2016 (Coles and Nabhan 2018). This decline was followed by an equally rapid recovery, with capitalization reaching 13.2 trillion ID ($11.2 billion) in June 2020, driven in part by foreign investors. The trend in the ISX index of stock prices has been less ambiguous. From a high of 998 in June 2015, the index has declined to 434 in June 2020 (CBI 2020a).

Stock market optimists point to the fact that the non-Iraqis are free to purchase stock through the ISX, except for restrictions on the percentage of bank shares that can be in foreign hands. Since the ISX is well organized, the poten-

Table 9.4 *Overview of Iraq Stock Exchange, June 2020*

	Market cap (ID billions)	Companies listed	Companies traded	No. of transactions
Telecom	6 549	2	1	23%
Banks	5 477	42	23	23%
Industry	622	21	8	44%
Tourism/hotels	333	10	5	3%
Agriculture	109	6	3	~0%
Services	93	10	6	6%
Money transfer	30	2	0	0%
Insurance	13	5	3	~0%
Investment	6	6	0	0%
Total	13 232 ($11.2 bn)	104	49	3 176 (100%)

Source: ISX (2020).

tial for increased foreign buying would seem to be substantial, except for the general regulatory hostility towards private businesses, opaque firm balance sheets and income statements, and the absence of credit rating agencies.

The rules of the ISX are conservative. A buyer must have 100 percent of the necessary funds in their account before making a bid to buy stock. A seller must have 100 percent of the necessary shares in their account before making an offer to sell stock. All trades are electronic and settled through the Iraq Deposit Center (IDC) on a same-day basis (T+0). Neither short selling (borrowing a stock and selling it with the hope of buying back the same stock at a lower price in the future) nor derivatives (such as stock options) are currently allowed. However, some brokers have expressed interest in increasing the settlement period to one or two days (T+1 or T+2), as well as allowing trade in stock options (puts and calls). These changes would allow buyers to effectively purchase stock before accumulating the necessary funds, or sellers to sell stock that they do not own on the date of the sale. The argument is that these changes would increase the activity and liquidity of the ISX by encouraging buyers to enter the market. However, it is unlikely that either of these changes will have a substantial impact on ISX activity in the absence of dealing with the binding constraint on investing in Iraqi equities: severe information asymmetry that strongly favors firm insiders. In addition, the conservative rules including T+0 settlements limit systematic risk (World Bank 2011a, p. 51). In view of the undermanned ISX staff assigned to monitor trades and the absence of clear relevant regulations, liberalization of the current trading rules would probably be destabilizing.

This information asymmetry in the ISX is caused by both the lack of reliable corporate data and an uncertain legal environment. As a result, corporate insiders – especially those who are politically connected – have access to more accurate information about corporate prospects that is not available at reasonable cost to outsiders. Company income statements and balance sheets are either unreliable or missing critical pieces of information. The asymmetry is partly a function of archaic Iraqi accounting standards. While the country is on the path to adopting the International Financial Accounting Standards, this adoption process will take several years at least.

Compounding the problem of obtaining accurate up-to-date business information from corporate balance sheets and income statements, is weak auditing. There is a severe shortage of trained accountants. In addition, there is little acceptance of the notion that auditors should provide an independent evaluation. Especially in the case of state-owned enterprises, balance sheets and income statements are more policy statements than attempts to accurately evaluate the status of the firms. The regulatory authority, the Iraq Securities Commission, is severely understaffed and apparently focuses its attention on receiving and filing the unaudited annual and quarterly company reports.

The Iraqi legal system provides weak protection for shareholders. The relevant laws are incomplete and contain numerous uncertainties. For example, there are no antitrust laws, and the World Bank ranks Iraq's bankruptcy system as tied for the worst in the world. Apparently, firms do not go bankrupt in Iraq; rather, their former managers loot them. In addition, the court system's backlog of cases is large, and resolution can take years (World Bank 2011a, p. 30).

The situation is exacerbated by the fact that members of the judiciary generally lack even a basic knowledge of commercial law. The final investor concern about judicial resolution of stock market disputes is the perception of widespread corruption. Judges and other court officials have been suspected of favoring more powerful or influential individuals or organizations. There are also some idiosyncratic issues that reduce liquidity in the ISX. There are a large number of holidays when trading does not occur. And the ISX management will periodically suspend trading when there is a substantial drop in stock prices.

The conclusion is grim. Until there is a substantial improvement in access to quality audited corporate income statements and balance sheets, a rewritten commercial code that sharply reduces uncertainties especially with respect to bankruptcy, and a substantial improvement in the quality and perceived integrity of the judicial system, investing in equities in the ISX is a game that only insiders – those with connections – can win.

INSURANCE

The exact size of the Iraqi insurance industry is unknown, although, like banking, it is dominated by state-owned entities. One way of measuring the scale of the Iraqi insurance industry is by looking at gross written premiums. Gross written premiums are an estimate of the revenues (premiums) expected over the life of an insurance contract. The three state-owned insurance companies are believed to have a total of 280–470 billion ID ($240–$400 million) in gross written premiums. The 18 private insurance companies are thought to do less than one-quarter as much business as the state-owned firms: between 70 and 95 billion ID ($60–$80 million) in gross written premiums. An estimated 15–25 percent of gross written premiums are paid out for reinsurance (World Bank 2011a, pp. 59–66).

The primary reason for the dominance of the state-owned insurance firms is their near monopoly on providing insurance for government entities. This monopoly exists despite the requirement of Article 18 of the Insurance Business Regulation Act that government entities make a public tender, or call for bids, whenever they seek to purchase insurance. In the rare cases where a public tender is actually made, few private firms win the contract to provide insurance.

The size and organization of private insurance companies varies greatly. At one extreme, there are five private insurance firms listed on the ISX with a combined market capitalization of 12.5 billion ID ($10.6 million). However, only three of these firms were traded in June 2020 (Rabee Research 2020a, p. 11). At the other extreme, some of the private insurance companies are very small, acting only as brokers (100 percent reinsurance) or maybe not operational. With respect to organization, there are five types currently in operation, three of which have unlimited liability which is not an accepted international practice in insurance.

There are two reasons given for the low level of insurance business in Iraq. First, no insurance is required either for operating an automobile or to cover employee workplace injuries. If both of these common requirements were put into effect, the demand for insurance would increase substantially. However, in the absence of more effective enforcement of Article 18 requiring public tenders, much of the increase will probably be captured by the state-owned insurance companies.

Second, there is the challenge of dealing with religious objections to insurance. A common view of Islamic scholars is that: "Conventional insurance is tainted with riba (interest) and ghara (gambling) and is therefore to be rejected" (Visser 2009, p. 102). Specifically, there are three objections: (1) insurance companies often invest premiums in interest-bearing assets that, as discussed above, are probably not consistent with Sharia law; (2) life insurance is a form of gambling: a person makes periodic payments with only a possibility of receiving a payoff; and (3) as discussed in Chapter 11, since there are strict rules in Islam on the division of an inheritance, Islamic scholars think that a policy purchaser should not be allowed to avoid these rules by designating a beneficiary for their life insurance.

Throughout the Islamic world, there have been attempts to create *takaful* or Sharia-compliant insurance organizations. There is not yet a widely accepted pattern for *takaful*. Some are cooperative forms of insurance that focus on mutual assistance, which is justified by the Islamic call for brotherhood. Essentially, members of a community each contribute to a common fund that is used to help any members of the community who suffer unexpected difficulties. These co-operative insurance organizations may have professional management and often hold little or zero year-to-year reserves. If premiums are insufficient in any year to meet commitments, the organization deals with the shortfall through some combination of reducing promised payments to those members in difficulty and requesting additional voluntary contributions from its other members. One of the biggest problems facing *takaful* is the absence of Sharia-compliant reinsurance.

One troubling development for those interested in reducing GoI influence in the economy is the substantial investment by state-owned insurance companies in the stock exchange. While data is difficult to obtain, it appears that state-owned insurance companies are major holders of equity in the manufacturing materials and services sectors of the private economy. This large – possibly majority – ownership of private firms by state-owned entities can be expected to lead to contradictory incentives. On the one hand, the state-owned insurance companies want the private firms to be as profitable as possible in order to enable the insurance companies to meet their claims but, on the other hand, GoI policy objectives might require state-owned entities to pressure private firms to engage in socially desirable actions that will reduce their profitability (World Bank 2011a, p. 56).

PENSIONS

Pensions in Iraq are in a state of flux. There are two separate funds, whose pay-as-you-go financing annually amounts to 4–5 percent of Iraq's GDP. This is one of the highest pension burdens in the region. The State Pension System (SPS) covers civil servants, the military, and survivors of martyrs; while the Social Security System (SSS) is responsible for pensions in the private sector. Civil servants tend to receive much more generous benefits (World Bank 2020a, Box 2, p. 8).

Both pension systems seek to replace 100 percent of the last salary received for all persons who earned less than twice the average income. An employee is eligible for a full pension at age 60 (age 50 for women in the private sector) after only 15 years of covered employment in the public sector or 20 years in the private sector. This very generous program mostly benefits government employees. Although public information on pensions is limited, it is thought that, as of December 2019, only about 4 percent of the non-governmental labor force – approximately 200 000 persons – were enrolled in private sector pensions. And there is not yet any social insurance for the part-time or the self-employed. The opportunity open to government employees to boost their last months' salary – and therefore their pension – allowed some to earn annual pensions that exceeded 100 percent of their last year's wage.

The disparity between the large number of government employees eligible for generous public sector system pensions, and the relatively few private sector employees eligible for less generous pensions from the SSS, frames the major challenge facing Iraqi pension reform. There are three options, all of which are considered unacceptable by important constituencies. First, the status quo – the current system – is viewed as fundamentally unfair to private sector employees. Not only do they receive less generous pensions, if they receive a pension at all, but they are also forced to contribute to the public sector pensions. This contribution is disguised by the fact that GoI revenues are primarily from oil exports rather than from the extremely low tax rate in Iraq. However, private sector employees indirectly bear part of the burden of public sector pensions, since oil revenues diverted to pay for these pensions could have been spent on infrastructure, education, health, security, or other public or merit goods.

The second option – expanding the coverage and retirement payments of private sector pensions to rough parity with those offered by the SPS – has widespread support. However, such an expansion will be very expensive – possibly 15–20 percent of GDP, even with $100 per barrel oil – with further increases as the population ages. Pension contributions are already inadequate to maintain long-term solvency and the public pension system is expected to move into deficit in 2025. When this occurs, an increasing portion of the pension expenditures will be paid out of the general budget, which will crowd out other needed expenditures. The proportion of the budget accounted for by pension costs will rapidly become unsupportable if, as expected, Iraq faces a decade or more of $60 per barrel oil.

In addition, if private pensions become mandatory, this will substantially increase the administrative and financial burden on private businesses; then this would also have an adverse impact on entrepreneurial activity. And as discussed in Chapter 11, generous public sector pensions contribute to the problem of potential entrepreneurs being diverted into government service rather than establishing and operating private firms.

The third option of increasing pension coverage of the private sector while reducing the generosity of public sector pensions is complex politically and financially. A 2006 Pension Law sought to combine the SPS and the SSS, but even optimistic estimates found it to be fiscally unsustainable; some estimated that pension costs under the 2006 Law would reach 60 percent of GDP. In 2016, a new draft social insurance law was submitted to the Council of Representatives but, as yet, it has not been approved. In general terms, this new law sets more reasonable limits on the payments into the pension system, eligibility standards, and expected pension payments (World Bank 2011b, p. 66). This progress was undone in November 2019 with a proposed amendment that increased public sector benefits, thereby further reducing fairness (IMF 2019b Box 2, p. 8). Unless there is agreement in the near future to rewrite the national social insurance laws of Iraq, then either a sharp cut in benefits or a large budget expenditure will be required.

Pensions do not face the same problems of Sharia compliance as insurance. The only serious restriction is that the investments in the pension fund must be *halal*: acceptable or good. There are definitional complications with determining which investments are *halal*. It is *haram* (that is, not *halal*) to invest in liquor stores, but what about investments in a trucking company that delivers liquor to stores?

OTHER FINANCIAL INTERMEDIARIES

While data on total assets is not available, the following five financial intermediaries are thought to be relatively small: (1) the Postal Savings Fund accepts deposits in its branches and invests them in a variety of sectors; (2) exchange companies are engaged in buying and selling foreign currency inside Iraq; most of the exchange companies are located in Baghdad; (3) the financial transfer companies are non-banks that specialize in transferring funds inside and outside Iraq, using accounts that they have established in Iraqi banks; (4) there are six financial investment companies listed on the ISX which funnel savings into equity, debt, and deposits; and (5) there are relatively new institutions that specialize in making small and medium-sized loans; the Iraq Company for Short-Term Loans was the first such firm established.

THREATS TO THE CULTURE OF CREDIT

Prior to Saddam's attempt to establish a socialist economy in Iraq, the country's merchants were renowned for their business savviness and their acceptance of the culture of credit. The culture of credit is the concept that loans should be used to expand productive capacity – not for consumption – and that loans should be paid back on schedule. Persons who live according to this culture can not only expect to be able to borrow larger sums in the future but also, as their reputation for using credit wisely spreads, have increased business opportunities. On the other hand, failure to repay loans on schedule leads both to inability to borrow in the future and sanctions.

During the 25 years of Saddam's rule, the culture of credit was undermined in two ways. First, loans were generally made not on the basis of creditworthiness and a solid business plan, but because of friendship, political imperatives, or corruption. Second, it was understood, even when not explicitly stated, that these loans would not have to be repaid. The banks would periodically simply mark interest as paid and add the value of the formerly overdue interest to the principal of the loan. On paper the banks were profitable, while in reality they were drowning in defaulted debt. Fortunately, there was only one generation between Saddam's attempt to establish a statist economy and its subsequent collapse. As a result, there still exists an older generation that understands the market.

SEQUENCING OF FINANCIAL SECTOR REFORM

Both the GoI and the International Monetary Fund (IMF) have a long list of reforms for Iraq's financial sector (see GoI 2020b, pp. 48–50). Although there is general agreement that the most important reforms are regulatory reform, recapitalization of the state-owned banks, and the liberalization of international capital flows, the sequencing of these reforms is critical. If state-owned banks are recapitalized before regulatory reform is achieved, then the experience of other post-socialist nations shows that the newly recapitalized banks will rapidly accumulate massive losses. Also, one of the lessons of the 1997 Asian financial crisis is that if the door to international capital movements is opened before there is a vigorous well-regulated domestic banking system, then there is a tendency towards financial instability including large-scale capital flight.

Based on these lessons, the GoI should sequence policy changes as follows. First, rationalize the regulatory environment of financial intermediation. Second, the management of the state-owned banks should be restructured followed by recapitalization. And finally, when both of these goals have been achieved, only then should the GoI seek to liberalize international capital flows.

First, Regulatory Change

Despite the long-recognized inefficiencies, reforms of financial intermediation in Iraq have been continually delayed. Many in the bureaucracy are happy with

Table 9.5   *Regulation of financial intermediation*

1st is best in world (190 countries in sample)	Iraq ranking 2020	Iraq ranking 2011	UAE ranking 2020
Registering property	121st	95th	10th
Getting credit	186th*	170th	48th
Protecting investors	111th	120th	13th
Enforcing contracts	147th	140th	9th
Resolving insolvency	168th*	183rd	80th

Note: *Iraq is tied for worst in the world.
Source: World Bank (2020d).

the current state of financial intermediation. State-owned banks, insurance companies, pensions funds, and so on, provide a variety of valuable services to the Baghdad bureaucracy. These range from providing individual loans to government officials with no expectation of repayment, to financing major SOE projects that lack any reasonable chance of return. In addition, the state-owned financial intermediaries provide a large number of jobs that can be distributed to favored individuals or members of favored groups.

But as much as members of the bureaucracy favor the current regulatory environment, the ability of Iraq's banking, equity, insurance, and pension intermediaries to facilitate strong long-term economic growth and development is severely constrained. Iraq's regulations are not just considered relatively hostile towards private business by Organisation for Economic Co-operation and Development (OECD) standards, but Iraq was last among the Arab countries according to "Doing Business 2020" (World Bank 2020d). Table 9.5 summarizes five indicators of the impact of regulations on financial intermediation in Iraq. The country with the most favorable business regulatory environment in 2020 is ranked 1st while the most hostile environment is ranked 190th.

The first indicator, registering property, refers to the procedures required to legally transfer title on immobile property. Unregistered property is very unlikely to be accepted as collateral for loans. The "getting credit" indicator includes the existence and efficient operation of credit information systems as well as protection of the rights of borrowers and lenders. Regulations to protect investors should clearly define the responsibilities of investors and firms, promote clear disclosure, require shareholder participation in major company decisions, and set rules for company insiders. Finally, an effective and quick system for resolving insolvencies – bankruptcies – should both ensure the survival of economically efficient companies and provide for the reallocation of the assets of inefficient ones. There have been no cases of a legal resolution of a bankrupt company in Iraq for the last 16 years.

In 2020, Iraq not only ranked in the bottom half of the 190 countries evaluated for these five indicators but, over the last decade, Iraq's relative position deteriorated with respect to four out of the five indicators and only one improved. In addition, Iraq ranked below the MENA country average – a region not known for quality financial regulation – for all five indicators. In fact, Iraq is tied for the worst in the world when it comes to getting credit and resolving business insolvencies.

Since financial intermediation in Iraq is dominated by the banking industry, the "getting credit" indicator is probably deserving of the most immediate attention. Fundamental to making loans is a reliable source of detailed credit information; that is, a credit registry. Currently, relatively few Iraqi firms or individuals are registered with a private or public credit registry for firms or individuals (IMF 2019b). This absence of credit history or even a reliable listing of current credit use by firms or individuals makes it very risky to make loans unless borrowers leave movable assets that can be stored at a bank as collateral. Creditors' legal rights are also very limited. Out of the ten items in the legal rights index, Iraq achieves only three. Combined with the regulatory hostility reflected in the other four indicators, it is difficult for banks to evaluate the risk of loans, for potential shareholders to judge companies, for insurance companies to estimate premiums, or for pension funds to safely invest. A longer discussion of the other costs of this regulatory hostility towards private business is included in Chapter 11.

With respect to creating a vibrant private financial sector, there are also technological gaps. The country lacks a secure modern communication system to support electronic funds transfer between different banks, core banking to connect checking accounts across multiple branches of the same bank, and debit and credit cards. Modernization of communications, by increasing the usefulness of private banks, should gradually reduce the dominance of cash transactions.

Regulatory reform in Iraq is more than a matter of having smart bureaucrats in Baghdad writing good regulations that are then approved by elected representatives. Actual execution of regulations is equally important. In many cases, current regulations appear adequate to address the problems of financial intermediation, but these regulations are ignored.

Second, Management Restructuring and Recapitalization of State-Owned Banks

Prior to recapitalization, it is important to reform the management of the state-owned banking system by adjusting managerial incentives. Otherwise, recapitalization will simply lead to further losses. The state-owned banks, especially Rafidain Bank, have acted more as fiscal agents of the GoI rather than traditional banks. In addition to making salary and pension payments for the GoI, the state-owned banks are involved with substantial policy lending to SOEs and other entities. As a result, managers of the state-owned banks have been evaluated and promoted more on their political and bureaucratic abilities than their ability to make good loans.

Changing the incentives for the management of state-owned banks will be difficult. It will require ceasing to use these banks as fiscal agents while, at the same time, encouraging them to making loans on commercial terms. Evidence from other post-socialist countries shows that a growing private banking sector tends to be associated with improved operation at state-owned banks in the same country. This is in part because improvements in the private banking sector put competitive pressure on state-owned banks, forcing them to become more efficient. Further complicating changing the managerial culture of the state-owned banks is ambiguous guidance from the CBI.

In general, the CBI must move away from directing policy lending and focus instead on exercising supervisory responsibilities to enforce sound banking practices and common regulations for all banks, state-owned or private. Progress has been made, but at a very slow rate.

While recapitalization of the state-owned banks should be delayed until the restructuring of managerial incentives is completed, the ongoing delay in recapitalization seems to have little to do with waiting for managerial changes. In 2006, the CBI and Ministry of Finance agreed to a plan to recapitalize these banks using GoI Treasury Bonds. However, the GoI has regularly announced delays in this recapitalization. The announced reasons for this continuing delay are the necessity of gathering detailed information on the banks' government accounts as well as the need for the CBI to prepare to effectively regulate the newly recapitalized institutions. However, there might also be bureaucratic pressure to delay the recapitalization process, for fear that the banks will no longer be so willing to fund projects supported by the various ministries. The delays in managerial reform and recapitalization are not costless, since every year that Iraq stumbles along with extremely inefficient financial intermediation results in slower economic development.

Third, Liberalization of International Capital Movements

Other post-socialist societies have benefitted by allowing domestic banks to borrow internationally, and encouraging foreign banks to purchase substantial ownership rights in domestic banks. In addition to providing new capital, foreign ownership tends to lead to greater efficiency and increased availability of new financial products. This increase in efficiency occurs not only in banks with foreign ownership but also by motivating greater efficiency in non-foreign-owned banks through increased competition.

There is concern that increased foreign ownership will exacerbate financial and economic instability, possibly by facilitating capital flight. Also, there is doubt that the CBI will have either the necessary knowledge or cooperation of foreign regulatory authorities to effectively regulate foreign banks. Regardless of the motivation, the CBI has made it difficult for foreign banks to either establish new wholly owned subsidiaries or buy substantial shares of existing Iraqi private banks.

In addition, the GoI is encouraging large-scale foreign investment and loans in order to finance a rapid expansion of the nation's infrastructure, especially in the petroleum sector. In many developing countries, increased international capital flows when the financial markets are both fragile and inflexible has led to acute financial instability. However, since this risk of financial instability is strongly influenced by the GoI's exchange rate policies, it is more appropriately discussed in Chapter 14, which focuses on monetary, and exchange rate policy.

10. Large industrial enterprises

> The previous ban on non-Arab foreign direct investment and the effects of three
> wars and a decade of international economic sanctions have meant that something
> like 90 percent of Iraq's industrial capacity – the SOE share – is seriously decapi-
> talized, asset-starved, obsolescent, inefficient, saddled with high production costs,
> over-staffed, and – as a result of looting – in a state of physical degradation.
>
> (World Bank 2004, p. i)

Industrial production in Iraq is dominated by state-owned enterprises (SOEs)
that tend to be high-cost, low-quality producers. Not only are they the leading
providers of essential services such as electricity and water but also SOEs
account for a large proportion of all consumer goods and industrial inputs.
Further reflecting their major impact on the country, SOEs are, collectively,
the largest employers after the national government. SOE total employment is
an estimated 600 000 persons (GoI 2020b, p. 10; al Mawlawi 2018).

Although the quality of data is poor, it is estimated that roughly 20 percent
of existing SOE factories are currently profitable or potentially profitable with
a reasonable amount of investment. Most of these profitable SOE factories are
in the petroleum or construction industries. Another 25–30 percent of the SOE
factories are empty shells; destroyed during conflict, severely mismanaged,
or looted to bare walls. However, employees of these wrecked SOEs continue
to be paid for showing up. The remaining half of SOEs are either considered
essential to national security – for example, weapons repair – or would require
large-scale investment, managerial restructuring, or a sharp workforce reduction in order to have any chance of achieving profitability.

This chapter analyzes large industrial establishments based on two sources
of data. The annual Industrial Large Establishment Statistics survey of the Iraq
Central Statistical Organization (CSO) provided data on 600 large establishments in 2018. The 2016 Government of Iraq (GoI) study, *Performance and
Fiscal Risks from Non-Financial State-Owned Enterprises in the Republic of
Iraq* (GoI 2016) provides more limited data on 157 SOEs and a more detailed
look at 136.

Table 10.1   Industrial establishments by sector

	2014	2015	2016	2017	2018
Number	616	600	566	551	600
Employees	134 818	129 024	109 574	111 374	114 762
Compensation (bn ID)	1 574	1 447	1 247	1 431	1 449
Average wages ('000 ID)	11 700	11 200	11 400	12 900	12 700
Value of production (bn ID)	4 271	5 469	4 969	5 998	7 191

Source: CSO (2019, Table 1, p. 4).

INDUSTRIAL OVERVIEW

In 2018, the number of large enterprises – those with 30 or more employees – was an estimated 1161. This total includes firms that were state-owned, private, public–private partnerships, and Iraqi–foreign partnerships. However, only 600 of these enterprises were actually operating in 2018; the others had suspended activities (CSO 2019, p. 3). While private firms accounted for almost 90 percent of all operating industrial enterprises, they produced only about 35 percent of total industrial output. As can be seen in Table 10.1, industrial firms employed an estimated 114 762 persons or between 1 and 2 percent of the country's labor force. Among the firms that reported data, workers received an average wage of about 12.7 million Iraqi dinars (ID) ($10,700) per year. The 2018 value of production was 7191 billion ID ($6.1 billion).

Primarily as a result of the war with ISIS, the number of firms, employees, and compensation declined from 2014 through 2016–17 before a strong recovery in 2018. However, the number of firms was still 3 percent lower in 2018 than in 2014, while the number of employees decreased by 15 percent over the same period. It is expected that the low oil prices of 2019 and 2020, combined with the 2020 Covid-19 epidemic, will slow the recovery of the industrial sector.

Government and public firms dominate total employment. While the average government or public firm had 1525 employees, the average private firm had only 43 employees. And average wages in private firms were half of those paid in government and public firms: about 7 million ID ($5900) a year. In fact, several hundred private sector employees – possibly owners – reported zero wages (CSO 2019, Table 4, p. 10).

There are a few sectors of the Iraqi economy where private firms are major players. Private sector firms and employment were mostly concentrated in two sectors: the manufacture of other non-metallic mineral products (International Standard Industrial Classification of All Economic Activities, ISIC 23) – this category includes refractory, structural clay, and cement products – and food products (ISIC 10 and 11). These sectors accounted for 92 percent of all private firms and 84 percent of private sector employment. While private firm employment also dominated wood products (ISIC 16), paper products (ISIC 17), and basic metal products (ISIC 24), employment in these sectors was low.

As expected, industrial firms were not evenly distributed among provinces. Baghdad alone accounted for 42 percent of national industrial employment. The other provinces with substantial industrial employment were Basrah, Babil, and Najaf, while at the other extreme fewer than 250 industrial jobs were reported for Anbar.

Surprisingly, the reported value added per employee is almost the same for public and private firms: 28.8 million ID ($24 000) a year. One possible explanation is that although employee wages in private firms are about half that of public firms, public firms tend to be monopolies or oligopolies while private firms tend to sell in more competitive markets. Therefore, public firms by charging higher monopoly or oligopoly prices have been able to achieve roughly the same value added per employee despite higher labor costs and less efficiency (CSO 2019, Tables 8 and 11).

Efforts over the last decade to increase the number of public–private partnerships and Iraqi-foreign owned firms have been unsuccessful. In 2018, there were only four mixed public–private firms with 597 employees, and only two Iraq-foreign owned firms with about 1300 employees (CSO 2019, Table 3, p. 8).

There are multiple reasons given for the failure to form more public–private partnerships or partnerships between Iraqi and foreign firms. After partnerships were agreed upon, the GoI often reduced subsidies to the public firm. And there is a tendency for ministries to retroactively change agreements for political reasons. For example, the Ministry of Industry and Minerals has sought to have foreigners investing in a project pay all of the salaries of its Iraqi partner, even for workers who are not employed on the project (US Department of State 2019, pp. 41–2). The small scale of many private firms, combined with few public–private partnerships and Iraqi–foreign firms, ensures the continued SOE control of industrial production.

ORIGIN OF SOES

SOE dominance of the Iraq economy is only about 40 years old. Saddam had at least three motivations for establishing SOEs after he took over the Iraqi government in 1979. First, consistent with the economic philosophy of the Arab Ba'athist Socialist Party, SOEs were seen as a way of accelerating economic development by forcing the adoption of what were then considered modern management techniques by firms large enough to internalize economies of scale. Existing private firms were combined into government-run organizations, while other SOEs were newly established in industries that the leadership of the Arab Ba'athist Socialist Party thought exemplified a modern industrial state. This "cargo cultism" led to the creation of industries such as truck and bus assembly where Iraq was at a severe competitive disadvantage. In addition, SOEs are subject to having administrative or production decisions made for political, not economic, reasons (GoI 2018, p. 94).

The second motivation for establishing SOEs was political. In common with other socialist leaders, Saddam found that SOEs provide the government with a means of both rewarding supporters and of punishing opposition. Directing a SOE provided a member of Saddam's family or other supporter with a well-remunerated position of status and political influence. The director of each SOE ensured that, whenever possible, jobs in their SOE were reserved for persons loyal to Saddam, the regime, and themselves. At the same time, SOE dominance of the economy made it more difficult, both financially and personally, for persons to oppose the regime. Any incipient opposition found it difficult to raise financial support, since most of the non-agricultural economy was controlled by the regime. In addition, individuals realized that if they were less than enthusiastic about Saddam's regime then they and their families would be excluded from employment.

Finally, it was believed that SOEs would allow the state to capture the profits that would otherwise have accrued to the private sector. Saddam wanted to ensure that a substantial portion of these funds would be diverted into the pockets of himself, his family, and his closest supporters. In other words, SOEs provided a means of facilitating large-scale corruption.

CURRENT STATUS OF SOES

It is challenging to get a clear understanding of the current status of Iraq's SOEs, for several reasons. First, while each SOE falls under the control of one of the Baghdad ministries, periodically a SOE or one of its factories will be shifted to another ministry or merged with or split from another SOE. In fact, while the number of SOEs is usually given as 192, actual counts have ranged from 176 to 195.

Second, each SOE is comprised of between one and 15 enterprises or factories, often of very different sizes. Some are quite small, with less than 100 employees, while others are huge. These enterprises are often located at a distance from their SOE headquarters and often act as separate entities with respect to technical issues. However, regardless of size or location, these enterprises are not registered as legal entities, they cannot sign contracts with customers, nor can they sign for bank loans.

Third, some of the SOEs are not really commercial entities as defined by the national SOE law Law No. 22 of 1997. Rather these SOEs act as fiscal agents for ministries with no commercial objectives (GoI 2016, p. 7).

Finally, the balance sheets and income statements of SOEs are non-existent, or works of fiction, or presented in a non-standard format. Even fundamental information is lacking. For example, most SOE firms cannot accurately report their number of employees. While SOEs have a rough estimate of the amount paid in salaries and benefits, the actual number of workers is generally unknown. This is in part explained by the fact that many SOEs have an unknown number of "ghost workers" who are collecting pay from the SOE but not showing up for work. For some SOEs, it is estimated that up to 25 percent of all employees are "ghosts", while two-thirds of the workers in some individual factories are ghost workers. Iraq's Special Borrowing Agreement with the International Monetary Fund (IMF) required a detailed census of all government employees to be completed by the end of 2006. However, 14 years later, the census is still incomplete due to the lack of cooperation, and in some cases the open hostility, of the SOEs. Large-scale overemployment in SOEs provides opportunities for enterprise managers to gain influence or, by auctioning off jobs, to accumulate substantial wealth.

Since their beginnings under Saddam, SOEs have received financial support from the GoI in multiple ways. First, SOEs receive direct subsidies in the form of cash transfers from the national budget. These subsidies continued after the 2003 fall of Saddam's regime and by 2010 amounted to approximately $2.5 billion or roughly 3 percent of total GoI budget expenditures (IMF 2011a, Table 3, p. 18). As a matter of scale, $2.5 billion was equal to almost 90 percent of GoI annual expenditures for all elementary and secondary education. Second, there are large indirect subsidies. SOEs receive free or inexpensive electricity, water, fuel, credit, and capital equipment. Third, transfer pricing aids non-oil SOEs. Since oil-related enterprises represent the only sector that has large annual cash surpluses, transfer pricing generally involves oil-related enterprises paying too much to SOEs that provide supplies to the oil industry, while receiving too little payment from SOEs that purchase fuel products. Fourth, SOEs have received bank "loans" from Rafidain and Rasheed banks where neither the SOE borrower nor the state bank lender expects the loans to be repaid. Fifth, the Board of Supreme Audit mandates government agencies to buy goods and services from SOEs unless the SOE price is more than 10 percent higher than the private sector competition (US State Department 2019, p. 44). Finally, there are regulatory subsidies. As will be discussed in detail in Chapter 11, Iraq has a regulatory environment that is extremely hostile to private business. This ensures that any private sector competition is limited, and creates captive markets for the high-cost, low-quality products and services produced by SOEs.

Table 10.2   Industrial establishments by sector

Ministry	Companies	Employees ('000)	Profitable companies in 2015
Industry and minerals	71	145.4	11
Electricity	24	83.0	1
Oil	18	143.6	8
Transportation	10	37.0	3
Construction and housing	8	13.7	2
Trade	7	10.5	1
Agriculture	7	4.3	1
Defense	6	20.5	0
Water resources	3	2.8	1
Communications	3	18.3	0
Total	157	479.1	28

Note: 19 SOEs associated with the ministries of Finance (12), Culture (4), Health (2), and Education (1) are not included in this table because of lack of data. Total SOEs is 176.
Source: GoI (2016, December, Figure 1, p. 9 and Table 7, p. 20).

While the absence of audited accounts combined with the organizational changes makes it difficult to provide a detailed analysis of the financial health of SOEs or their component enterprises, the little information that is available points to one conclusion. Despite large indirect subsidies, favorable transfer pricing, and limits on private sector competition, most Iraqi SOEs have lost money every year since the mid-1980s.

For example, Table 10.2 illustrates the profit/loss status of 157 out of the total of 176 SOEs. Only 28 firms, 18 percent, were profitable in 2015. Data for SOEs associated with the ministries of Finance, Culture, Health, and Education are not available (GoI 2016, p. 12). However, it is believed that most of the SOEs associated with these ministries were also unprofitable.

In addition to excessive employment, the distances between factories further complicate the difficulty of profitably managing SOEs. Factory locations were often determined more by political considerations than those of economic efficiency. In view of the degraded transportation and communication infrastructure that will be discussed in Chapter 12, physical distance not only increases costs of production but also inhibits coordination. The primary argument made in favor of geographically distributed SOEs is to provide industrial employment outside of the large metropolitan areas of Baghdad, Basrah, and Mosul. However, in many cases the impact on local employment is minimal. SOE employees are bused in from the major metropolitan areas to work at the outlying SOE plants or, at least, to collect their paychecks.

As discussed above, with the exception of the oil sector, there is little international interest in developing joint projects with Iraqi SOEs. The few that have occurred appear to be in three areas: cement production, fertilizer, and energy generation. Among foreign companies interested in partnerships with Iraqi SOEs, Chinese interest has been strongest, although there have been French and British initiatives. Generally, there will be well-publicized agreements that lead to no or meager results.

Many of the medium-sized and small investments originate from Iraq's neighbors and may include substantial "false foreigner" participation. False foreigners are Iraqi émigrés who invest in Iraq. These investors and their partners probably have a more accurate idea of which investments have the best risk-to-reward ratio. Also, it is believed that the GoI will be more transparent and less arbitrary with international investors than with its own nationals.

SHOULD IRAQ REFORM ITS SOEs?

The establishment of SOEs in oil exporting and other natural resource "rentier" states, such as Iraq, is not rare. In three – Saudi Arabia, Qatar, and the United Arab Emirates (UAE) – SOEs appear to be relatively well managed (Hertog 2010, pp. 262–3). In Iraq, the results are mixed. On the favorable side, Iraq's SOEs are the primary providers of most domestic producer and consumer goods and services. In addition, since SOEs provide a large number of reasonably well-paying jobs, maintaining SOEs is considered a practical means of reducing political instability in post-conflict Iraq. There is also a privatization-related fear that the elimination of the favored status of SOE will lead not to a shift to private sector production in Iraq, but rather to a sharp increase in imports, and therefore increased dependency on foreign suppliers.

The arguments against continuing to heavily subsidize SOEs extend beyond the substantial budget burden. First, since SOEs in Iraq tend to be low-quality, high-cost monopoly or oligopoly producers, they reduce the efficiency of both upstream and downstream business entities. It is not just a matter of the quality of the products and services provided; suppliers to SOEs and consumers of their products realize that the SOEs are, by their very nature, politically powerful. Any commercial disputes will most likely be resolved in favor of the SOE. As a result, private suppliers to SOEs and private purchasers of their goods and services must suffer not only from quality issues, but also from acute uncertainty about whether any agreement with an SOE will be executed as contracted.

Second, heavily subsidized SOEs tend to crowd out private sector competitors and stall market development both directly and indirectly. Directly, private producers find it difficult to compete with heavily subsidized SOE products and services. Indirectly, each SOE is associated with a ministry which, in order to minimize its subsidy costs, often uses available regulatory authority to prevent private sectors from successfully competing with its SOE.

Third, SOE dominance in Iraq reduces the country's ability to compete in global markets. Foreign purchasers do not have to tolerate the quality issues and unpredictability associated with trading with an Iraqi SOE. Iraqi consumers have shown a willingness to pay a premium for the predictable quality of imported goods, ranging from tractors to *halal* frozen chickens. Whether state ownership acts as a barrier to foreign direct investment is more ambiguous. Having the GoI as a partner can both ensure political support for the enterprise, and also reduce profit opportunities if one's government partner is corrupt.

Fourth, proponents argue that excessive SOE employment reduces political instability by providing jobs for unskilled young men. However, expanded SOE employment may actually be destabilizing. It is not uncommon for the SOE of a ministry to be "captured" by a religious sect, a party, or a tribe as a means of providing funding for the organization as well as jobs for its members. Thus, SOE employment can be destabilizing by supporting, with government funds, members of political parties or other groups that may be in opposition to the rule of law.

Fifth, SOEs tend to be less concerned with the environment. World Bank reports on studies in Bangladesh, India, Indonesia, Thailand, and Brazil point to a direct relationship between SOE dominance and pollution levels (World Bank 1995, pp. 38–41). The report concludes that: "SOE are better placed to evade pollution regulators than their private counterparts" (World Bank 1995, p. 40). This is not just a function of the fact that SOEs tend to have outdated infrastructure, or that SOE are concentrated in industries that tend to be more polluting. Even after adjusting for infrastructure age, SOEs tend to pollute more than private firms in the same industries.

Finally, there is an increasing consensus that, in addition to the effects of microeconomic inefficiencies discussed above, a large dependency on SOEs has an adverse impact on a country's long-term macroeconomic growth. In part, this results from divergence of scarce government spending from education or infrastructure investment to SOE subsidies. In addition, government ministers face contradictory demands when making ministerial decisions that could help the economy but also injure their associated SOE. Galal et al. (1994) estimated that for nations such as Iraq where SOEs accounted for 10 percent or more of gross domestic product (GDP), cutting the SOE percentage of GDP in half would raise real GDP growth by 1 percent per year for the foreseeable future.

Table 10.3   SOE reform options

	Type of reform	Ownership	Management
1	Status quo	State	State
2	Improved state management	State	State
3	Managerial contracts	State	State/private
4	Private–public partnership	State/private	State/private
5	Regulated privatization of natural monopolies	Private	State/private
6	Complete privatization	Private	Private

HOW CAN IRAQ REFORM ITS SOES?

The GoI has stated its intention to reform its SOEs. In accordance with the country's Industrial Strategy 2030, the GoI has four goals for SOEs. They must be autonomous, able to cover all expenses. Privatized SOE must be profitable. They must meet international business standards. Finally, the market value of SOEs, whether privatized or continuing with private ownership, must be higher than their 2018 value (GoI 2018, p. 94). This SOE strategy of privatization was confirmed in even stronger terms in the government's 2020 White Paper (GoI 2020b). In this document, the GoI stated that the only "Successful Companies" are those SOEs that are totally or partially privatized. Non-privatized SOE will be classified as either "Troubled" or "Failed" (ibid., p. 60).

All too often the debate over SOE reform in Iraq is treated as a choice between two extremes. One must choose either the status quo or the radical reform of privatizing all SOEs. But this is a simplistic view, since there are at least six options for Iraqi SOE reform. The options are: (1) the status quo; (2) better state management of SOEs; (3) managerial contracts to allow private management of SOEs; (4) public–private partnerships; (5) the privatization of natural monopolies but with detailed state regulation; and (6) complete privatization. As can be seen from Table 10.3, three of these options assume continued state ownership, one proposes joint private–state ownership, and the remaining two options are forms of privatization. It should be noted that the GoI has a more limited view of privatization. Under this view, reform options 3 through 6 are all considered to be "privatization" (GoI 2018, p. 94).

One issue that requires careful consideration is the precise standard for successful SOE reform. There are four standards commonly discussed in Iraq. First, maintenance of employment levels is often cited as an important goal. In fact, this is probably the most important criterion in Iraq: any SOE reform program that is expected to result in an increase in unemployment is probably a non-starter. The White Paper proposes to resolve this difficulty by transferring any surplus SOE labor to public works projects (GoI 2020b, pp. 61–2). Of course, this raises the difficulty of inappropriate skill sets; for example, the excess worker at an SOE might be an office clerk, but a public works project might need a skilled carpenter.

Second, successful reform would result in the recovery of a stagnant industry. In this case, reform would lead to the production of better-quality goods and services at lower real costs.

Third, successful reform should contribute to the country's economic development. This could result from setting free resources including labor that are "trapped" by moribund SOEs.

Finally, since most SOEs are a burden on the national budget, successful reform would be one that reduces or eliminates this burden. The reduction in the SOE-related budget burden would occur through some combination of the reduction or elimination of the substantial direct and indirect subsidies that most SOE receive, and the transition of some SOE firms into profitable enterprises that could transfer funds to the GoI. The White Paper calls for the Ministry of Finance (MoF) to phase out direct financial subsidies by 2024 (GoI 2020b, p. 60).

Unfortunately, the first two definitions of success – preservation of traditional industries and associated employment – are often inconsistent with the latter two definitions: acceleration of economic growth and reduced burden on government budgets.

But there are also political limitations on SOE reform even if it has a desirable impact on economic development. Government ministers and SOE directors seem willing to accept SOE reform as long as there is no change in the geographical distribution of factories, and the authority of the ministries. Of course, these restrictions make substantial reform very difficult to achieve.

The situation is similar to that of the People's Republic of China in the 1980s in that the motivation for SOE preservation is both political and economic. Within Iraq's current coalition government, the 34 ministries are divided up among the various political parties, reflecting election results and the party leadership's ability in political negotiation. Some of the ministries are important because of their critical roles in the nation. These would include the Finance, Foreign Affairs, and Oil ministries as well as those of Defense and Interior (police). However, some of the other important ministries are valued only because they control SOEs. These SOEs provide the ministers with large budgets and status as well as the capability to distribute government jobs to loyal supporters. If a ministry loses its SOE to privatization then there will be fewer prizes – desirable ministries – to divide among coalition members and important members of the opposition. This would complicate government coalition formation in Iraq, since there would be fewer desirable ministries to be divided as political spoils. SOE privatization could be politically destabilizing.

In fact, only the first four options in Table 10.3 – status quo through private–public partnership – are seen as viable alternatives for Iraqi SOE reform in the foreseeable future.

The argument against even a partial privatization of production, export, and sales of petroleum and natural gas is primarily political. As was discussed in Chapter 7, oil is generally viewed as a national legacy of all of the people of Iraq. It would be political suicide for a leader to propose the privatization of an Iraqi oil-related firm. As a result, the 18 oil-related SOEs will probably remain under state ownership and management for the foreseeable future.

The other group of SOEs that is most likely exempt from substantial reform is – strangely enough – the 25–30 percent of SOEs that either were destroyed in the multiple conflicts of the last 20 years or are uncompetitive at any possible set of input and output prices. Reform is unlikely for these "empty shell" firms because any reform short of liquidation will be unlikely to provide any benefit. These firms should be viewed as a form of welfare: the employees continue to receive their paychecks despite the lack of any need for their labor. Even if these empty shell SOE factories were reopened, actual direct job creation might be small. Most of the direct spending on equipment and spare parts (60 percent) for revitalized SOE factories went for imports, not domestically produced items.

But there is a danger of overstating the arguments in favor of the status quo. There are factories where the advantages of substantial reform are clear. The Iraq Stock Exchange (ISX) estimated that as many as 296 factories could be at least partially privatized in order to sell stock on the ISX (World Bank 2011a, pp. 52–3). Some ministries appear to be willing to experiment with either substantial change in SOE operation or even privatization. The 2018–2022 National Development Plan identified at least 34 SOEs controlled by 11 ministries that should be "restructured, partially/fully privatized, and/or incorporated" within four years (GoI 2018, p. 96) But how can the management of existing SOEs be improved either to prepare them for partial privatization or, at least, to reduce the burden on both the national economy and the national budget? The answers to these questions might be best discussed under three headings: increased transparency, management reform, and joint venture versus privatization.

Increased Transparency

Increased transparency combined with reduced ambiguity of SOE managerial incentives is not only a necessary precursor to privatization but also should lead to increased efficiency in SOEs that will not be privatized. The most important component of increased SOE transparency is to replace indirect subsidies by direct ones in order to allow the more accurate targeting of GoI assistance.

Unraveling the knot of indirect subsidies is difficult, especially since many SOEs are either customers of or suppliers to other SOEs. The "prices" of goods and services involved in these inter-SOE transactions diverge greatly from those that would balance demand and supply. This non-market transfer pricing enables Iraq's oil and financial sectors to subsidize the rest of the SOEs. In addition, it is even difficult to obtain accurate information on the volumes of electricity, water, other utilities, and credit utilized by individual SOE factories. In many cases, metering or other measurement methods are not even attempted. Even when estimates of the volume of these utilities and credits are collected, this information is treated as confidential within each SOE factory and generally not shared with the parent SOE, the national government, elected officials, or – least of all – the public.

In addition, even when reasonably accurate estimates of the volume of subsidized utilities are available, there is the question of the proper prices. Official prices are based more on politics than on market clearing. As will be discussed in Chapter 12, these official prices are often below the average costs of production and sometimes they are less than marginal costs.

If SOEs were a relatively small part of the Iraqi economy, determining the correct transfer prices would not be as difficult. Officials could observe the input and output prices received and paid by firms in the private sector, and use these observations to estimate the correct relative prices for transactions involving SOEs. However, since SOEs dominate the Iraqi economy, there is a shortage of domestic private sector price observations for officials to extrapolate from. In addition, there is a simultaneity problem. For example, for the electrical industry to estimate its real shadow price of electricity generation, it requires information on fuel costs. However, for the fuel refineries to estimate their costs of fuel production, they require knowledge of the price of electricity.

The pricing problem is exacerbated by Iraq's extremely hostile regulatory environment for private businesses, as will be discussed in Chapter 11. This hostility towards private businesses has two effects. It reduces the scope of the private sector, and it forces most of the private businesses that do exist in Iraq into the underground economy to avoid the hostile regulations and bribe demands from officials. This not only reinforces SOE dominance of most Iraqi markets but also limits information on relative market prices for SOE planning.

Therefore, to increase the transparency of indirect SOE subsidies, it is necessary not only to do a better job of estimating the volume of utilities and credit received but also to attempt to value these subsidized inputs, either by attempting to estimate real shadow prices or by using relative prices from a revived private sector market. It is expected that increased transparency will lead to a deterioration of the income statements of most SOE factories, with the exception of those in oil and gas.

Improved SOE transparency will probably face ministerial opposition because it will reveal a more realistic estimate of the true budget impact of SOE inefficiencies. However, by converting hidden subsidies into explicit budget expenditures, this reform may strengthen the public's desire for further reform. The next step would be to improve SOE management.

Management Reform of SOE

Current management of firms in Iraqi SOEs suffers from ambiguous incentives. The managers are assigned multiple – often contradictory – goals that are evaluated according to bureaucratic, not market, standards. For example, loyalty will probably be more valued than efficiency. Incentive ambiguity can be reduced in several ways. Large firms can be unbundled so that managers are not required either to manage very diverse products lines or to manage product lines that compete with each other. Rationalization of private sector regulation and the resulting growth of private sector competition will not only motivate SOE managers to do a better job but also enable state-owned firms to sell their output and buy their inputs in larger markets. SOEs could also be restructured so as to provide factory directors with greater autonomy, to reduce the number of decisions that must be referred to the SOE headquarters or the ministry in Baghdad (see the detailed discussion in World Bank 1995, Chapter 5).

Another approach that is often advocated is that the SOE sign a formal agreement or contract with an individual manager explicitly laying out the SOE firm's goals and means as well as providing financial or other incentives for successfully achieving these goals. When the individual contracted to lead a firm is a government official, it is a performance contract; and when an SOE contracts with a private sector firm or individual to administer a state-owned firm, then it is referred to as a management contract. Both are widely utilized. In 1995, the World Bank found over 550 performance contracts in 32 countries and 202 management contracts in 49 countries being utilized (World Bank 1995, pp. 107–8). While the research is difficult to interpret, it appears that performance contracts with government officials generally fail to achieve the desired results, while management contracts with private sector persons are generally successful.

The failures of performance contracts – contracts between a government and a manager who is a public employee – to result in a substantial improvement in SOE management are caused by asymmetrical information, the lack of meaningful rewards for success, and lack of trust in government's commitment to its agreement. Public sector managers of an SOE factory know much more about their firm's activities than an outsider. They can use this information asymmetry to negotiate a performance contract that provides them with complex but easily achievable – soft – targets. These complex soft targets ensure that the manager will be evaluated as a success even if little real progress is made.

In addition, one would expect that a contract would have the greatest motivational impact on the residual claimant: the claimant who is allowed to keep whatever is left after other contractual parties have been paid. But most performance contracts state that the government, not the manager, is the residual claimant. The additional reward to the public official who successfully manages a SOE may be zero or a very small fraction of the gains from their managerial success. Maybe the manager will receive an earlier promotion to the next level of the bureaucracy. But, if they are honest, the manager is unlikely to become rich from success or devastated by failure. Finally, since in a performance contract one element of the government is contracting with another, it is difficult for the manager to have confidence that any commitments made by the government will be executed as contracted. Typically, the government will make a wide range of promises to the manager, concerning everything from managerial autonomy to delivery dates for required inputs, as well as commit to specific prices for expected outputs. However, both parties realize that these promises are likely to be modified in response to political or bureaucratic changes (World Bank 1995, pp. 120–133).

Management contracts – contracts between a government and a private party to operate the SOE firm for a fee – have been more successful. In a study of 158 management contracts in 50 countries, the World Bank found that two-thirds of the firms experienced improvements in both profitability and productivity (World Bank 1995, pp. 137–8). Unlike performance contracts for government managers, management contracts for private sector managers appear to have a better chance of overcoming the problems of asymmetrical information, adequate risk and rewards, and trust in government's commitments.

Management contracts tend to have fewer and more transparent performance standards. The private managers are granted greater autonomy, and their financial rewards are more closely connected to firm success. Finally, management contracts provide options for the government to improve the credibility of its commitments. The government can offer long-term contracts with the possibility of renewal. These contracts motivate the manager to take a long-term view of investment and training, since the manager expects to be able to benefit from any resulting improvements. In addition, the government signals its serious commitment by engaging in the expensive search for the right manager, by making announcements that are perceived as politically costly, for example allowing layoffs or price increases, or by making substantial investments (World Bank 1995, pp. 140–48).

It should be noted that, in addition to the exact terms of the management contracts, Iraq's economic environment would probably have a substantial

impact on whether there is a measurable improvement in SOE firm performance. Management contracts tend to be more successful when the SOE firm
that has implemented a management contract is in a competitive industry:
when there are other SOE firms or private sector firms competing for the same
customers. In Iraq, this would argue that management contracts will be more
successful if there was a simultaneous effort to encourage private businesses.

Management contracts tend to be more successful when the managers have
broad authority to hire, train, and fire workers without government interference. This will result in some of these SOE firms experiencing a drop in the
firm's wage bill, the total amount paid to employees. However, most reformed
SOE firms actually expand their wage bill. Either they have fewer workers but
pay them more, or the increase in labor productivity – as a result of improved
management – leads to increases in both production and employment. Despite
these optimistic results for the wage bill, it is expected that allowing SOE
private sector managers the necessary authority to prune unnecessary employment, especially "ghost" workers, would be extremely difficult in Iraq. Even
in joint ventures, the GoI has made it very clear not only that workers are not
to be fired, regardless of performance, but also, in some proposed joint venture
agreements, that if a worker quits or retires they are to be replaced, with the
selection of the new employee arranged by the Iraq partner.

Research also shows that management contracts tend to have more favorable results when foreign finance is involved. This is most likely because the
presence of such finance reflects a more serious commitment to the contract
on the part of the domestic government. In other words, it is believed that
governments are less willing to break an agreement with a foreign financial
institution, government, or international organization than one with its own
people. If the foreign interests have an equity agreement in an Iraqi firm, then it
is a joint venture, which is a form of privatization in that the government, while
still the dominant partner, gives up some control over an SOE firm.

Joint Ventures or Privatization

Which SOEs are likely to meet these demanding standards, especially that
of profitability? In Iraq in early 2006, T.W. Curran developed five "success"
factors and five "failure" factors that could be used to determine whether
an SOE joint venture or privatization would succeed. Curran then evaluated
each of almost 200 SOEs according to each of these factors with a ranking of
1–4, with 4 most likely associated with joint venture or privatization success.
A summary of the results is shown in Table 10.4.

Table 10.4   Successful joint venture or privatization scores

Score	Likelihood of joint venture or privatization success	Number of SOEs	Examples
35–40	Excellent	6	Cement, oil production
31–34	Good	58	Refining, fertilizer, electricity generation
21–30	Fair	126	Textiles, software
10–20	Poor	5	Missiles, PDS food imports

Source: Curran (2006).

Success factors

A joint venture or privatization of an Iraqi SOE is more likely to be successful if it possesses one or more of the following characteristics. First, natural resource-based firms that utilize petroleum, phosphate, or other inputs that Iraq possesses in generous amounts.

Second, SOEs that produce products that have high transportation costs relative to their value. Due to their high transportation costs, such products tend to face reduced competition from imports. A prime example would be cement for construction.

Third, SOEs that produce goods or services which are in demand from Iraqi consumers will have a greater likelihood of success than firms that sell most of their output to other SOEs or the government. This is somewhat controversial, since some argue that another SOE or the government would provide a guaranteed market for the output of a new joint venture or privatized SOE, which would make it unnecessary for the firm to engage in cutthroat market competition. However, as discussed above, SOEs tend to be untrustworthy business partners.

The fourth favorable characteristic is the degree to which an SOE's good or service is used in construction. Iraq will experience large construction expenditures over the next decades, for three overlapping reasons. As a result of four decades of conflict and mismanagement, reconstruction requirements are great. Also, a large proportion of the population is ill-housed. Finally, the GoI's plans for expansion of oil and gas production, refining, and exports will require substantial downstream and upstream infrastructure construction.

The fifth favorable characteristic is whether an SOE is actually operating; currently producing a product, even in an inefficient manner. If an SOE factory is just an empty shell, a successful joint venture or privatization is unlikely.

Failure factors

First, any firm that relies on a complex supply chain will find it very difficult to survive in Iraq's current environment. As will be discussed in the transportation sections of Chapter 12, Iraq currently lacks reliable, reasonably inexpensive transportation. For example, Iraq's substantial promise as a producer of fresh and frozen agricultural products is severely constrained by the lack of a reliable shipping cold-chain.

Second, similar reasoning supports a pessimistic view concerning the eventual profitability of firms engaged in complex or time-sensitive manufacturing. Unexpected disruptions in the quantity or quality of electricity and other common inputs, in transportation, and unpredictable application of government regulations, generally results in production delays and halts, inventory losses, and sometimes damaged manufacturing equipment.

A third factor is associated with Iraq's relatively small population of about 40 million persons. As a result, products with strong economies of scale such as automobiles will require access to an export market in order to make a profit. However, as will be discussed in Chapter 13, until recently Iraq suffered from a severe case of the "Dutch Disease". The dominance of oil exports resulted in a strong Iraqi dinar that increased the foreign price of Iraqi exports.

Fourth, since one of the purposes of establishing a joint venture or privatization is to reduce the budget burden, production of a good or service that is largely dependent on highly subsidized inputs such as fuel, electricity, and water will find it difficult to survive privatization.

The final joint venture or privatization failure factor is that it is unlikely that Iraq will be able to successfully achieve a sustainable share in any market which is already dominated by low-cost, high-quality international suppliers. For example, in view of the dominance of South Asian producers, it is very unlikely that an Iraqi firm will succeed in profitably making cotton clothing.

Likelihood of successful joint venture or privatization

Each success factor was scored 1–4, with 1 representing a poor showing. Failure factors were evaluated inversely, that is, scored 1–4 but with a 4 showing an almost complete absence of the failure factor. Using this scoring system, a score of 40 represented an ideal candidate for a successful joint venture or privatization while a score of 10 represented a very poor candidate (Table 10.4).

The validity of this joint venture or privatization metric shown in Table 10.4 is supported by the initial seven joint ventures proposals as well as the GoI 2018 privatization proposal. The first joint ventures were five cement plants (Sinjar, Fallujah, Kubaysa, Kirkuk, and Al Qaim) and one each for fertilizer (Baiji) and electrical equipment (Diyala) (SIGIR 2009a, Appendix A, p. 24). At the time of writing this book, there have been no privatizations. However, the GoI has identified multiple SOEs associated with 11 SOEs for partial or full privatization. Consistent with Curran's success and failure factors, most of the SOEs identified for privatization are in the energy, construction, transportation, and certain manufacturing sectors (GoI 2018, Table 18, p. 96).

However, almost two-thirds of Iraq's SOEs have only a "fair" or "poor" chance of being part of a successful joint venture or privatization, although some SOE firms might be attractive as "shells" for foreign investors: providing an Iraqi partner which is already authorized to do business in Iraq. If the GoI decides to actually privatize one or more SOE factories, then there is a growing body of research on privatization that might guide the GoI to a successful privatization.

SOE privatization: lessons learned

Of course, there have been many attempts to privatize SOEs in other countries that might provide useful lessons for Iraq. Since the conclusions of this growing field of research are wide-ranging, only a few of the most important will be mentioned here. (For more extensive discussions, see Galal et al. 1994; Megginson and Netter 2001; World Bank 1995). Probably most relevant to Iraq's situation are the experiences of other post-socialist states in Eastern Europe and Asia. The primary conclusion is somewhat cynical. The likelihood of successful privatization is very low unless government leaders benefit either professionally or personally; that a SOE privatization will benefit the entire economy is insufficient. For example, much has been made of the regulatory and other changes that led to a rapid acceleration of Chinese economic growth. However, it should be noted that the People's Republic of China did not reform its SOEs; it outgrew them (Qian 2003).

The specific form of privatization is critical: privatization alone does not guarantee more efficient performance because some SOEs are harder to privatize successfully than others. Even with "successful" privatizations, improvements in efficiency tend to occur slowly (Stallings and Peres 2000). Also, successful privatizations impact the quality of life in a country as well as the quantity of goods and services. For example, privatization of water services tends to lead to reduced child mortality as water quality improves (Galiani et al. 2006). Finally, as discussed above, SOE privatization tends to have a favorable impact on the environment, since the state has difficulty policing itself (Hettige et al. 1996). The challenges facing privatization are many, and the best method of privatization is industry-specific.

As noted in Table 10.3, there is an option where the state gives up all ownership rights but continues to substantially influence the management of a former SOE firm. If the former SOE firm is a natural monopoly, the most common transition is to a heavily regulated private firm. Whether this transition is successful or not is dependent on the economic and political environment. To the extent that there is a vibrant private sector economy and the government's commitments are credible, then a regulated private natural monopoly tends to be more efficient than an SOE. However, if either of these requirements are missing then there is little evidence that a transition to a regulated private sector firm is beneficial. But believing that a transition to managerial contracts, joint ventures, regulated natural monopolies, or privatization is more efficient than the status quo is a necessary but not a sufficient condition for successful SOE reform. There must also be a will to reform.

WILL IRAQ REFORM ITS SOES?

According to a World Bank study (World Bank 1995, pp. 233–6), there are two indicators that government leaders will support substantial SOE reform: a recent change in a country's governing regime or coalition, and an economic crisis. A regime change or economic crisis will tend to decrease – at least for a period – political and economic dependency on SOEs. In addition, if the SOEs, their leadership, or their workers are associated with the former regime or coalition then it is easier for the new regime or coalition to reform or even eliminate them. A crisis has a similar impact in that, during a crisis situation, the government may be more willing to try a dramatic restructuring of the economy.

There have been three fairly dramatic regime changes in Iraq in the last two decades. In 2003, Saddam's regime fell to the US-led coalition and Iraq entered a period of occupation. This continued until the constitutional referendum of December 2005 and the succeeding national and regional elections that established a new Iraqi government. Periods of occupation by US-led coalitions were accompanied by radical SOE reform in Japan and Germany after World War II. In fact, the initial reconstruction program in Iraq proposed widespread privatization of SOEs (Gunter 2007). However, an internal US government debate on the merits of radical SOE reform, combined with an upsurge in violence, prevented substantial SOE reform prior to the referendum on the new Constitution. Most recently, street protests beginning in October 2019 led to the resignation of the Prime Minister and early elections.

The new Iraqi government has publicly taken a measured approach of examining the benefits of various types of SOE reform on a factory-by-factory basis. In private conversations, Iraqi officials, with the possible exceptions of those of the Ministry of Industry and Minerals and the Ministry of Finance, have little interest in any substantial changes in the country's SOEs. With all of the other issues demanding the Prime Minister's attention, it would seem that privatization of SOEs would have had a low priority except for the economic crisis.

For at least the next decade, the health of the Iraqi economy and budget will be determined primarily by the value of petroleum exports. With the collapse in oil prices that began in 2014, it is increasingly recognized that direct and indirect subsidies to SOEs are a large drain on the shrunken government revenues. This recognition has led to increased interest in SOE reform.

The shift in GoI attitudes towards privatization is evident from the difference in the treatment of SOEs in the periodic documents on development strategy. In the National Development Strategies for the years 2005–07 and 2007–10, the restructuring and privatization of SOEs was a major component of the GoI strategy for "revitalizing the private sector" (GoI 2005b, pp. 25–6; GoI 2007, pp. 51–2). However, the National Development Plan: 2010–2014 is more pessimistic. It notes that the "economic, financial, legal, and administrative measures necessary to restructure public institutions" are lacking. As a result, it counsels delay until the proper foundation in all of these areas can be prepared (GoI 2010, p. 172). The most recent National Development Plan: 2018–2022 recognizes the difficulties in reforming SOEs, but lays out a specific goal of 15 privatizations by 2025 (GoI 2018, pp. 94–7).

In view of the difficult political and economic situations, it is likely that Iraq's long-term SOE strategy will be a modified version of the Chinese model. Such a policy would have four major components. First, where possible, engage Iraqi SOE factories in joint ventures with foreign investors. Second, increase the efficiency of remaining SOEs by implementing managerial contracts. Third, liberalize the regulatory environment that impedes growth of Iraq's private sector. Finally, allow the private sector to gradually become Iraq's major source of national income and employment by outgrowing SOEs.

11. Entrepreneurship in post-conflict Iraq

Economic attitudes and mentalities of the propertied urban classes in Babylonia [sixth century BC] can be described in terms of two basic (although necessarily idealized) models: a rentier type and an entrepreneur type. Rentiers seek to obtain a reliable income from mostly inherited positions and resources with little risk, by exploiting prebends [temple benefices] and landed property. Entrepreneurs tend to engage in highly profitable but also risky venture businesses in a competitive environment.

(Cornelia Wunsch 2010, p. 45)

Iraq faces a severe unemployment problem because of its rapidly growing population. Any year in which the economy creates less than 340 000 new jobs will lead to further growth in the pool of unemployed young men, with associated political instability. As discussed in previous chapters, the public sector is limited in its capacity to absorb each year's addition to the labor force since government services are already severely overmanned. Almost 50 percent of Iraq's labor force is currently directly or indirectly on the public payroll.

Prior to 2003, with the exception of non-grain agriculture, the private sector was suppressed as part of the Arab Ba'athist Socialist Party's philosophy. Since then, private sector growth has occurred mostly in the informal sectors of the economy. The quality of employment data is very low, but an estimated 20 percent of the labor force was employed in the informal economy in 2019. As expected, most firms in the informal economy are very small and in the service sectors.

In early 2020, the United Nations (UN) agency, the International Organization for Migration, surveyed 456 micro and small enterprises. Some of the results are summarized in Table 11.1. Wholesale and retail firms had the highest estimated average annual revenue of 152 million Iraqi dinar (ID) ($129 000), with each firm having about four employees. Food and agriculture firms had an estimated average annual revenue of 141 million ID ($119 700), with each firm having about 8.5 employees. Firms in the services area had both the lowest average annual revenue and the fewest employees. The low revenue per employee of these micro and small firms are consistent with very low wages.

As shown in the last column of Table 11.1, the micro and small firms in the sample have limited access to finance. Only 4 percent of these firms borrowed in the formal financial system, with an average loan of 13 million ID ($11 500). While 26 percent of these firms were able to borrow informally, the average loan was only 6 million ID ($5000). There was substantial variance among

Table 11.1   Micro and small enterprises

	Yearly revenue	Employees per firm	Revenue/employee	Credit: formal/ informal
Construction and manufacturing	$38 000	5.0	$7 600	0%/22%
Food and agriculture	$119 700	8.5	$14 100	12%/29%
Services	$23 600	3.2	$7 400	0%/21%
Wholesale and retail	$129 000	4.0	$32 300	5%/31%
Entire sample	$80 200	5.4	$14 900	4%/26%

Source: International Organization for Migration (IOM) (2020).

sectors: 12 percent of food and agriculture firms were able to borrow formally, although with an average loan size of 10 million ($8400); at the other extreme, construction and manufacturing firms as well as services firms reported no formal borrowing. As a result of this limited access to finance through the formal sector, micro and small firms are forced to rely on retained earnings, family support, and suppliers' credit to meet their needs for day-to-day operating capital. There is little capital available for expansion. As a result, not only is the private sector small, but it is also relatively inefficient.

In economic terms, Iraq is well inside its production possibility frontier, and due to innovations and technological change in the rest of the world Iraq is falling further behind as that frontier is moving out at a rapid rate. To reduce unemployment and achieve higher living standards for its people, Iraq needs to expand its non-oil economy. The creation and growth of micro, small and medium-sized enterprises (MSMEs) has the best potential to create the jobs needed. While the situation is complex, it appears that the binding constraints on the expansion of private sector MSMEs is a mismatch between the demand for and supply of entrepreneurs in Iraq, combined with very high transaction costs for entrepreneurial acts.

While there has been relatively little empirical research on entrepreneurship in Iraq, the Center for International Private Enterprise (CIPE 2008a, 2008b, 2011) surveys and the high-quality analysis of Desai (2009) have laid a strong foundation for future work.

WHAT DO ENTREPRENEURS DO?

A common misconception is that every Iraqi small businessman or woman is an entrepreneur. This error has led to government policy initiatives that either have little effect on economic growth or may even retard it. Not every small business owner is an entrepreneur. Rather, "Entrepreneurs are individuals who, in an uncertain environment, recognize opportunities that most fail to see and create ventures to profit by exploiting these opportunities" (Gunter 2012, p. 387).

It is widely accepted that entrepreneurs specialize in business-related "judgmental decisions" where there is no obviously correct answer and information is costly (Casson 1987, p. 151; see also Casson et al. 2006, pp. 3–4). In other words, entrepreneurs make business decisions based on insights in an uncertain environment. Uncertainty is different from risk. Risk occurs when there are well-defined possible outcomes, a sufficient sample size, and an accepted statistical procedure (Knight [1921] 1971). If one or more of these characteristics is missing, a possible outcome is uncertain.

The entrepreneur faces a challenge when they seek to profit from their judgmental decision or insight. Usually, the entrepreneur's ability to immediately sell the insight is severely limited because the insight is either too simple or too complex. If too simple, the mere description or demonstration of the insight to a prospective purchaser will allow its theft. At the other extreme, it is difficult to convince another that the insight will eventually lead to a substantial return on investment. As a result, the most common method for an entrepreneur to benefit from an entrepreneurial insight is to establish a new firm.

Especially in a developing, transitional economy such as Iraq, it is useful to distinguish between Kirznerian and Schumpeterian entrepreneurship. The former primarily involves arbitrage or speculation. Error, conflict, natural disaster, or unexpected regulatory changes create disequilibria in one or more markets. The Kirznerian entrepreneur – motivated by profit – creates new ways of buying in one existing market and selling in another market. These entrepreneurial acts of arbitrage or speculation resolve market disequilibria as well as reduce profit opportunities for others who later attempt to perform the same arbitrage or speculative act (Kirzner 1979, p. 92). In other words, a successful Kirznerian entrepreneur moves an economy from inside its existing production possibility frontier towards its efficient limit.

The Schumpeterian entrepreneur is also motivated by profit. However, this entrepreneur seeks to earn a profit through innovation even in markets that are initially in equilibria. They perceive new possibilities for products, markets, manufacturing processes, and so on, that are not apparent to others. If the Schumpeterian entrepreneur is successful, creative destruction will occur. As new processes, products, or markets are created, old products rapidly lose market share, former manufacturing processes are unexpectedly rendered obsolete, and so on (Schumpeter 1911, pp. 214, 228). A successful Schumpeterian entrepreneur shifts an economy's production possibility frontier outwards.

Figure 11.1 illustrates the interrelationship of innovative Schumpeterian and arbitrage/speculative Kirznerian entrepreneurs. The demand for entrepreneurial acts, from which the demand for entrepreneurs is derived, is a function of change. A state that is experiencing no change (no change in consumer preferences, production methods, trade patterns, demographics, and so on) will have little demand for entrepreneurs, since uncertainty is low and any profits for filling gaps in this stationary state economy have long since been competed away (Kuran 2010, p. 64).

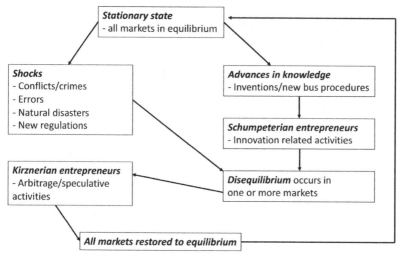

Source: Gunter (2012, Figure 1).

Figure 11.1   Kirznerian and Schumpeterian entrepreneurship

Of course, such a stationary state is hypothetical. Shocks or advances in knowledge constantly disturb Iraq's and every other economy. Shocks cause disequilibria in multiple markets that provide arbitrage/speculative profit opportunities for Kirznerian entrepreneurs. Advances in knowledge create innovative profit opportunities for Schumpeterian entrepreneurs that lead to the disruption of existing markets as new markets, new products, or new processes, rendering existing practices obsolete. Schumpeterian entrepreneurship creates disequilibria, resulting in opportunities for arbitrage/speculation on the part of Kirznerian entrepreneurs.

Currently, Iraq has a desperate need for Kirznerian entrepreneurs to fill the gaps – resolve disequilibria – in its economy. However, at the country's current stage of economic development, the need for Schumpeterian entrepreneurs is not so great.

Aside from their effects on market equilibria, Kirznerian and Schumpeterian entrepreneurs also differ in their characteristics. Successful Schumpeterian entrepreneurs tend to have more formal education, create larger entities, and require more finance. And they tend to be more visible. Kirznerian entrepreneurs are more likely to establish small – possibly single-person – firms, tend to have limited education, and their need for finance can often be met from informal sources such as family.

There has been some progress in determining the characteristics of entrepreneurship in developing countries. Most relevant to a discussion of entrepreneurship in Iraq are the following.

At a low level of economic development, much entrepreneurial activity is invisible to an outsider or an official. Among the reasons for this invisibility is that entrepreneurs seek to avoid taxation, demands for bribes, or hostile regulations by establishing themselves in the informal economy.

Entrepreneurial activity can be either opportunity-driven or necessity-driven. In the former case, someone becomes an entrepreneur as a matter of choice; while in the latter case, a person becomes an entrepreneur because they have no other options. Acs (2006; see also Estrin et al. 2006, pp. 712–14) notes that opportunity-driven entrepreneurial activity has a greater impact on economic development. It should be noted that most entrepreneurship in Arab countries – including Iraq – is necessity-driven (Noland and Pack 2007, p. 245).

Entrepreneurs in developing countries tend to spread their attention over several related businesses as a form of diversification (Lingelbach et al. 2005, p. 3; Stevenson 2010, p. 174). It is almost a cliché that when an Iraqi entrepreneur decides to give you a business card he has to shuffle through the cards of his multiple businesses in order to give you the correct one.

In many developing countries, the personal and professional rewards to unproductive rent-seeking are greater than those of productive entrepreneurship. There may be higher profits from persuading the government to put your competition out of business than to attempt to produce and sell a better product.

These characteristics constrain the types of policies that will be successful in creating a more favorable environment for entrepreneurship in Iraq. It is difficult or impossible for government agencies to efficiently identify individual entrepreneurs to target assistance or encouragement. Therefore, as discussed below, effective policies must be built on a knowledge of the general demand and supply of entrepreneurs in Iraq. Hopefully, this will identify policies that can either produce complementary inputs or reduce the transaction costs of entrepreneurship.

DEMAND FOR ENTREPRENEURS

Iraq is experiencing a series of strong shocks. The major causes are oil-funded economic growth, the transition from a socialist to a market economy, and the recovery from severe domestic and international conflict. These shocks create markct disequilibria and opportunities for Kirznerian – arbitrage – entrepreneurs.

Natural Resource Dominance

The Government of Iraq (GoI) is planning to increase petroleum output and exports over the next decade. As discussed in Chapter 7, this will require massive foreign and domestic investment in the discovery, exploitation, refining, and transportation of oil and the accompanying natural gas. The rapid expansion of the country's energy sector is expected to severely disrupt the non-oil economy with respect to consumer goods and services as well as employment.

Transitional Economy

Iraq is also engaged in a transition from centrally planned economic development – which long predates the Arab Ba'athist Socialist Party regime – to a more market-oriented or capitalistic economy. This transition will require changes in both the macroeconomic environment – exchange rate, inflation, budget expenditures – and the microeconomic environments in Iraq. Based on the Eastern European experience, one would expect a shift from industry to services, from domestic to international markets, and from intermediate to final goods (Estrin et al. 2006, p. 698). International organizations such as the International Monetary Fund (IMF) as well as the US government tend to focus on macroeconomic stabilization, not improving microeconomic efficiency, as the most important policy initiative.

Unlike most Eastern European transitions, macroeconomic stability was achieved relatively quickly in Iraq (Estrin et al. 2006, p. 699). The currency is stable, the central bank has been willing to act aggressively to contain inflation, and the ratio of national debt to gross domestic product (GDP) is fairly low as a result of debt forgiveness and GDP growth. However, the microeconomic transition has barely begun. The large industrial firms discussed in the previous chapter are products of the statist period, and continue to expend a great deal of managerial energy to avoid substantial change. Therefore, existing firms are more likely to be part of the problem than the solution when it comes to economic liberalization and employment creation (Estrin et al. 2006, p. 693).

Post-Conflict Society

A third source of shocks that contributes to the demand for entrepreneurship in Iraq is the rebuilding of a war- and sanctions-devastated economy. The GoI, the USA, and the many international organizations that are involved in the post-conflict rebuilding of Iraq have focused on an intellectually simple effort to return the economy to its antebellum state. In other words, these organizations have concentrated on returning infrastructure, state-owned enterprises (SOEs), and the general economic environment of Iraq to its condition in some previous, more favorable, period.

But much of this rebuilding ignores the dramatic social, political, and economic changes in Iraq and the world since Saddam first imposed a socialist template on the country. During the long war with Iran, the UN sanctions period following Saddam's invasion of Kuwait, the US-led coalition invasion of Iraq in 2003, and the long-drawn out fights with the al Qaeda and ISIS insurgencies, there have been substantial changes not only in the product and employment preferences of the Iraqi people but also in the people themselves. They have different skills and knowledge, social circumstances have changed, and people have different aspirations. The world outside Iraq has changed as well since the country began its 45-year (1958–2003) experiment with socialism. As a result, Iraqi attempts to return to some halcyon antebellum world are likely to lead to a lower quality of life. Iraq's businesses, along with the rest of its society, must adapt to a new labor force, new demands, new sources of supply, and new technologies. This requires entrepreneurship.

Furthermore, markets in Iraq are fragmented both geographically and economically. Products that are usually sold together elsewhere in the world, in Iraq are sold separately or on different schedules. Prices and supplies of many food products and utilities are controlled by various government agencies. The SOEs contribute to the uncertainty in many markets. Some SOEs continue to produce high-cost, low-quality goods and services. Others are not currently producing any products but are attempting to obtain government loans or grants to restart production. This increases the uncertainty facing private firms. Demand for a good or service might be currently strong, but if SOEs re-enter the markets with heavily subsidized products then private firms may find it difficult to compete. All of these structural and institutional changes have greatly enhanced the demand for entrepreneurs. But given that there is a large demand for entrepreneurs in Iraq, will there be a sufficient supply?

SUPPLY OF ENTREPRENEURS

Baumol (1990, p. 894), Schumpeter (1911), and others see the supply of entrepreneurs as determined primarily by culture and history, and therefore exogenously determined. Although it is important to note that entrepreneurs are not homogeneous. As Desai et al. (2010, p. 3) point out, entrepreneurs possess differing amounts of skills, discount factors, access to markets and networks. Therefore, the discussion of the supply of entrepreneurs in Iraq will consider not only the impact of history and culture but also the impact of rent-seeking, and of two other characteristics that tend to be associated with successful entrepreneurship: appropriate human capital and the capacity to make credible financial commitments (Harris 1970, pp. 348–9).

Are the Iraqis an entrepreneurial people? Since entrepreneurship tends to be associated with self-employment, a Gallup Organization survey on preference for self-employment provides some depressing results. Iraq, along with Egypt, Yemen, and Syria, has the lowest self-employment preference among young people in Middle East and North Africa (MENA) countries (Stevenson 2010, p. 185). Why do not more young people in Iraq seek to be entrepreneurs? Countries that have recent experience with a market economy tend to have a more favorable entrepreneurial environment (Stevenson 2010, p. 174). This is partly caused by the fact that entrepreneurs tend to come from families with previous entrepreneurial traditions (Estrin et al. 2006, p. 714). One might contrast the strong reaction to economic liberalization of the People's Republic of China with its roughly 35 years (1949–82) of exclusive socialism to the more troubled reaction of the USSR/Russia which banished market activities for about 75 years (1914–89). Iraq had about 45 years of socialism (1958–2003), which means that only the oldest family members remember relatively free markets. Another determinate of the supply of entrepreneurship and entrepreneurs is the general attitudes in a society.

Cultural Environment for Entrepreneurship

While there is a continuing debate on which cultural characteristics are essential for entrepreneurship to flourish, the following characteristics receive a great deal of attention. Entrepreneurs tend to be young, and youthful societies tend to be more willing to try innovations (Estrin et al. 2006, p. 707). Therefore, Iraq's low average age should contribute favorably to its entrepreneurship environment. Entrepreneurship-friendly societies or cultures tend to focus more on the individual than the community, have less respect for authority or age, are more focused on financial rewards than position and status, and accept that any rewards from accomplishment should accrue primarily to the individual and not be mostly diverted to the members of an extended family, tribe, or the state.

According to the controversial Hofstede cultural analysis, the average Iraqi does not have the characteristics desirable for entrepreneurship. The average Iraqi tends to score low on individualism, with a strong commitment to improving the welfare of a group, usually an extended family, rather than themselves. Also, Iraqis tend to have high "power distance", which is the belief that there is a hierarchy of power in which everyone has a place. Subordinates expect to be told what to do, and the best boss is a generous autocrat. Finally, according to the Hofstede analysis, Iraqis tend to have great respect for tradition, combined with an emphasis on getting quick results (Hofstede Insights 2020).

It is possible that younger Iraqis are less influenced by these cultural characteristics. At The Station, an entrepreneurial incubator in Baghdad, a number of young people are working on establishing their own businesses. Based on conversations, their entrepreneurship – whether Kirznerian or Schumpeterian – focuses on using modern technology to meet real needs. The Station is not alone in this effort: The Ark is also dedicated to getting university graduates started in their own businesses. But are these efforts to create a more favorable environment for entrepreneurship consistent with Islamic beliefs and practices?

About 97 percent of the Iraqi people are Muslim, and most older Iraqis state that religion is an important influence on their lives. Therefore, the supply of entrepreneurs in Iraq is, in part, a function of whether Islam provides a favorable environment for entrepreneurship. Some researchers look for evidence in religious writings, but these are difficult to interpret or contradictory (Kuran 2010, Chapter 3; Lewis 2002; Patai 2002). A more productive approach would be to examine how Islam and other cultural factors have shaped the institutions that affect present-day entrepreneurship in Iraq.

Since the Prophet Mohammed was a merchant, it is no surprise that Islam was initially very entrepreneurial. In fact, as late as the tenth century, almost three-quarters of all Islamic religious scholars earned a living from business (Kuran 2010, p. 66). In outstanding examples of successful entrepreneurship, Muslim traders crossed the world and founded trading posts in Africa, India, and China. Baghdad became a major world trading center where merchants bought and sold hundreds of products from the far reaches of the three continents. The city was also renowned for manufacturing, especially in textiles and metal products. Cross-national economic comparisons are always challenging, but the limited data available shows that Baghdad may have been the richest, most populous, most economically dynamic city in the world in the first centuries of the second millennium. Supported by an extensive system of irrigation, what is now Iraq supported a population estimated at more than 2 million persons, approximately the same population as the United Kingdom during the same period.

Baghdad's golden age of business, along with its scientific and artistic creativity, came to an end in AD 1258 when the Mongols sacked Baghdad. Most of the population was slaughtered, including the last Caliph of the Abbasid Caliphate, and the city was devastated (Previté-Orton 1952, p. 754). Attempts to rebuild the city were frustrated over the next 40 years by further vicious raids that not only smashed the urban areas in the Tigris and Euphrates valleys but also destroyed large parts of the ancient irrigation systems that had fed a large and growing population. It was not until about six centuries later, in the late nineteenth century, that Iraq finally returned to the population and per capita real income levels that the land had achieved before AD 1258 (Ansary 2009, pp. 87–8; Maddison 2007, p. 192).

Whether the destruction of Baghdad, combined with later military setbacks in Western Europe (the final *reconquista* of Spain in 1492, the failed Siege of Vienna in 1529, and the defeat at the Battle of Lepanto in 1571), shook the confidence of the Islamic leadership is difficult to determine. After all, the three major Islamic states continued to win battles on their periphery and dominate the social and political life at the core of the Islamic world. However, in retrospect, Islam's period of geographic expansion came to an end, to be replaced by a period of growing European influence in Islamic states (Ansary 2009, pp. 217–46).

According to Bernard Lewis (2002, pp. 151–60), there was a gradual acceptance that the Islamic states were losing power and influence compared to the growing and innovative states in the West. This led to a search for the reason behind this reversal of power. One response was that believers had fallen away from the pure Islam of the first half-millennium after Mohammed. Therefore, the remedy was to throw out any innovations that were not firmly grounded in the Qur'an or Sunnah. The Sunnah are the words and practices of Mohammed which together form the core of Islamic law – Sharia law (see Ansary 2009, Chapter 13, pp. 247–68).

This centuries-long movement towards a more conservative, less creative perspective on a variety of issues had an adverse impact on attitudes towards business in general, and entrepreneurs in particular. It was necessary to ground business innovations in Islamic precedents. This had two consequences that still influence attitudes towards entrepreneurship in Islamic countries such as Iraq. First, business institutions were developed that favored personal over impersonal exchange. Second, the practice of justifying innovations in centuries-old writings concealed the significant innovations of the period.

Douglass C. North (1990, pp. 34–5) developed a framework for the study of exchange that might be useful. The most restrictive type of exchange has characterized trade through most of history. Such trade is limited to the "personalized exchange involving small-scale production and local trade". In the absence of what most would recognize as established norms, rules, or laws, trade-related agreements are "enforced" by repetitive transactions and shared cultural values among persons who know each other well.

Over time, there arose a second form of exchange that allowed transactions with persons who were not members of the same family, tribe, or village. These transactions permit more complex agreements over greater time and distance. However, such exchanges are subject to demanding conditions. They require that the parties to such an exchange have "kinship ties, bonding, exchanging hostages or merchant codes of contract. Frequently, the exchange is set within a context of elaborate rituals and religious precepts to constrain the participants" (North 1990, pp. 34–5). An individual or firm who fails to live up to such an agreement can expect exclusion from future trade opportunities, but also this trader will run the risk of being shunned or spurned by their community. These non-formal methods of enforcing exchange agreements are a complicated but effective entrepreneurial response to the lack of formal third-party enforcement. (See Estrin et al. 2006, pp. 716–19 for a further discussion of "coping strategies"). However, the transaction costs of negotiating such non-formal agreements are high.

North's (1990) third form of exchange is impersonal exchange with third-party enforcement. This third party, usually a government, enforces agreements with reasonable predictability, fairness, and cost. This allows the possibility of mutually beneficial transactions among domestic or even international strangers. The expenditure of effort to build and maintain the elaborate relationships required by the second form of exchange is unnecessary with effective third-party enforcement.

Islamic business culture is consistent with North's first and second forms of exchange but is generally inconsistent with the third. The national governments of Islamic states have tended to avoid creating the institutions that are intended to directly support private business activity, with the possible exception of building grand bazaars in major cities. The motivation for such bazaars might have been more to facilitate the regulation and taxation of commerce than to encourage it (Kuran 2010, p. 77).

However, as long as most non-Islamic trading peoples also primarily dealt in North's (1990) first two forms of exchange, Islamic traders were not at a disadvantage. But when institutions with reasonably efficient third-party enforcement of contracts appeared in the West, Islamic businessmen started to lose the competitive race (Kuran 2010, pp. 63, 71). It is not that any Middle Eastern country lacks laws to govern sophisticated transactions among strangers; these institutions of a modern economy are almost universal. However, they are not a natural outgrowth of pre-existing domestic business institutions; rather, they were either imposed by colonial powers or adopted in an attempt to prevent the dominance of foreign traders and firms. This reduced both their effectiveness and their perceived legitimacy.

Modern commercial laws in Islamic states operate somewhat in an institutional vacuum that makes contracting and enforcing transactions both more difficult and more expensive than in other cultures. This not only reduces the incentives to engage in many entrepreneurial efforts, but also allows foreign entrepreneurs to compete more effectively. Examples of institutions that complicate the contracting and enforcement of agreements with strangers are seen in the laws of inheritance and corporation.

In the West, a successful businessperson could will their business to one or more of their children either directly or by establishing a corporation and ensuring that their preferred heir or heirs would have a controlling interest. Sharia-compliant inheritance laws restrict the ability of a wealth holder to distribute their assets by a will since these inheritance laws set out specific shares for each family member. The impact on entrepreneurial activity is significant. Under Islamic inheritance laws, it was very difficult to pass on a family business. If there were multiple children then it is likely that a family business will be dismantled upon the death of the founder (Kuran 2010, p. 69).

Nor can this problem be avoided by establishing a corporation, a legal person, to continue the entrepreneur's business. Islamic law respects only flesh-and-blood persons; corporations (as understood in the West) cannot be established (Kuran 2010, p. 69). Since Islamic law recognizes only flesh-and-blood persons and yet enforces strict inheritance laws, most private businesses, however successful, rarely last longer than a single generation.

The adverse impact on entrepreneurship of Sharia laws such as those of concerning inheritance and corporations should not be exaggerated. It is possible to develop Sharia-compliant alternatives for incorporating innovations, or *bid'ahs*. A *bid'ah* is any type of innovation in Islam. If the Qur'an and Sunnah outline a way of life that cannot be improved upon then any *bid'ah* is suspicious and should be examined carefully before it is considered acceptable. Since the Qur'an provides guidance for an extensive range of government and business activities, judging the acceptability of a business innovation can be both complex and controversial. Among the products whose adoption was delayed because of concerns that the innovations were unacceptable were coffee, the printing press, and in 1960s Saudi Arabia, television (Kuran 2010, pp. 72–3).

As a result, entrepreneurs are often forced to spend extensive amounts of time and influence to negotiate an acceptable religious decision on a business-related *bid'ah*. Or complex procedures may be necessary in order to construct a relatively simple business transaction in a manner that is religiously acceptable. For example, the prohibition against interest may be overcome by the use of *murabaha*, where one party (a bank, perhaps), who has at least

constructive ownership of an item, announces the actual cost and any mark-up
to the buyer (see Chapter 9; and Schoon 2008, pp. 32–3). Developing, coordinating, and executing such acceptable alternatives to otherwise forbidden
activities may substantially increase the difficulty of bringing an innovation to
market. Kuran (2010, p. 81) compared dealing with *bid'ah* to the difficulties
of dealing with environmental regulations in the USA. Such regulations rarely
explicitly forbid an activity but, however worthy their purpose, they tend to
make it more complex and expensive to produce a new product or adopt a new
way of doing business.

While Islam dominates the religious environment of Iraq, the participation
of other religious sects is not insignificant. When over 1600 small businessmen
were asked for their religious affiliation, 6 percent of those that responded
stated that they were either Christian or another non-Shi'a, non-Sunni religious
affiliation. This 6 percent response is twice their estimated representation in
the population at large. It should be noted, however, that almost 22 percent
of those surveyed refused to answer the question about religious affiliation
(CIPE 2008a, Table 30). Jews are not represented in the sample. Contemporary
records from 1908 estimate that almost one-third of the population of Baghdad
was Jewish. As late as 1947, there were an estimated 117 000 Jews in Iraq, 2.6
percent of the total population, almost half of which lived in Baghdad (Issawi
1988, p. 124). In fact, this Jewish community of Baghdad was formerly considered to be a fruitful source of entrepreneurs in Iraq, and was responsible for
a variety of business initiatives. However, growing hostility towards Iraqi Jews
since the late 1940s drove them into exile (Tripp 2000, pp. 123–6, 196).

Socialism and Entrepreneurship

The changes in Iraq's entrepreneurial environment brought about by the rise
to power of the Arab Ba'athist Socialist Party in 1968 are more a matter
of degree than of kind. Since at least the Hashemite Monarchy (1941–58),
non-agricultural economic activities were dominated by a relatively small
number of Iraqis who gained economic power through familial and friendship
relations with the national political leadership. These patron–client relationships ensured political power by providing a mechanism to reward loyalty to
the regime with status and economic rewards (Tripp 2000, p. 197).

Creating and maintaining patron–client relations was one of the motivations for the establishment of SOEs. SOEs generally oppose new firms in
their industries, not only to reduce competition but also since new firms may
support further economic reform (McMillan and Woodruff 2002, p. 153).

In Iraq, this opposition takes several forms: harsh regulation of private business (discussed below); harassing private businesses where SOEs are important suppliers to or buyers from private firms; and, as was discussed in Chapter

4, through massive corruption. As a result, many small firms choose to stay in the informal (underground) economy, with all of its associated inefficiencies. In a 2011 survey, only 44 percent of all businesses met the legal requirement to register with the GoI. There was a wide divergence among industries, with the highest rate of registration in the construction industry (51 percent), since registration is necessary to bid on government contracts (CIPE 2011, pp. 18–19). These percentages may exaggerate the actual level of compliance, since unregistered entities are probably less likely to participate in surveys.

Unregistered or informal firms account for an estimated 20 percent of Iraq's employment. Since informal firms are unable to access the legal system, they tend to engage in North's (1990) first two forms of exchange. Surviving in the informal economy with its lack of formal regulation is inherently an entrepreneurial act. However, small firms in the informal economy often survive more by their agility at navigating the various layers of governmental authority rather than their ability to recognize purely economic opportunities (Estrin et al. 2006, p. 699). If this pool of active entrepreneurs could be freed from the high transaction costs associated with the informal economy, the potential for real economic growth and employment are significant.

The fact that both Islam and socialism tend to discourage entrepreneurship should not be taken to mean that they support similar economic policies. For example, Muhammad Baqir al-Sadr, a renowned twentieth-century Iraqi cleric, wrote a major economic work where he used eight times more space in attacking socialism than capitalism (Kuran 2004, pp. 98–9). His view – which is widely held among Iraqi religious leaders – is that there are three alternative economic structures: socialism, capitalism, and Islamic economics.

Wage-Takers, Entrepreneurs, and Rent-Seekers

As Baumol (1990, p. 894) pointed out, if the supply of entrepreneurs is almost entirely a function of culture and history then it is a "counsel of despair" since nothing can be done to increase the supply of entrepreneurs. However, he states that it is possible that a potential entrepreneur will engage in unproductive or even destructive entrepreneurship, as opposed to the productive type that leads to economic development. Therefore, it might be possible, by creating the proper incentives and institutions, to persuade potential entrepreneurs to engage in welfare-increasing initiatives (Baumol 1990, pp. 897–8).

It is useful in the case of Iraq to divide career opportunities into three categories: wage-takers, entrepreneurs, and rent-seekers. Wage-takers are satisfied with jobs with minimum compensation uncertainty. While such employment can exist in both the private and the public sectors, in Iraq public sector employment provides the most favorable environment for wage-takers. On average, employment with a government ministry or SOE provides 40–60 percent higher wages than equivalent jobs in the small formal private sector. In addition, public sector jobs in Iraq provide stronger job security, better benefits, a pension, and a more relaxed pace of work. Any improvement in the number or relative compensation of public sector employment in Iraq can be expected to reduce the pool of potential entrepreneurs. Even for those potential entrepreneurs who avoid the temptation of a government job, there is still the problem of diversion from entrepreneurship to rent seeking.

Rent-seekers are persons who seek to make a living by capturing economic rents rather than by engaging in voluntary trade (Krueger 1974; Tullock 1967, pp. 224–32). For example, a rent-seeker could bribe a government official to restrict auto imports to a single port controlled by the rent-seeker. This monopoly misallocates resources, reduces aggregate welfare, and redistributes income away from consumers in favor of the monopolist (see Tullock 2005a, p. 9). Rent-seeking can also take the form of legal activities such as lobbying the government to award a contract to a company that is not objectively the high-quality, low-cost producer.

Rent-seeking impacts the pool of Iraqi potential entrepreneurs in two ways. First, if potential entrepreneurs were attempting to maximize their wealth or status then one would expect them to choose to be either entrepreneurs or rent-seekers, based on whichever of these had the greatest potential return. Second, some rent-seekers could work at diverting entrepreneurial profits into their own pockets. This would discourage potential entrepreneurs by reducing profits. The implications for public policy are clear, although difficult to execute. If incentives and institutions can be created that discourage rent-seeking without placing a burden on productive entrepreneurship, an economy would gain in two ways: reduced damage to the economy from unproductive or destructive entrepreneurship, and an acceleration of economic development from more persons engaged in productive entrepreneurship. Changing these incentives and institutions will be a much more rapid process than waiting for cultural change (Baumol 1990, pp. 916–19).

Incentives in Iraq may be shifted in the direction of productive entrepreneurship by increasing the supplies of complementary inputs in the entrepreneurship process. These would include providing the potential entrepreneur with the necessary knowledge or skills, and ensuring that they are able to provide adequate financial guarantees. Without appropriate education and ability to provide financial guarantees, the likelihood of entrepreneurial success is low.

Education of Entrepreneurs

Private sector firms in Iraq tend to be very small compared to other countries. As stated in Chapter 10, a business with 30 employees or more is considered large. According to the 2020 International Organization for Migration (IOM) study, the average micro or small business in Iraq has an average of only 5.4 employees. According to another study, about 83 percent of private businesses in Iraq were sole proprietorships, 13 percent were family-owned, and 2 percent were non-family partnerships or domestic corporations (CIPE 2011, Charts A1 and A3, pp. 43–5).

With respect to education, a study of entrepreneurs in Egypt found a wide distribution of educated workers among MSMEs. For example, while 20 percent of the entrepreneurs in micro enterprises possessed a university degree, 13 percent were barely literate or illiterate (El-Gamal et al. 2000, p. 13). Harris (1970, pp. 352, 354) found similar results in Nigeria, where many successful entrepreneurs were illiterate or not very educated. Using the distinction discussed above, while Schumpeterian entrepreneurship in the Organisation for Economic Co-operation and Development (OECD) countries tends to require substantial formal education, this is not true for Kirznerian entrepreneurs in countries at Iraq's stage of economic development. Successful Kirznerian entrepreneurship at the MSME level requires relatively low levels of formal schooling.

However, without being able to read, write, and calculate, the odds are strongly against becoming a successful entrepreneur. Unfortunately, as discussed in Chapter 3, the proportion of the Iraqi population that is either illiterate or failed to complete elementary education is high. Aside from the social and political benefits, it can be expected that universal elementary education will lead to more persons attempting entrepreneurship, and a greater likelihood of success for those who make the attempt.

Aside from basic general education, research has shown that the best education for potential entrepreneurs is working for or being associated with an independent businessman (Stevenson 2010, p. 178). As Lingelbach et al. (2005, p. 7) states: "Entrepreneurship is a lonely profession rendered more difficult without the benefit of mentorship and apprenticeship". Of course, there may be a self-selection bias; potential entrepreneurs may seek out current entrepreneurs to work for. However, the tendency for family members of independent businessmen to become businessmen themselves points to a learning process in addition to self-selection. It is not clear whether the learning involves specific business skills (for example, how to keep accurate records), the inculcation of a business attitude, or both. In contrast, having worked for a government agency or SOE appears to reduce the likelihood of becoming a successful businessman. Since an estimated 50 percent of the Iraqi labor force works directly or indirectly for the government, this reduces the learning-by-doing opportunities for potential entrepreneurs.

Ability to give Financial Guarantees

If a potential entrepreneur is unable to provide financial guarantees, then their likelihood of success is low (Harris 1970, p. 351). Without finance, the entrepreneur must try to convince others to advance resources without a reasonable assurance of payment based on limited information. In other words, they must try to convince others to become entrepreneurs like them without providing so much information that their prospective partners will steal their innovation.

The adverse impact on in the Iraqi economy of the serious problems in the banking, bond and stock markets were discussed in Chapter 9. However, the difficulty of obtaining finance is a special burden to potential entrepreneurs. "Getting credit" and "protecting investors" are two of the ten subcategories that are used to estimate the World Bank's (2020e) "Ease of Doing Business". In 2020, Iraq ranked 186th in getting credit (tied for the worst ranking in the world). Not only is the regulatory environment bad for getting credit in Iraq, but also it is deteriorating. As can be seen in Table 11.2, Iraq's "getting credit" rank was 180th in 2015. The low ranking with respect to getting credit was caused by weak legal rights for creditors, and an almost complete absence of either private or public credit information on firms or individuals. The relatively small number of non-SOE bank loans are made primarily to government workers, bank employees, and influential politicians, not to entrepreneurs. With respect to protecting minority investors, Iraq ranked 111th in 2020, slightly improved from 2015. The poor "protecting minority investors" ranking is based on low rankings in three metrics: the extent of disclosure of significant firm transactions, the extent of director liability, and the ease of shareholder lawsuits.

Of course, bank loans and equity sales are not the only sources of finance for an Iraqi entrepreneur. Many entrepreneurs raise needed finance by borrowing from family members or obtaining credit from suppliers or buyers. A decade-old nationwide survey of all Iraqi households showed that the primary source of loans was family members in Iraq or abroad. These family loans accounted for over 81 percent of all loans. Trade credit from merchants was the second most popular source, at about 11 percent. State-owned or private banks supplied about 6 percent of total loans, with moneylenders, employers, and non-governmental organizations accounting for the rest (COSIT 2008, Table 10-2, p. 756; see also CIPE 2011, Chart A6, p. 47).

The dominance of family savings and retained earnings as financing sources has an adverse impact on potential Iraqi entrepreneurs. Not only is finance for future operations limited by previous business profitability, but also achieving any significant scale of production will probably be a gradual process as each stage of production must finance succeeding stages. Gradual growth makes an entrepreneur more vulnerable to rent-seeking by regulatory authorities.

Table 11.2   *Ease of doing business in Iraq*

Category	Iraq 2020	Iraq 2015	MENA range
Ease of doing business	172nd	160th	16 to 187
Starting a business	154th	144th	17 to 173
Dealing with construction permits	103rd	145th	3 to 186
Getting electricity	131st	102nd	1 to 187
Registering property	121st	116th	10 to 187
Getting credit	186th**	180th	25 to 186
Protecting minority investors	111th	114th	3 to 183
Paying taxes	131st	57th	1 to 158
Trading across borders	181st	178th	48 to 188
Enforcing contracts	147th	119th	9 to 166
Resolving insolvency	168th**	189th**	44 to 168

Note: ** Tied for worst in world.
Source: World Bank (2020e), "Ease of Doing Business" Index.

Just like newborn animals in the wild, a firm is most vulnerable to regulatory predators when it is small. As noted by Parkinson (1957) in one of the more controversial chapters of his classic work, since bureaucracies tend to operate at a very slow pace, rapid growth will reduce the period of vulnerability to regulatory exploitation. Once a firm becomes large enough, its vulnerability to regulatory predation falls, since a large firm can demand (or pay for) political protection from the grasping bureaucracy (Parkinson 1957, pp. 92–4). On the other hand, a slow-growing private firm gives regulators an extended opportunity to perceive, evaluate, plan, and execute the successful exploitation of the firm.

External Sources of Entrepreneurs

There are external sources of entrepreneurs that could also help meet the demand. Since the rise of Saddam, large numbers of Iraqis have fled the country. The UN estimates that almost 2 million persons (7 percent of the population) have fled, including 40 percent of the country's middle class (Oxfam International 2007, p. 3). While most are living lives of poverty in neighboring countries, a substantial number reached the West.

Over the last four decades, many of these former Iraqis have become entrepreneurs either by choice or by necessity in their new home countries, or have obtained the skills and access to financing essential for successful entrepreneurship. The return of this Iraqi diaspora would be especially valuable since they can be expected to have a realistic view of the Iraqi business environment.

TRANSACTION COSTS FACING ENTREPRENEURS

Transaction costs are the costs involved in negotiating, executing, and enforcing an agreement to trade, a good, or service other than the costs to the seller of producing the product and to the buyer to consume it.

Rule of Law

The scope for entrepreneurship is a function of the size of the market, since a larger market allows for increased specialization. Since the days of Adam Smith (1776, p. 9) it has been recognized that increased specialization facilitates invention and innovation and therefore entrepreneurship. However, the size of a market is determined not only by geography, culture, and technology, but also by the rule of law.

As discussed above, private sector trade in Iraq is comes in three types: the first two are local personalized exchange and impersonal exchange enforced by non-formal means. The third type of exchange – trade between strangers with third-party enforcement – is limited. The primary cause of this limitation is the lack of a commercial code that provides for predictable, fair, and inexpensive resolution of commercial disputes. This has severely constrained the size of the market for many goods and services. Smaller markets tend to mean lower entrepreneurial profits and, therefore, fewer entrepreneurs.

The current Iraq commercial code severely limits transactions among private strangers, as a complex ubiquitous bureaucracy attempts to manage most economic activity. This is a common situation in transition economies. Business regulations were developed during a previous planned-economy regime and are inappropriate to a more market-oriented economy (Stevenson 2010, pp. 190–91). Relatively simple business activities require substantial paperwork and the approval of multiple levels of the Iraqi government. Like all sophisticated bureaucracies, these agencies take a great deal of time to approve routine requests, but tend to stall completely when faced with any kind of innovation. If under common law "all is allowed except what is specifically forbidden", then in Iraq "all is forbidden except what is specifically allowed".

These complex business regulations also tend to provide large rewards for rent-seekers. As was discussed in Chapter 4, the Iraqi economy is structured so as to discourage private business and facilitate very profitable rent-seeking.

Returning to Table 11.2, of the 190 nations in the 2020 survey, Iraq ranked 172nd overall, which revealed it to have one of the most hostile regulatory environments for business not only in the MENA countries but also in the world. Of the ten subcategories evaluated by the World Bank, Iraq scored particularly poorly in starting a business (154th out of the 190 countries evaluated), getting credit (186th, tied for worst in world), trading across borders (181st), and resolving insolvency (168th, again tied for worst in the world). Iraq not only ranks in the bottom quintile in most subcategories, but its relative standing has deteriorated over the last five years: its overall ranking was 160th in 2015. The Iraqi regulatory environment for private business is bad and getting worse.

Particularly damaging to any attempt to encourage entrepreneurship in Iraq are the regulatory barriers to starting a new business, getting credit, and closing a business. The required procedures are complex, difficult, and expensive. As a result, these regulations act as severe "taxes" on entrepreneurship that either discourage entrepreneurs from attempting to create a new business or encourage them to operate in the informal or underground economy with all of the associated inefficiencies that were so vividly described by De Soto (1989, 2000).

As an illustration, consider starting a new business in Iraq. The World Bank "starting a new business" ranking is based on a survey to determine the procedures needed to start a small or medium-sized business legally, without paying bribes. In Iraq, there are eight separate procedures (nine for women) involving multiple government offices. If the potential entrepreneur knows exactly what to do and is unwilling to pay bribes then it will take 26 days (27 for women) and about 24 percent of the average Iraqi annual income (about 2.4 million ID or $2000) to register a new business. Contrast this to registering a similar business in the state of Delaware in the USA, which can be accomplished online in about half an hour and costs $139.

Getting credit is made difficult because of the absence of most standard legal protections for creditors. On a ranking of 0–12, Iraq's legal rights index is 0. And only 1.3 percent of the population are listed on a credit registry, compared to a MENA average of 15.8 percent. In addition, there is no collateral registry. This absence further reduces the ability of firms to borrow.

Trading across international borders is also very difficult for an Iraqi entrepreneur. Iraq ranked 181st out of the 190 nations surveyed when it came to the difficulty of importing or exporting a dry container of ordinary legally manufactured goods to or from a business located on the periphery of the country's largest business city. For example, what are the costs of shipping a container from the dockside in Basrah to the suburbs of Baghdad? Not only does it take about 74 eight-hour days to complete the paperwork and obtain the necessary approvals, but also the cost per container of border compliance, including providing the necessary documentation, is an estimated 1.5 million ID ($1150). Exporting is equally complex and expensive: 38 days and 3.4 million ID ($2900) per container. These costs include fees, document preparation charges, terminal handling charges, and inland transport. The fees do not include tariffs, duties, or bribes. For comparison purposes, a similar shipment from Saudi Arabia would take only six days to process and cost about $400. In Iraq, only high-value goods that can be ordered with a long lead time will be legally traded. Other products either will not be traded or, more likely, will be smuggled.

Finally, Iraq is tied for the worst ranking in the world with respect to resolving insolvency, which considering the competition from some other extremely dysfunctional countries is an impressive achievement. For example, in a country in the top-third of surveyed nations such as the MENA state Djibouti (ranked 44th in the world), closing a business takes about one and a half years and stakeholders recover an average of 44 percent. In Iraq, the average recovery is zero since no firm has successfully completed a formal bankruptcy. The impact of this dysfunctional bankruptcy law on entrepreneurs is twofold. Since there is little chance that a creditor will be able to recover even a fraction of their investment if a business fails, they will hesitate to lend. Also, without an effective bankruptcy law, it is difficult for an entrepreneur to recover from a failed endeavor. If an entrepreneur attempts to establish a new business, the unsatisfied creditors from previous endeavors will seek to attach the assets of the new firm.

Not only are Iraqi business regulations generally hostile to private business, but also they are difficult for even an educated businessman to understand. Across the six cities in a CIPE survey, only 16 percent of businessmen thought that the regulations were understandable, while 31 percent responded in the negative (CIPE 2008b, Table 18, p. 7). This creates unnecessary uncertainty that further discourages potential entrepreneurs.

A related issue is whether Iraqi businessmen are confident that, after expending substantial time and money complying with incredibly complex laws and regulations, the authorities will carry out their end of the "bargain" (Estrin et al. 2006, p. 705). This confidence is apparently low in Iraq. Security aside, when Iraqi businessmen were asked to name their top three items that had the most adverse impact on the growth of Iraqi business, the results were surprising. The top two factors cited nationwide were, as expected, corruption (volunteered by 61 percent of all businesses surveyed) and the weakness of Iraqi infrastructure (cited by 48 percent). However, in third place was "not applying laws and regulations" which was cited by 41 percent. This law and regulation factor has been interpreted as revealing a lack of confidence that even if correct procedures are followed, the authorities will do what they said they would do. According to the survey, this has a greater adverse impact on businesses then difficulties in obtaining finance, high fees, or the problems of obsolete equipment (CIPE 2011, Table 1.1, p. 17).

The complexity and obscurity of business regulations in Iraq are not unloved artifacts of the Saddam regime. Rather, the expense and time required for a businessman to comply with the regulations increases the incentives to pay bribes to avoid the regulations.

Corruption

Whether corruption (the abuse of public power for private benefit) amounts to a significant transaction cost in Iraq is sometimes debated. As an Iraqi businessman told me: "In the West, firms pay lawyers to ensure that transactions are executed in the intended manner while, in the Middle East, firms pay bribes for the same purpose". If, as in Iraq, the legal system is opaque, inefficient, and unpredictable, then paying bribes may be not only more efficient but also cheaper.

This argument that, in view of the ineffectiveness of the Iraqi legal system, businessmen should use a "second-best" corruption solution has at least two weaknesses. First, corruption is a very expensive way of facilitating business transactions. About 37 percent of all firms reported receiving at least one bribe request (World Bank 2020c, Table 5.2). And the size of the bribes are substantial. In a 2011 survey of business owners in nine Iraqi provinces, the CIPE reported that over half of the businesses stated that corruption accounted for more than 20 percent of their total costs of doing business (CIPE 2011, Chart 5.1, p. 29). This is consistent with an earlier survey which revealed that between 12 percent (Kirkuk) and 26 percent (Basrah) of firms reported that corruption accounted for 40 percent or more of total costs (CIPE 2008a, Table 27).

In addition, the costs of corruption tend to worsen over time. There are strong incentives for government officials to continually increase the complexity and expense of complying with regulations in order to encourage businessmen to pay more substantial bribes. Thus, widespread corruption not only increases the cost of doing business but, as discussed above, it discourages potential entrepreneurs.

Transportation Costs

Especially in developing countries, transportation costs can be a substantial contributor to overall transaction costs. Just as the absence of various institutions can reduce expected entrepreneurial profits by restricting the size of the market in which the entrepreneur can trade, so can absent or low-quality infrastructure have the same effect. For example, an Iraqi agribusiness entrepreneur saw an opportunity in producing high-quality chickens using modern feed and veterinarian care, but was unable to bring his product to market because of the lack of rapid road transport. Much of his stock was dying before it reached the urban market.

Of course, this problem could be seen as another entrepreneurial opportunity: to provide secure cold-chain truck transportation so that a quality product can be provided to the market despite bad roads. However, much infrastructure is non-excludable and, therefore, requires government or charitable provision. This topic is discussed at greater length in Chapter 12. The takeaway is that poor or non-existent infrastructure increases transportation costs and therefore reduces potential entrepreneurial profits.

In summary, there is currently a great need for an increase in private sector employment in Iraq in an environment of rapid change, and entrepreneurs can play an important role in creating this employment. This leads to the critical question of which policy initiatives should be adopted to increase the likelihood of successful entrepreneurship.

WHAT CAN BE DONE?

There is a strong temptation to provide a laundry list of the many initiatives that individually or collectively have facilitated entrepreneurship in other countries (see, for example, Stevenson 2010, pp. 197–201). In fact, the GoI (2020b) in its 80-page 2020 White Paper was unable to resist this temptation. However, any such list, to avoid doing more damage than good, must be carefully crafted to meet the specific conditions of Iraq. First, as discussed in Chapter 4, Iraq suffers from corruption at every level of the government. Any policy recommendation intended to facilitate entrepreneurship that can possibly be distorted to elicit bribes will be so distorted. For example, it is believed that the 2009 GoI program to provide microloans to potential entrepreneurs had very little impact since most of the funds were diverted to ministerial clients.

Second, ministerial capacity is limited not only by a general lack of managerial experience but also by the strict hierarchy in ministerial decision-making. Even relatively minor decisions must proceed up the bureaucracy, step by step, with each layer attaching carefully crafted comments until it arrives at the top. As a result, the minister and his few trusted aides struggle to deal with a flood of requests for decisions. Routine decisions are long delayed; innovative requests involving new programs often cause the bureaucratic decision process to seize up entirely.

Finally, cooperation among ministries is more the exception than the rule. As discussed in Chapters 7 and Chapter 12, the ministries of Oil and Electricity are often unable to coordinate either the supply of fuel to the major big city electrical generation plants or the supply of electricity to the three largest refineries. As a result, Iraq continues to suffer from shortages of both electricity and refined fuels. The moral is that any pro-entrepreneurship policy that requires inter-ministerial cooperation or even cuts across different ministries' areas of responsibility will be difficult to execute with any efficiency.

Because of corruption, limited managerial capacity, and the difficulties of inter-ministerial coordination, it is best to focus on a few initiatives with a reasonable chance of success rather than spread limited legislative and managerial resources over a large number of policies. I think that strong arguments can be made that corruption, regulatory hostility, and low levels of education are the most serious problems. The possible causes and cures of corruption were analyzed in Chapter 4, so I will focus on the other two constraints.

Regulatory Hostility

Along with corruption, the greatest barrier to Iraqi entrepreneurship is regulatory hostility towards private business, especially with respect to starting a new business, getting credit, trading across borders, and closing a business. Since 2004, the World Bank and the US government have strongly encouraged the GoI to rationalize its Saddam-era business regulations. However, little progress has been made. One reason is that in Iraq, as in the rest of the MENA countries, the government has focused more on encouraging large-scale foreign investment than on efforts to encourage the creation of domestic small businesses (Stevenson 2010, p. 173).

With respect to rationalizing Iraq's commercial code to provide a more favorable business environment, there are several options. The GoI is currently engaged in a long-term process to revise its commercial code. But in view of the rising unemployment and underemployment rates, Iraq does not have the time for a gradual, decade-long process. One alternative to speed up the process of rationalization is to adopt the commercial code of another state such as that of the United Arab Emirates (UAE), an Islamic state with a much more favorable private business environment. Another option would be to establish a special economic zone (SEZ) to experiment with various initiatives that could then be applied nationwide.

Regardless of the process, re-writing Iraq's commercial code to reduce or eliminate three specific barriers to entrepreneurship should lead to a substantial private sector expansion and reduced unemployment. These changes are to eliminate paid-in minimum capital requirements, develop a credit reporting system, and introduce better procedures for resolving insolvencies.

Eliminate paid-in minimum capital

Paid-in minimum capital requirements were developed to enable the authorities to control who is allowed to start a business, and as a way of providing security to creditors (World Bank 2020, pp. 42–3). The first justification motivates firms either to pay bribes in order to get permission to operate without meeting the capital requirements, or to avoid the capital requirements by remaining in the informal economy with all of the associated inefficiencies. As a guarantee for creditors, capital requirements have generally failed. Firms could borrow the necessary capital required for registration and withdraw it as soon as registration is approved. Clearly, the existence of a paid-in minimum capital requirement is a barrier to formally starting a business, and this requirement especially affects potential female entrepreneurs, who tend to have less capital.

Currently, a prospective Iraqi business owner is expected to deposit the initial capital at a commercial bank and obtain a confirmation receipt, which must be filed with the company registration application. Although the capital will be blocked in the bank account under the name of the company, it can be withdrawn immediately upon the issuance of the certificate of registration. The required initial capital is equal to about 15 percent of an Iraqi's annual income. For comparison, the typical MENA state requires 5 percent. In fact, 12 MENA countries are among the 120 countries worldwide that require zero paid-in capital (World Bank 2020d, p. 44). Iraq has made some progress over the last decade in reducing this requirement. In 2011, the paid-in capital requirement was equal to about 44 percent of average per capita income (World Bank 2011c, Table 2, p. 9).

Even a sharply reduced minimum paid-in capital has adverse effects. Based on World Bank research on a sample of 93 countries, "The higher the paid-in capital requirement for business start-ups, the lower the business entry rate in the economy" (World Bank 2020d, Figure 3.2, p. 45). Since Iraq's paid-in minimum capital requirement fails to achieve its stated goals and is a barrier to entrepreneurship, it should be eliminated.

Developing a more inclusive credit reporting system

Credit bureaus or registries gather information on individuals and firms in an attempt to judge their creditworthiness. This information can include data on borrowing, bill paying habits, income, and so on. Credit bureaus and registries facilitate business by reducing the problem of asymmetrical information where one party to an agreement has more or better information than another party to the same agreement. Effective credit bureaus and registries tend to lead to more lending, since banks and others have a better awareness of the borrower's capacity to repay a loan; that is, there is a lower probability of default.

While Iraq now has a credit registry, it is woefully inadequate. The quality of the data is low, for several reasons. Banks and other institutions are unwilling to share information on their individual or business customers. The registry only has recent data; there is a dearth of credit information on individuals or firms from even two years ago. Creditors are unable to access this data online. Finally, borrowers do not have the right to access their data to ensure its accuracy.

In addition, coverage is limited. In 2019, the registry had credit information on only about 282 000 persons – 1.3 percent of the population – and 8600 firms. According to the World Bank's "Ease of Doing Business" survey, Iraq's "getting credit" score is 0 out of a maximum of 8 points. The average for the MENA is 5.3 (World Bank 2020e, pp. 28–32).

A study showed that firms are 9 percent less likely to identify the difficulty of obtaining finance as a major barrier when credit bureaus or registries exist. The creation of an effective credit reporting bureau or registry tends to increase the amount of private credit in an economy. Not only are more funds available to the private sector, but also, since financial intermediaries are better able to judge default risk, there is tendency for lower interest rates and reduced collateral requirements (World Bank 2020e, Figure 3.4, p. 48). The effect will be uneven. In Iraq, as discussed in the previous chapter, SOEs and other public firms have little difficulty obtaining credit on reasonable, even generous, terms. It is small private firms that have great difficulty borrowing. Creating an effective bureau or registry is expected to have a substantial favorable impact on small firm borrowing and associated employment creation.

Establishing an effective credit bureau or registry is not complex. Among the MENA countries, Iran, Jordan, and Qatar have all established credit bureaus or registries rated by the World Bank as 8 out of 8. While simple bureaus or registries are limited to distributing both positive and negative data on individuals and firms, most now offer credit scoring services, data on utility credit, and all services are available online.

Introduce better procedures for resolving insolvencies

When a firm becomes insolvent, ideally there are two options. If the firm would be viable with a lower level of debt, new management, or other structural changes; it can be reorganized. If the reorganization is successful, the financial health of the firm is restored, and it continues to operate as a going concern. If reorganization is not an option or the firm is evaluated to be not viable even with partial debt forgiveness, the firm can be liquidated, and any assets distributed among its creditors. The exit of this bankrupt firm from the market frees up its labor, capital, and other resources to be used more efficiently in producing another good or service.

In Iraq, there are two serious problems with efficiently resolving insolvencies. First, similar to the situation in one-third of the countries in the world, reorganization is not an option under current Iraqi law. In fact, there is no accepted way for creditors even to propose a reorganization. Second, although there are regulations related to liquidation, there have been no foreclosures or liquidations in at least three decades. Private firms, especially small private firms, view the formal liquidation process as being a waste of time and money since it is extremely unlikely that the previous owners will be left with anything of value after the courts are finished. Without a formal process of resolving insolvencies, insolvent private firms in Iraq are looted and/or abandoned by their owners. In contrast, public firms become zombie companies: they continue to operate because of a continuous transfusion of funds from state banks. It is for this reason that Iraq is currently tied for last place in the World Bank (2020e) "Doing Business" survey rankings for resolving insolvency.

A revised insolvency regime would give priority to reorganizing insolvent firms over liquidation. An example of such a revised insolvency regime would be India, which despite a court system famous for its lethargy has approved reorganization plans in about 20 percent of all cases, and when liquidation is necessary, the recovery rate is about 72 cents on the dollar (World Bank 2020e, "Removing Obstacles to Entrepreneurship" pp. 54–5). In addition to giving priority to reorganization, it is critical that Iraq reduce the time and expense of resolving insolvency. The average for the MENA is 2.7 years at a cost of 14 percent of the value of the firm (World Bank 2020d, p. 56). As long as resolving an insolvency in Iraq is a long, drawn-out process at great expense, it will continue to be avoided.

Improving the regulatory environment for entrepreneurship
Improving Iraq's regulatory environment will be a daunting task. It will provide massive corruption opportunities, task ministerial managerial capacity to the limit, and require extensive inter-ministerial cooperation; the odds are poor that such a re-write could be completed and implemented in less than a decade even with strong support from the Prime Minister's office. But Iraq cannot afford to wait a decade for the increased employment and market efficiency that greater entrepreneurship is expected to produce (Gunter 2009a). Aside from the economic impact, a substantial rise in the pool of unemployed young men threatens political stability. A more radical approach should be considered: the wholesale adoption of an existing set of business regulations of another country.

Adoption of the existing regulations of OECD countries would require difficult and controversial translations into Arabic and Kurdish, and are thought to be too non-sectarian for Iraq, with its constitution that emphasizes the role of Islam. Kuwait's World Bank (2020e) "Ease of Doing Business" ranking is 83rd (out of 190) compared to Iraq's ranking of 172nd. However, recent history between these two countries makes the Iraqi adoption of Kuwait's code extremely unlikely. Possibly the best choice for adoption is the business regulations of the UAE. In fact, the GoI has sent delegations to Dubai, one of the seven Emirates that compose the UAE, to discuss methods of facilitating business (*Khaleej Times Online* 2012). The UAE does not have the language or religious issues of the Western nations. Its "Ease of Doing Business" classification (World Bank, 2020e) ranks 16th.

In the critical category of starting a new business, the UAE provides a much better environment for entrepreneurs than Iraq's current regulatory morass.

There are fewer procedures, that require much less time to complete: three days in the UAE, compared to 26 days in Iraq. With respect to the three areas of reform that can be expected to have the greatest impact on expanding entrepreneurship, the UAE scores substantially better than Iraq. Required paid-in capital is zero in the UAE compared to 14.6 percent in Iraq. Credit bureaus and registries cover 51 percent of the population, compared to 1.3 percent in Iraq. Finally, the recovery rate upon insolvency is 27.7 cents on the dollar (World Bank 2020d); not a great rate, but an improvement on the zero recovery rate in Iraq.

Another alternative for accelerating the rationalization of Iraq's commercial code would be to either convert existing free trade areas into SEZs, or establish a new SEZ. A detailed SEZ proposal for Iraq is beyond the scope of this book. However, there are some general characteristics that should be considered. Because the government is already running a large fiscal deficit, building new infrastructure to support the SEZ is probably out of the picture. Therefore, the SEZ should be located where there are existing road, airport, electricity, and water infrastructures. Rather than develop a new commercial code, the SEZ might begin by adopting existing regulatory systems. For example, the SEZ might adopt the UAE commercial code and the US labor code. To simultaneously provide political oversight by the elected Iraqi government while limiting excessive interference and corruption, the SEZ might be managed by a foreign company from, say, Germany or Switzerland on a long-term contract with the Iraqi government.

But an SEZ is a means to an end: reducing the regulatory hostility towards the private sector. Currently, as a result of corruption and regulatory hostility, most small businessmen and women of Iraq are forced to keep their businesses in the underground or informal economy. As expected, operating in the informal economy is extremely inefficient, and yet such businesses employ an estimated 20 percent of the labor force. If they could be freed from the adverse conditions described above and allowed to operate openly, then one could expect rapid growth in the private sector, accompanied by a rise in productive employment to help absorb the 0.3 million Iraqis looking for their first job each year.

Primary, Intermediate, and Secondary Education

Education increases the likelihood that an entrepreneur will be successful (Baumol et al. 2007, p. 153). Most Kirznerian entrepreneurship does not require high levels of formal education or training. In most cases, a solid primary and intermediate education generally provides an adequate foundation. Unfortunately, as shown in Table 3.2, an estimated 14 percent of Iraqi males and 31 percent of Iraqi females are illiterate, while almost one-third of males and half of females did not have a primary education. And these data are probably overly optimistic, since they do not incorporate ISIS's deliberate destruction of schools and forcing many young people into refugee camps with extremely limited education opportunities. As recommended in Chapter 3, GoI education policies should focus on achieving 100 percent attendance at elementary school for boys and girls aged 6–11 years, as well as eradicating illiteracy among older Iraqis. This should increase the potential pool of entrepreneurs.

Reducing corruption, regulatory reform, and extending education in the direction discussed above face several challenges. First, asking the ministries to "do less" with respect to regulating private businesses will be difficult, since a reduction in ministerial responsibilities will be seen as leading to reduced power and status as well as reduced opportunities to extract bribes. Second, easing the entry of new firms is necessary, but the degree of success in bringing about a substantial entrepreneurial-led expansion of private sector employment will depend on whether the GoI intentionally or unintentionally offsets improvements in the regulatory environment with adverse budget changes (Stevenson 2010, p. 172). Finally, if oil prices are "too low" or "too high" over the next decade then regulatory reform will most likely stall. If the fall in oil prices that began in late 2014 continues then – in a repeat of 2006 – the GoI will probably respond to the resulting severe budget constraint by defunding all new initiatives in order to reserve spending for the compensation of current and former government employees. However, in the unlikely event that oil prices return to the $100 per barrel level, then there will be little sense of urgency in encouraging entrepreneurship and the expansion of Iraq's private sector. The GoI can respond to demands for employment by further increasing the already bloated government payrolls.

What is needed is the political will to push change through a bureaucracy distinguished by its inertia. Elected and appointed officials may discover this political will through necessity. In view of the rapidly growing labor force in Iraq, creating private sector jobs through entrepreneurship might be the only viable means of avoiding severe political instability.

12. Infrastructure and essential services

> This is an enterprise which at first sight appears gigantic, but its accomplishment, in the opinion of those who are acquainted with the localities, and have examined its capabilities, is considered comparatively easy. An Imperial firman has lately been granted to an English Company ... for the construction of a railway to cross Turkey in Asia, by the Valley of the Euphrates.
>
> (Count Edward de Warren 1966, p. 138)

Almost every history of modern Iraq states or implies that Iraq is an artificial construction (for example, Catherwood 2004). Much is made of the fact that the British constructed the country from three separately governed Ottoman provinces: Basrah, Baghdad, and Mosul. However, these "artificial construction" arguments generally focus on ethnic and political differences among the regions and ignore the trade and routes that have tied the country together for millennia.

Traditional trade routes in Iraq reflect geography which favors North–South movement. The riverbeds of the Euphrates and the Tigris not only provided water and power but also were the home of most of the population throughout history. Between the cities of Baghdad and Basrah and the port of Al Faw on the Persian Gulf, these rivers were navigable for fairly large vessels, while roads and railroads ran parallel to the Euphrates to middle Syria and along the Tigris through Mosul to northern Syria near the Syrian–Turkish border.

In fact, before the completion of the Suez Canal in 1869, the Basrah–Baghdad–Mosul route was a critical route for world trade. The products of the East converged on Basrah and proceeded up the Tigris and the Euphrates in large boats to Baghdad. There they traded for the products of the West brought overland from Antioch on the Mediterranean or from what is now Turkey through Mosul (Rousseau 1966, pp. 135–6). Steam navigation on the Tigris and the Euphrates and railroads connected the major cities and carried large amounts of goods and passengers that tied the three cities together in complex trade transactions (Issawi 1966, pp. 137–53, 179–85). Therefore transportation – like the power and communication infrastructure of Iraq – is more than a means for providing goods and essential services to the population; it is also a means of strengthening the bonds among the Iraqi people.

Since the complete destruction of Baghdad by the Mongols in 1258 CE, a succession of governments in what is now Iraq have gradually created and maintained an extensive infrastructure to provide essential services such as water, power, and transportation. But in the last four decades, civil conflict, the long Iraq–Iran War (1980–88), the wars with US-led coalitions in 1990–91 and 2003, the associated United Nations (UN) sanctions, and the more recent fights with the al Qaeda and ISIS insurgencies have resulted in a sharp deterioration of national infrastructure from both direct damage and foregone maintenance. While the engineering problems are challenging enough, the incentives issues are even more difficult. Saddam bought loyalty by providing most public goods and services for free or at very low cost. These subsidies were massive, and were financed by crude oil exports and large-scale foreign borrowing. Since these fees were generally insufficient to cover maintenance costs, much less the capital costs required for any expansion, infrastructure improvements and expansions were driven not by demand but rather by the willingness of the Baghdad bureaucracy to expend funds. Both during Saddam's regime and in the period following his fall from power, these expenditure decisions appear to be made more for political than economic reasons.

In this chapter the focus will be on electricity generation, potable water, airports, water transportation, railroads and roads, and communication. The equally important infrastructure associated with exploration, production, and transportation of Iraq's large oil and gas resources was discussed in Chapter 7; and dams and the extensive irrigation system were treated in Chapter 8.

ELECTRICITY GENERATION

Providing reliable quality round-the-clock electricity is the most critical infrastructure challenge facing Iraq. Despite billions of dollars in maintenance and investment, as well as the hard work of many Iraqis, the gap between the demand and supply of electricity is greater than ever. This has resulted in outages that not only have stalled economic development but also have worked a genuine hardship on the population. Without air-conditioning, summer temperatures in Central Iraq of 43°C (110°F) are almost unbearable and constitute a public health hazard for the very young, the very old, and the sick.

Electricity in Iraq faces problems of both quantity and quality. There are scheduled and unscheduled blackouts in almost all Iraqi cities. In 2018, it was estimated that the quantity demanded of electricity was about 158 terawatts (1 terawatt is 1 trillion watts), while the actual quantity supplied to the grid was about one-third less: 103 terawatts (Tabaqchali 2020a).

Despite its huge energy resources, Iraq imported about 12 terrawatts of electricity primarily from Iran and from floating Turkish power plants in the Persian Gulf (EIA 2019). As a result of both economic and political issues, the electricity imports from Iran show a great deal of year-to-year variance. For example, in the summer of 2018, Iran withheld 1.4 gigawatts (billion watts) of electricity exports to Iraq because of failure to pay for previous transmissions (Ashwarya 2020). With respect to electrical quality, for the reasons discussed below, electricity from the grid is often substantially below the 50 hertz frequency promised. This can damage consumers' electrical devices.

In 2016, installed capacity was about 26 000 gigawatts although, as will discussed below, the available capacity is lower than installed and the actual peak generation is lower still. Gas turbines accounted for almost two-thirds of the installed capacity. The efficiency of such turbines is an estimated 30–40 percent, although if the exhaust heat from a gas turbine is used to provide heat for a steam powered turbine – a combined cycle plant – then the efficiency can exceed 60 percent. One of the advantages of gas turbines is that they can rapidly be turned on or off. Steam powered turbines accounted for about a quarter of installed capacity. The remaining 10 percent of installed capacity is about evenly divided between diesel engines and hydroelectricity (World Bank 2019b, p. 40).

While large quantities of natural gas are wastefully flared off in Iraq, there are still shortages of gas for gas turbine electricity generation. In some years, as much as half of the fuel for these turbines was gasoline, crude oil, and heavy fuel oil (HFO). HFO is in excess supply because Iraqi refineries are inefficient and, as a result, produce a lower proportion of valuable lighter fuels and a higher proportion of sludge-like HFO than refineries in other countries. In fact, some refineries in Jordan and Turkey use Iraqi HFO instead of crude oil as feedstock. However, exporting HFO is difficult since it must be heated before it can be pumped, and it contaminates the pipelines, railroad tanker cars, or trucks in which it is carried. In fact, inability to dispose of HFO has led to shutdowns of Iraqi refineries. Therefore, HFO used for electricity generation in Iraq is adopted as a second-best solution. However, the use of gasoline, crude oil, and HFO to fuel electricity generation units has four disadvantages: they are more expensive than gas, less efficient, more polluting, and they increase the maintenance burden as inappropriate fuels increase wear on generation equipment.

There is a close relationship between electricity generation and refineries. Unexpected electrical blackouts shut down refineries, while failure to produce and deliver the proper quality and quantity of fuel leads to either a reduction in electricity generation or increases in maintenance costs. This symbiotic relationship requires close coordination between the ministries of Electricity and Oil that unfortunately does not exist. It has been proposed that the two ministries be merged into a new Ministry of Energy in order to improve this vital coordination, but political implications will probably prevent such a rationalization (Gunter et al. 2020, p. 12).

By 2009, Iraq's electricity supply finally exceeded pre-2003 levels. However, the estimated gap between demand and supply continued to grow.

This gap was taken as a clarion call for increased investment in electrical infrastructure. However, such statements ignore the challenges of estimating true electricity demand as well as the difficulty of increasing productivity in power generation.

Prices and Physics

The first problem is that the official price of electricity is not only less than the estimated cost of electrical production, but also substantially below estimates of the market clearing price. As a result, the demand for electricity was not constrained by the cost of the electricity as much as by incomes and the availability of goods and services that use electricity.

When there was an increase in incomes, especially of civil servants with their strong job security, this led to an increase in the purchases of air conditioners and other electricity-using appliances, resulting in an increased demand for electricity (Allawi 2007, p. 257). While there are several estimates of the demand for electricity, in the absence of a realistic market price for electricity these estimates – including the 158 terawatts estimate referenced above – are almost meaningless. Since the marginal cost to the user of additional electricity consumption is very low, any increase in the supply of electricity leads to the purchase of more electricity-using goods and services. As a result, the gap between amount demanded and amount supplied never closes. This problem with artificially low electrical prices was recognized fairly early in the reconstruction process by both military and civilian authorities (see, for example, Robinson 2008, p. 173). But inertia and leadership overload prevented any serious attempt to charge reasonable electrical charges until late 2015. As a result, electricity tariff increases have been inadequate.

Prior to 2015, electricity tariffs averaged 1.7 cents per kilowatt/hour (kWh) or less. In January 2016, the Ministry of Electricity introduced a new tariff schedule with the highest tariff, 15 cents/kWh, for government entities. There were lower tariffs for commercial, industrial, and agricultural users, with the lowest tariff, 3 cents/kWh, reserved for households. The average electricity tariff for the five categories was 8.11 cents/kWh. Even if 100 percent of these tariffs had been collected, receipts would have covered only about two-thirds of the estimated average cost of electricity generation of about 13 cents/kWh. However, in response to widespread protests, the pricing system was modified several times, each time reducing tariffs. By 2018, as a result of these reductions, the highest tariff for government electricity use was only 10 cents/kWh and the average tariff for the five categories was 5.16 cents/kWh (World Bank 2019b, Figure A1.4, p. 42). While the final 2018 tariff was more than three times higher than in 2015, the unwillingness of the Government of Iraq (GoI) to charge a tariff sufficient to cover costs is troubling. There are at least two additional unresolved issues concerning electricity pricing.

First, metering is not available in many areas, and is considered unreliable. In the absence of reliable metering, many commercial and households refuse to pay even their low electricity bills. Second, there is a perverse impact on the Ministry of Electricity. There is little incentive for the Ministry of Electricity to connect and maintain electrical connections to low-income districts, since each new customer imposes a cost on the Ministry of Electricity but provides little or no revenue. Since the price system has little effect on restraining demand, the Ministry of Electricity is forced to continue an unending effort to increase supply. Some of the methods used to increase the short-term supply of electricity have adverse long-term implications.

Like most of Europe and the Middle East, Iraq operates on a frequency of 50 hertz (hz) standard. (The USA operates on a 60 hz standard, which means that most electrical equipment is incompatible). Part of the difficulty of supplying reliable electricity is the physics involved. Because there is no reasonably economical way of storing electricity, a continuous flow of electricity must be maintained between where it is generated at the power plant, and where it is consumed: the load. Since electricity travels at almost the speed of light, it must be simultaneously generated, transmitted, and consumed. In other words, at every instant of time, the amount of electricity generated must equal the amount consumed, within a fairly small margin of error.

If the amount generated exceeds that consumed, the excess must be immediately shed. This shedding is accomplished by rapidly bringing additional load onto the line (Jurewitz 1987, pp. 10–11). On the other hand, if the quantity of electricity consumed exceeds that generated, then there are really only two possibilities. If the differential is small, a temporary decrease in frequency will allow continued operation while additional generation capacity is added to the grid, possibly by starting gas turbines. However, an excessive drop in the electrical frequency will damage equipment. If additional generation capacity cannot be immediately added, then to avoid a large frequency drop there must be a rapid reduction in load by dropping consumers, resulting in unexpected power outages. On average, Iraqis can count on receiving only about 15 hours a day of electricity. Of course, there are patterns to electricity usage in Iraq that allow some scheduling of generation. For example, in the summer the demand for power hits a maximum in late afternoon, while on Fridays industrial demand is down. Ideally, system operators can adjust capacity to prevent a drop in frequency or a power outage.

Unfortunately, the power grid is dysfunctional in Iraq. The daily power outages are a result of both mismanagement and criminal and insurgent acts. With respect to management of the electric grid, grid operators have sought to increase the available power by normally operating at between 49.0 and 49.5

hz instead of the 50 hz standard. However, operating at lower than standard frequencies reduces the system's capability to deal with unexpected spikes in electrical demand. For example, if the system was running at 50 hz then, in response to an unexpected increase in demand, the grid could temporarily drop to 49.7 hz until additional generation capacity was brought online. However, if the grid is already running at 49.0 hz then a further cut in frequency runs the risk of damaging electrical equipment. As a result, grid operators have no choice but to drop load: to cease supplying electricity to as many consumers as necessary to reduce demand until it equals the available supply. Therefore, while operating at lower than standard frequency does allow a small increase in power, it also degrades system stability and increases the likelihood of unannounced outages (Meese 2007, p. 1).

Electricity cannot be directed to take a particular path. Rather, it travels over all available paths between generators and users, with the power flow on each path being in inverse proportion to the impedance of each line (Ohm's Law). As the output of different generators changes to match consumption and there are changes in the location of new users, flows on transmission lines are constantly changing. Therefore, not only must the electrical grid maintain sufficient generation capability to meet the maximum load at any time, but also there must be sufficient capacity on transmission lines so that each can safely accept rapid increases in electrical flows. Due in part to the destruction of power lines during both the 2005–07 war with al Qaeda and the 2014–17 war with ISIS, major electrical lines in Iraq lack this necessary reserve transmission capacity. As a result, about 40 percent of the generated electricity is lost between the generation plants and the company or household consumer.

Politics of Power

Electrical generation and distribution during the pre-2003 Ba'athist regime were dominated by two political considerations. First, central control over electricity was maintained through the use of a few large power generating units. Second, electricity was directed to favor particular regions and cities. For example, in the summer of 2002, Baghdad had electricity 24 hours per day, seven days per week, while other parts of the country received no electricity even though there were electrical generation units or transmission lines within their boundaries (Allawi 2007, p. 257).

Despite efforts by the US-led coalition to minimize damage to the electrical grid following the 2003 invasion, widespread looting crippled the system. In addition, de-Ba'athification and "brain drain" to surrounding countries, combined with the adverse effect of years of isolation during the sanctions period, had resulted in a severe shortage of trained Iraqis to maintain and operate the electrical system. The situation was exacerbated by not only a shortage of the refined fuels but also an inability to deliver the correct fuels to the power generation plants in a timely manner. Finally, the insurgencies targeted the electrical grid in 2005–07 and again in 2014–17. Attacking the grid was a powerful tactic because there had been a widespread expectation that the 2003 US-led invasion would be followed by an increase in both the supply and the reliability of electricity. Therefore, the insurgents found attacks on the electrical grid to be a relatively risk-free way of undermining confidence in the GoI and coalition forces (SIGIR 2009b, pp. 144–5). Also, regions and cities that had received little or no electricity in the final days of Saddam's regime demanded a more equitable distribution. Attempts to achieve such a distribution resulted in energy outages in Baghdad and other major cities, even as national electrical generation slowly climbed back to pre-invasion levels. These shortages are exacerbated by illegal diversion of electricity. At the local level, these diversions of electricity lead to "rats' nests" of illegal wire taps on almost every city block as individuals and small entities attempt to obtain electricity without payment. In addition to the impact on electrical supply, these connections by non-professionals lead to fires and periodic accidental electrocutions.

However, the more serious diversions occur at the regional or provincial level. Throughout Iraq there are relays that are built to "trip" when the frequency drops to dangerous levels, producing a local or regional blackout in order to protect the national grid. To protect their local consumers, grid operators in some regions of Iraq disabled the relays. As a result, if there is an excessive drop in frequency, blackouts will occur in other regions where the relays are not disabled. As expected, this motivates the operators in other regions to also disable their relays. Of course, if enough relays are disabled nationwide then there is the increased potential for either a drop in frequency sufficient to cause severe equipment damage, or unexpected nationwide blackouts.

For the reasons given above, it is difficult to determine exactly who is receiving electricity that is either generated in Iraq or imported. According to one source, households and agricultural users account for almost 88 percent of all customers, but they provide less than a third of total revenue. On the other hand, public, commercial, and industrial users are only 12 percent of customers but provide over two-thirds of revenue. As a result, the Ministry of Electricity seems to give priority to providing service to these large users rather than to the general public (World Bank 2019b, Figure A.2.1, p. 50). Even within each general category of user there are favored consumers. For example, Baghdad has an emergency grid intended for essential services such as hospitals, police stations, and so on. During periods of shortage, the regular grid may go black, but the emergency grid will probably still have power. However, a variety of individuals and entities have used their influence to obtain a connection to the emergency grid for their residences or businesses. There is even a tractor factory in Iskandariyah – a town south of Baghdad – that was connected to the emergency grid. In addition to fairness, the addition of many non-emergency establishments increases the likelihood that the emergency grid itself will go black. Of course, as long as there is a gap between the demand and supply of electricity, its distribution will be controversial. So how can Iraq increase its supply of electricity?

Factors Limiting Electricity Generation

By mid-2010, the nameplate capacity of national electrical generation – the maximum output under ideal conditions – exceeded estimated demand. But conditions are never ideal in Iraq. A better standard is feasible capacity, defined as the estimated maximum output in view of actual conditions at power plants. In other words, if: (1) no power plants were offline for scheduled or unscheduled maintenance; and (2) sufficient fuel (for combustion turbine, thermal power, and diesel plants) or water (for hydroelectric plants) were available, then Iraq should have the feasible capacity to produce enough power to meet peak demand without electrical imports. However, actual peak production in Iraq is only about 70 percent of feasible capacity, for a variety of reasons. These include unplanned maintenance outages, shortages of fuel or attempts to use substandard fuel, low water at the hydroelectric plants, and a variety of miscellaneous factors. The gap between estimated maximum output and actual production is especially large for gas turbines, which account for about half of total installed capacity but are utilized at about 37 percent. This low rate of utilization is caused in part by the use of gasoline, crude oil, and HFO, which increases periods of unscheduled maintenance (World Bank 2019b, Figure A1.2.b, p. 40).

In its simplest form, Iraq has five possible ways of reducing the gap between the demand and supply of electricity:

- Increase the price of electricity.
- Increase nameplate generation capacity.
- Increase efficiency of existing power plants.
- Increase imports of electricity from Turkey and Iran.
- Increase private (non-grid) generation of electricity.

The January 2016 increase in electricity tariffs actually led to an almost 50 percent decrease in revenue, even though electricity consumption rose (World Bank 2019b, Figure A1.1, p. 38). Apparently, many Iraqi firms and households simply ceased to pay their electricity bills, or reduced metered usage by illegally tapping power lines, contributing to the tangle of wires hanging above many streets and alleys. This collapse in revenues was one of the motivations for the tariff reductions in 2017 and 2018.

To summarize the tariff problem. About 40 percent of generated electricity is lost before it is delivered to government agencies, firms, or households for consumption. These government agencies, firms, and households pay only about 50 percent of the amount billed. The amount billed covers only about 40 percent of the actual cost of electricity used. As a result, electricity tariff revenues pay for only 5–10 percent of the total cost of electricity generation in Iraq.

One option would be to extend the use of smart meters, accompanied by more aggressive collection efforts. To prevent hardship among households that would have difficulty paying higher tariffs, the GoI might provide an electricity subsidy for low-income households. After the collection effort has improved, the GoI should increase the tariff to cover the average cost of generation. However, more aggressive collection of higher tariffs will be extremely unpopular, and politicians may consider it political suicide to revisit this issue for at least several years.

Increasing nameplate generation capacity is by far the most popular option among politicians and bureaucrats. Massive construction projects, especially with a large imported component, provide much greater opportunities for graft and favoritism than efforts to improve the efficiency of existing capacity. However, adding substantial capacity is often fraught with engineering challenges. For example, delivering new generators from the Syrian border to a Baghdad power plant required the reinforcement of bridges along the route (Robinson 2008, p. 327). The cost of increasing nameplate capacity is substantial. The Ministry of Electricity has proposed spending $25 billion over the next five years to replace the generation capacity destroyed by ISIS as well as prepare the generation transmission capacity needed to meet the expected doubling of electricity demand to 45–50 gigawatts by 2030 (Fairbanks 2019; Gunter 2018b). And even in the unlikely event that the necessary expenditures are authorized, it can take 18–24 months to bring new generation capacity online, 2–6 years to upgrade existing transmission lines, and 4–10 years to build new lines.

Unfortunately, despite the great cost and effort, the actual increase in electrical generation from new capacity is much less than expected. For example, the GoI has shown a bias towards purchasing and installing gas turbines, because they can quickly be brought online. In addition, such turbines can be fueled with processed natural gas which is in generous supply in Iraq. However, in 2019, almost two-thirds of Iraq natural gas, valued at about $5.2 billion, was wasted: it was flared off for safety reasons (Ashwarya 2020). During the same period, Iraq imported large quantities of natural gas from Iran.

Yet, in addition to the large burden on the national budget, this imported power requires highly skilled labor and management. Unlike the existing system of generators that burn HFO, gas turbines require fuel that meets specific tolerances, and trained maintenance personnel. Iraq has neither.

The most controversial option is to accept that Iraq does not have a shortage of capacity, but rather it is incredibly inefficient in its use of existing capacity. If Iraq were able to increase the efficiency of existing power generation, a substantial increase in electrical supply would be possible without the massive investment under consideration. For example, if the grid operators were able to minimize unscheduled maintenance outages and – with the cooperation of the Ministry of Oil – ensure that the correct fuel was available in sufficient quantities for existing generators, then the existing electrical infrastructure could produce enough additional electricity to close half the current gap between demand and supply.

Increasing the efficiency of electrical generation is almost always treated as an engineering challenge. But a more productive approach might be to focus on incentives. Currently, managers and workers at electrical generation plants are not rewarded for long-term increases in efficiency, nor financially punished for unnecessary drops in the quality or quantity of electricity. In the absence of such incentives, politically motivated efficiency drives can be expected to have only a short-term impact on electrical production, often paid for with a long-term degradation of equipment.

The fourth option is to increase electricity imports from Turkey and Iran. While there is large year-to-year variation, Iraq imports about 12 terawatts of electricity from Iran. There are also Turkish floating electrical generation plants in the Persian Gulf that provide electricity to the Basrah area. There are complex political and economic issues associated with these imports. Exporting electricity allows Iran to earn additional foreign exchange in the face of the sanctions related to its nuclear program. At the same time, Iraq is engaged in complex negotiations with both Turkey and Iran concerning support for rebel groups and, in the case of Turkey, water flows in the Euphrates and the Tigris. Iraq's continuing dependency on imported electricity weakens its negotiating position with these countries and has led to difficult conversations with the United States over possible sanctions violations. Also, importing electricity seems to conflict with Iraq's comparative advantage. If Iraq could efficiently capture the large quantities of natural gas currently flared, it could not only provide for domestic energy production but also potentially become a regional low-cost electricity exporter.

Finally, due to the uncertainty of electrical supply from the grid, many Iraqis have turned to private electrical generation often using black market fuel for the generators. As can be seen in Table 12.1, this is especially true in Baghdad, where about 45 percent use either community or private generators as their primary electrical source. In the rest of Iraq, only 14 percent rely on community or private generators as their primary source of electricity, but more than half use community or private generators when the grid is down. The estimated price of buying electricity from community or private suppliers is 40 cents/

Table 12.1   Electrical supply in Baghdad and other provinces (%)

Source of electricity	Public generator (the grid)	Community generator	Private generator	No source
Baghdad primary	55	33	12	0
Baghdad secondary	28	18	37	17
Other provinces: primary	87	10	4	0
Other provinces: secondary	11	35	26	29

Source: COSIT (2008, Table 2-28, pp. 124–5).

kWh, eight times greater than the grid tariff. And the community or private generators insist on being paid, or no electricity.

WATER AND AIR TRANSPORTATION

Ports

There has been substantial river transport in Iraq for millennia. Beginning in 1839, the use of steamers not only led to a sharp decrease in the cost of shipping from Basrah to Baghdad by both the Euphrates and the Tigris rivers, but also reduced transportation times dramatically. By the 1860s, shipping times had fallen to two to three days from Baghdad to Basrah, and four to five days on the more difficult upstream voyage from Basrah to Baghdad (Issawi 1966, pp. 146–7). Until the railroad construction boom after World War I, river transport dominated Iraq's domestic and international trade.

Currently pipelines, railroads, roads, and air transport carry most of Iraq's current trade, and the future for river transport is limited due to falling water levels in both rivers but especially in the Euphrates. However, the ports of Iraq will continue to be the major routes of imports and exports, especially for imported commodities such as wheat and exports such as petroleum products and fertilizer that have a low value to weight ratio. But to restore substantial port throughout will require at least four expensive initiatives: clearing wrecked vessels from the navigation channels, restoration of bridges that were destroyed during conflict and "temporarily" replaced by floating or low-clearance bridges, dredging the channels, and a substantial upgrade of the port facilities.

Iraq's access to the Persian Gulf is severely constrained by geography and politics. The Euphrates and Tigris rivers join north of Basrah city to form the Shatt al-Arab waterway, which then flows into the Gulf. From Abu al Khasib to the city of Al Faw on the Gulf, a distance of about 100 km (60 miles), the Shatt al-Arab waterway forms the border between Iraq and Iran. This border was heavily fought over during the Iran–Iraq War in the 1980s, and still suffers from periodic cross-border military raids and the traffic of well-armed smugglers in consumer goods and antiquities. The distance from where the Shatt al-Arab waterway enters the Gulf to the border with Kuwait is only about 60 km (37 miles).

Iraq has four commercial ports (Umm Qasr, Khor al Zubair, Abu Falous, and al Maqal), two oil exporting ports, and four platforms for exporting oil crammed into this relatively small area of Basrah province. These four ports and 48 docks have a design offloading capacity of 17.5 million tons of cargo annually (GoI 2018, p. 146).

The anchorages for large oil tankers to take on cargoes of Iraqi oil are just off Al Faw on the Shatt al-Arab waterway. The largest cargo port is Umm Qasr, which is Iraq's only deep-water port and accounts for over half of the country's port capacity. Umm Qasr is near the border with Kuwait on the Khawr Az Zubayr river, which flows north and is connected by a canal to Basrah city and the Shatt al-Arab waterway. Umm Qasr is actually three ports, Umm Qasr South, Mid, and North, with Umm Qasr North the most modern. Unlike most of Iraq's transportation infrastructure, Umm Qasr North is not a state-owned enterprise (SOE) but is managed by a private entity, Basra Gateway Terminal. It has the reputation of being the best-managed port, and can handle both containers and general cargo. It is capable of berthing post-panamax vessels (vessels too large to transit the Panama Canal) up to 300 meters long. Ashore, the port has extensive dry and cold storage facilities.

The port of Khor Al Zubair is on the same river as Umm Qasr about halfway between Basrah city and the Gulf. The much smaller ports of al Magal and Abu Flous are near Basrah city on the Shatt al-Arab waterway. The current depth of the channels and alongside the port docks is at most 10 meters too shallow for the largest container and cargo ships.

The gain to the Iraq economy of modernizing its seaports would be large. However, expanding the throughput of the port system of Iraq faces five challenges. First, with the exception of Umm Qasr North, Iraqi ports are currently not capable of competing on price and available services, resulting in many shippers bypassing Iraqi ports to offload in Kuwait, Saudi Arabia, or Iran. Second, there is severe truck and other road traffic congestion in and near the port facilities. Recent attempts to reduce road congestion by authorizing a private contractor to control entry has suffered a typical Iraqi problem: various ministries have issued waivers that allow their trucks to both ignore entry control and refuse to pay the fees that are the contractor's revenue. Third, dredging the waterways to allow large vessels to access these ports must be continuous, but is expensive. The fall in oil prices and the demand for the GoI to spend more on post-ISIS reconstruction means that the necessary funds are not available. Fourth, there are security issues outside of the immediate port area.

Finally, as a result of almost 40 years of conflict, there are over 200 vessels sunk in or near the port of Umm Qasr, and hundreds more in the channels from Umm Qasr north to Basrah city. Not only do these vessels limit access to the ports, but also the sunken ships are believed to be leaking hazardous chemicals from munitions, pesticides, refined fuels, and unknown toxins. Although the GoI with the support of the government of Japan and the UN have begun to remove some of these wrecks, it will be a long process. As a matter of priorities, the GoI is focusing its initial efforts on removing those wrecks that block the main channels, rather than those believed to be leaking dangerous chemicals. This priority has international implications since chemicals from wrecks in Iraqi waters may contaminate the water drawn into desalinization plants in Kuwait, with a possible adverse impact on that nation's drinking water.

With channels cleared of wrecks and dredged deep enough to allow large container vessels to dock, the North and South Railroad Lines upgraded, and customs officials corruption controlled, Iraq has a good chance of returning to its century-old role as the preferred route for shipments from Asia to Turkey, Syria, and Jordan. The National Development Plan: 2018–2022 contains a detailed strategy for achieving this long-term goal (GoI 2018, pp. 146–7). However, previous development strategies also contained plans, although less detailed, for port and rail restorations and upgrades, but these plans were not executed, primarily due to a lack of funding (GoI 2005b, p. 25; GoI 2007, pp. 49–50). It remains to be seen whether the newest plan will also fail from lack of investment.

Airports

Since 2008, Iraq has experienced a dramatic increase in international air travel. While the total number of airports with paved runways has remained almost unchanged, the number of airports with runways longer than 3 km, that can accommodate the largest civilian aircraft, has increased from two to 20. There are five major international airports: in Baghdad, Basrah, Najaf, Arbil, and Sulaymaniyah cities. The other major airport in Mosul was destroyed during the ISIS occupation and has yet to completely reopen. Baghdad International Airport (BIAP) has gone from 14 regularly scheduled non-governmental civilian international flights per week in 2007 to over 150 departures a week in mid-2020. In addition, there is a growing volume of airfreight shipments. Najaf International Airport (NJF), converted from a military airbase in 2008, continues to expand in order to accommodate large numbers of religious tourists. International businessmen and women, many in the energy industry, account for many of the passengers flying into Basrah.

Arbil and Sulaymaniyah airports are located, of course, in the Kurdish Regional Government (KRG). Following the 2017 vote for KRG independence, Baghdad closed the KRG airports to international travel for five and a half months. This greatly complicated business, since traveling straight to KRG airports does not require an Iraqi visa, while transiting through Baghdad does. Although the KRG airports have been reopened to international flights, neither passenger nor freight traffic has recovered to pre-2017 levels.

One of the most serious challenges facing Iraq's air industry is the requirement to develop and maintain an up-to-date air traffic control system with the necessary radar, communication, and coordination systems. The difficulty is not in obtaining the necessary equipment, but rather in training the personnel to perform these critical jobs. Of course air travel, while quick, is expensive, especially for bulky products. For intra-Iraq shipments of such goods, reliance must be placed on railroads and motor traffic.

GROUND TRANSPORTATION

Railroads

After water transport, rail is by far the most efficient means of transporting bulk goods such as agricultural products and fertilizer. It is estimated that, compared to truck transport, railroads can reduce transport costs by up to 90 percent (Easterly 2006, p. 280). Iraq's almost 2400 km (1440 miles) of rail lines generally follow the riverbeds of the Euphrates and Tigris rivers. The nation's railroads are almost a century old, beginning with the Baghdad North line, from Rabiya on the Syrian border to Mosul to Baghdad, which was completed by a German construction firm in 1918. This was followed by the Baghdad South line, from Baghdad to Basrah, which was completed by the British in 1920. The newest lines are the Baghdad West and Traverse lines which were completed in 1987. The Baghdad West line was originally built to carry phosphate from its sources in western Iraq to Baghdad and then to Basrah for export. In addition to poor maintenance and the effects of conflict, rails that are too light for modern railroad operations limit the two older lines' capacities. Table 12.2 lists the four major rail routes, their length, and a summary of their current state of repair.

Iraqi Republic Railways (IRR) is an SOE with 109 stations, 31 locomotives, and 1685 units of rolling stock. IRR is organized into 11 operating divisions, which is excessive for the length of track (US Embassy 2015, pp. 3–5). In addition, the Iraq railroad system also suffers from outdated and poorly maintained engines and other rolling stock, as well as an outdated communication system. The wide mix of engines and other equipment reflects a long history of politically driven railroad investment. For example, immediately after the invasion

Table 12.2   Major rail lines

Line	Major cities	Length	Status
Baghdad South	Baghdad, Basrah, Umm Qasr (Port on Persian Gulf)	609 km (380 miles)	Operational – some sections have substandard rail
Religious	Musayyib, Kerbala	25 km (16 miles)	Operational
Baghdad North	Baghdad, Mosul, Rabiya (Syrian border)	524 km (325 miles)	Not operational – security situation
Baghdad West	Baghdad, Fallujah, Al Qaim, Akashat (Syrian border)	520 km (320 miles)	Not operational – security situation
Traverse	Haqlaniya, Bayji, Kirkuk	252 km (160 miles)	Not operational – security situation

Source: GoI (2018, Table 25, p. 150).

in 2003, IRR listed engines from Japan, China, Turkey, Czech Republic, Russia, and France. Most of these engines were not operational as a result of the inability to obtain parts, the lack of maintenance, or the post-invasion looting. Track is poorly maintained, which limits the average speed to about 60 kmh (40 mph) and has contributed to several accidents.

Almost 99 percent of all rail passengers were riding the Baghdad South line between Baghdad and Basrah. Cargo ton-kilometers have increased dramatically over the last several years and again the Baghdad South line was the busiest, carrying most of the total railroad tonnage. The Baghdad West line between Baghdad and Al Qaim on the Iraq–Syria border and the Baghdad North line between Baghdad and Mosul were previously the second- and third-busiest lines before their destruction by ISIS.

Among the engineering challenges facing IRR is completing the construction and integration of a microwave communication system that will allow the centralized dispatch, control, and tracking of trains throughout Iraq. Next most important is upgrading the track on existing lines to world standards, including parallel tracks, which will allow faster – 130–200 kMh (80–120 mph) – longer, and heavier trains. Priority should be given, first, to upgrading the Baghdad South line from the Persian Gulf port of Umm Qasr to Baghdad; and second, to upgrading the Baghdad North line from Baghdad to Mosul to the Syrian border at Rabiya. Integrated with port improvements, the upgrading of the Baghdad North and South lines to world standards would have several advantages for Iraq. The railroad would provide lower-cost transport of bulk imports (grain) and exports (phosphate-based fertilizer). If substantial container capacity could be developed then there is potential for substantial carrying trade from the Persian Gulf to Turkey, Syria, and Jordan. Another advantage is political.

An efficient north–south rail system would further tie together the residents of Iraq's three largest cities with trade and travel.

However, lack of physical investment in communication and rails are not the binding constraints on IRR productivity. Like most of the other SOEs, IRR suffers from excessive employment: over 10 000 employees inefficiently running a system that in other countries would require from one-quarter to one-tenth as many employees. As is true with the other SOEs, the employees cannot be fired for failure to perform their assigned tasks to expected standards, or even have their pay docked for failure to show up for work.

Despite or, possibly because of, the massive overmanning of IRR, it is extremely bureaucratic. Eleven operating divisions, each with its own division head, for a 2400 km system is excessive. Based on the experience of other developing countries, at most four operating divisions would be more than adequate. Combined with over 20 non-operating departments, the large number of operating divisions means that there are over 30 division and department chiefs reporting directly to the Director General of IRR. Modern management theory recommends that the number of reporting subordinates be six or less. In addition, many of the employees lack the necessary skill set for their jobs. Rather than provide the necessary training or education, it appears that IRR compensates for unskilled employees by requiring a series of checks to be performed by upper-level management before even relatively minor decisions are made. For example, despite the improved national rail communication system, dispatchers in Baghdad still hesitate to actually dispatch trains. Rather, stationmasters generally authorize each train to proceed to the next station, where it must wait for permission of the next stationmaster to continue.

In addition, cargoes are often transported based on ministerial-level decisions driven by relationships and mutual favors, rather than the impact on railroad revenues or costs. As a result, many trains are severely delayed, carry products that are not those of the highest net revenue, or have empty cars. The ministries of Trade and Oil have a strong bias in favor of using truck transport rather than rail for transporting bulk goods, despite the greater expense. This preference is partially motivated by the fact that they do not want to be dependent on IRR. Investment in railroad infrastructure can be expected to have little long-term impact on railroad efficiency if there is no rationalization of managerial organization and incentives.

Roads

Like its railroads, much of Iraq's 42 600 km (26 000 miles) of paved roads follow the Tigris and the Euphrates rivers. There has been little expansion of the road network over the last several decades, with only about 30 percent (11 300 km or 7000 miles) of the roads designed for high-volume traffic.

Road usage has increased sharply since 2003, for two reasons. The suspension of rail freight services forced shippers to put their cargoes on trucks. Also, there has been a large increase in the number of passenger cars in Iraq; many smuggled into the country from Iran. Since there is no legal requirement for automobile insurance or for drivers' permits, the roads are increasingly filled with bad drivers. Compounding the problem of increased traffic, the existing road network is badly located and suffering from much deferred maintenance.

Driven by Saddam's military strategy, the road network reflects military necessity rather than economic priorities. The major roads run from Baghdad west to the Syrian and Jordanian borders, and from Baghdad south to Basrah. While there is a divided highway from Baghdad to Mosul, there are no first-class roads running either north to the three Kurdish provinces (or Turkey) or east to Iran. Again, reflecting military needs, one of the two first-class roads from Baghdad to Basrah runs east of the Tigris through the cities of Al Kut and Al Amarah. This road runs roughly parallel to the Iraq–Iran border about 40 km (25 miles) away, and was designed to allow the rapid transfer of military forces along the border.

As in many developing countries, the critical road transportation constraint in Iraq is not building roads, but maintaining them (Easterly 2006, p. 165; Szirmai 2005, p. 622). Road maintenance would seem to be relatively inexpensive in Iraq. In the lower two-thirds of the nation, temperatures rarely, if ever, fall below freezing, which excludes a major cause of road deterioration. Also, in the lower two-thirds of the nation the land is relatively flat, which simplifies roadway engineering, although there are almost 700 major bridges and another 600 minor bridges required not only for the two rivers but also for the large number of irrigation canals. However, the absence of traffic control and weigh stations – only 15 stations nationwide in 2015 – has led to a large proportion of overweight trucks which has caused rapid deterioration of major trucking routes.

Bad roads impede economic development in nations such as Iraq with substantial agricultural sectors (Easterly 2006, p. 22). Bad roads increase the cost of transporting agricultural products by an estimated five times, and the increased time required to bring crops to market tends to reduce their quality. Since there is no cold-chain truck transportation network in Iraq, a substantial portion of agricultural products are bruised or spoiled before they reach urban markets. In fact, in Baghdad, Basrah, and other urban areas, foreign imported vegetables and meats are often cheaper and of higher quality than those produced in the same or neighboring provinces.

Of course, the inadequate road network in Iraq has broader implications. Bad roads isolate villages from medical care. Teachers, especially female teachers, are reluctant to accept teaching positions in inaccessible villages. In addition, almost all railroad crossings of roads are at ground level, which can result in long delays and rail car accidents. As expected, poor-quality roads tend to be more dangerous and have contributed to the doubling of road traffic fatalities in only three years (COSIT 2012, Table 6/11; Salem 2016). While it is widely accepted that Iraq would greatly benefit from better maintenance or upgrades of existing roads, incentives make this difficult to achieve. Most roads are public goods: neither rival in consumption nor excludable at reasonable cost. Therefore, most roads in Iraq are financed by the national government, but built and maintained by contractors chosen by provincial or local governments.

Unfortunately, in Iraq, the road construction and maintenance business is riddled with corruption. A large proportion of the money committed to road maintenance or construction is diverted into other uses or private accounts. The construction firms are generally chosen on the basis of political or family connections as well as a willingness to pay bribes. This leads to a bias in favor of building new roads instead of maintaining old ones, since building new roads provides more corruption opportunities as well as photo opportunities for governmental officials. The quality of road construction and maintenance is low because of the substitution of lower-quality materials for those specified in contracts. This results in more rapid roadway deterioration and additional contracts for restoration.

There are no easy answers to the problems associated with upgrading Iraq's road system. ISIS's destruction of the roads and bridges was devastating. For example, in Mosul city, ISIS wrecked all five Tigris bridges and an estimated 43 percent of urban roads (World Bank 2018c, pp. 89). The continuing security challenges in the border areas of the country, combined with the division of responsibility and, therefore, accountability among national, provincial, and local entities, makes progress difficult. The lesson from the rest of the developing world is that the effectiveness of local democracy determines the quality and quantity of roads. If decisions about road construction and maintenance are made in the capital city and funding comes from the national Treasury, then quality tends to be low. However, when residents not only pay taxes to care for roads but also play a major role in selecting local government officials, then roads tend to be of better quality. In view of the dominance of the Iraqi central government in planning and funding the country's road network, Iraq can expect to suffer from a poor road network for the foreseeable future.

COMMUNICATION

The rapid growth of telephone access was one of the very few unambiguous successes of the early reconstruction period. Access to a phone service increased from approximately 1 million (one per 27 Iraqis) pre-invasion to approximately 38.2 million (equivalent to 98 percent of the population) in 2018. About 50 percent of these phones are "smart" cellphones that could potentially be used for e-banking and "e-government". There are three major operators in the mobile phone sector, with pre-paid plans being the most popular. In developing countries, cellphone ownership tends to be a very productive investment. This is especially true in countries such as Iraq that still lack widespread reliable mail, or package delivery services. Cellphones are not only used by Iraq's private sector to coordinate their businesses, but they are also important to government agencies, SOEs, and even the ministries of Interior and Defense. Internet access has improved sharply, with almost half the population having regular access.

There has also been a sharp increase in the general public's access to media. Within one year of the US-led invasion, the number of television stations increased to 21, while the number of radio stations rose to over 80. In both urban and rural areas, satellite dishes have sprouted on almost every roof. It is estimated that about 70 percent of all families have access to television, while radio reception is almost universal. Few of these TV and radio stations are government stations; most are associated with one of the religious, political, or tribal groups. There is a wide range of programming, from Iraqi soap operas to foreign films and shows, to religious services.

Whether the transition from the government-controlled media of the Saddam era to the current media "free-for-all" has been entirely favorable is widely debated. It may provide unifying themes across tribal, political, and religious divides. For example, some soap operas seem to have wide national viewership. Other analysts think that the association of media outlets with various groups tends to divide the population and encourage instability. Another connection between increased access to diverse media and political instability is found in Huntington's (1968) hypothesis that was discussed in Chapter 5 of this book. According to this hypothesis, increased media access leads to the creation of new aspirations or desires for a better life as Iraqis learn how people in the developed world live. Unless these new aspirations are at least partially matched by economic development, then social frustration will increase. Political instability is not inevitable as a result of this frustration; increased geographic or professional mobility, combined with flexible political institutions, may ameliorate social frustration.

ENGINEERING EFFICIENCY VERSUS ECONOMIC EFFICIENCY

While the various types of infrastructure all suffer from specific problems, there are some common themes. There is a tendency among both Iraqis and foreigners to see solutions to infrastructure problems primarily in terms of engineering proposals and budget expenditures. This is only partially true;

without improved incentives, progress will be expensive, difficult, and temporary.

For example, in Baghdad, I sat in on a presentation concerning a relatively new water purification plant that was suffering from deterioration in water quality combined with a decrease in throughput. Apparently, the plant was suffering from poor maintenance that resulted in clogged pipes and pump damage. The presentation focused, first, on the most efficient way of clearing the pipes and replacing the pumps; and second, on how this repair could be financed. I raised the issue of incentives. The Iraqi supervisors and workers at the purification plant were very knowledgeable about the plant's operation. They knew what maintenance had to be performed in order to keep the plant operating at acceptable levels of efficiency. Therefore, the fact that this main-tenance had not been performed probably meant that they lacked incentives to do it. Without changing incentives, the same problem would reoccur every few years. I was informed that any discussion of incentives had political implica-tions and would have to wait for a political consensus.

When expanding or maintaining Iraq's infrastructure is under consideration, questions of incentives should not be an afterthought. Iraq is not Sweden. The Baghdad bureaucracy is less professional, less well trained, and – to be blunt – less honest than that of any of the Organisation for Economic Co-operation and Development (OECD) countries.

In Iraq, almost all of the infrastructure construction and maintenance costs throughout the 15 non-KRG provinces are paid for by the national govern-ment. For example, with respect to the provision of clean drinking water in urban areas, the national government builds the reservoirs, purification plants, and the system of pumps and pipes that will bring adequate quantities of clean drinking water to each consumer. Then an agency takes responsibility for operation and maintenance of the system, possibly charging a small fee – less than the average cost – for the water. This causes several problems.

First, there is a tendency for a significant portion of the funds dedicated for any area of infrastructure to be spent by the ministerial bureaucracy in Baghdad on salaries and benefits for the ministerial employees. When the national ministry makes the decisions on which water system to build and who shall build it, the primary motivation is often not the welfare of the ultimate water consumer, but rather how building the necessary infrastructure can increase the influence or wealth of the ministry and its officials. When construction material is purchased, it will often be from entities that have relationships with ministerial officials. Hiring is driven more by connections than competency. Promotions are based on the ability to successfully navigate a complex web of political relationships, rather than on engineering or other competency. Maintaining well-paid employment of ministerial employees is generally more important than consumers receiving adequate clean water.

Second, since the consumer pays little or nothing for the water, their needs can safely be ignored. The only exceptions are if the consumer has political connections or is willing to offer a bribe. The bureaucracy has little incentive to provide services to anyone who lacks political or other connections, since they will not receive increased funds for providing the service to a particular customer. As a result, consumers must pay bribes to receive basic essential services.

Thus, the national financing of both the fixed and variable costs of Iraqi infrastructure leads to perverse incentives on the part of both the bureaucracy and the consumer. Since Baghdad ministries pay the agencies responsible for providing the "last mile" of public goods from the earnings from crude oil exports (and not from the payments by the customers), they have little incentive to meet customer needs. At the same time, since customers pay little or nothing for the services provided by the infrastructure, they have little incentive to either avoid waste or make rational decisions among several inputs. When there is electricity available from the grid, there is no reason not to use all electrical appliances at maximum capacity. On a warm day, the air conditioner in a family shop will be running at maximum cooling with the shop door open to encourage customers to enter.

What is the alternative to the traditional means of providing public infrastructure in Iraq? A second-best solution would be to charge the consumer a price equal to the average cost of providing essential services. This system rationalizes incentives. Since consumers will be able to withhold payment unless the promised goods or services are provided, agencies managing infrastructure will focus less on the Baghdad ministries and more on actually providing the goods or services demanded. The adverse impact on the living standards of low-income Iraqis could be offset by a GoI transfer payment to poor families to offset the cost of utilities. Then, if the government is unable to provide reliable potable water or electricity, the consumer can purchase them from a water truck or community generator.

This is a very unpopular option in Iraq, where surveys have shown that consumers of essential services are not interested in paying anything for these services, but expect the government to efficiently provide almost unlimited essential services for free. Politicians across Iraq feed this desire for efficient free essential services, arguing that if they are elected or appointed to ministerial positions then things will work the way they are supposed to. Of course, once in power, these politicians face the same perverse incentives as their predecessors. And without innovation that provides rational incentives, it is unlikely that infrastructure will improve without a great deal of wasteful expenditure.

13. International trade and finance

From Basra dates, rice, and sometimes wheat and barley are shipped to Musqat, Surat,
and the Gulf of Cambay. The writing pens used by Persians and Turks are made out of
reeds that grow east of the Shatt al-Arab; a large amount is also sent to India. Horses
bred by the Arab tribes living west of Basra and Baghdad are highly valued in India.
A large number is shipped each year to Surat and Gujarat.
(G.A. Olivier 1988, p. 179)

Iraq's international trade and capital flows exhibit four salient characteristics. First, reflecting its long history as a major trading route between Europe and Asia; the Iraqi economy is dominated by international merchandise trade. Combined with its $66.8 billion in imports, Iraq's $89.6 billion in exports produced an estimated 2019 foreign trade to gross domestic product (GDP) ratio of 70 percent. There are few countries in the world at any level of economic development whose economies are so dominated by international merchandise trade. And, since the data excludes smuggling, the official statistics understate the actual value of trade.

Second, Iraq is more dependent on the export of a single natural resource than any other country in the world. Oil accounted for almost 97 percent of the country's 2019 merchandise exports.

Third, Iraq's balance of payments experiences large year-to-year changes. When oil prices are high, as in 2017 and 2018, the country reports a large surplus in its balance of trade, but when the price of oil is low such as in 2020, the country runs a trade deficit. In addition, its capital balance also changes sharply year to year, most likely because of changed perceptions of political instability. Before the 2020 collapse in oil prices, Iraq had accumulated a large holding of international reserves, an estimated $68.0 billion. Finally, there is evidence of large-scale capital flight from Iraq, with possibly more than $20 billion moved illegally out of the country between 2015 and 2019.

MERCHANDISE TRADE

Data Quality

The quality of data on both the volume and the value of Iraq's merchandise trade is very low. Smuggling across the borders with Iran, Syria, and Turkey mean that the official statistics capture only a fraction of all cross-border transactions. In fact, international organizations generally disregard the trade data reported by Government of Iraq (GoI), preferring to rely instead on adjusted counterpart data.

Counterpart data is used as follows. In theory, Iraq's reported exports to, say, the United Kingdom should equal that country's reported imports from Iraq, adjusted for the difference between the price of exports that are usually listed as fob (free on board – insurance and freight costs are not included) and imports that are usually recorded as cif (including cost, insurance, and freight). If the adjustment for the difference between fob and cif is made, then one can use counterpart data to estimate Iraq's exports and imports. However, in Iraq's case, counterpart data is of limited value since two of its major regional trading partners – Turkey and Iran – also have difficulty with accurately reporting their exports and imports.

With respect to unreported merchandise trade, probably the most disturbing is the extensive trade in stolen or looted historical artifacts. Some of these objects of both artistic and historical interest are obtained from Iraqi museums or government storage facilities either by theft or, more commonly, by bribing the guards. However, many of the items are obtained directly from the many archeological sites in southern Iraq that lack security. As a result, not only do artifacts that illustrate the incredible history of Iraq disappear into the international black market in art, but also tomb raiders, in their eagerness to quickly find items of value, often damage the sites. The most-used smuggling routes are between Iraq and Iran in the south, where looted Iraqi artifacts are sent east using the same routes that are used for black market consumer goods from Iran moving west.

Exports

In 2018, Iraq exported $91.9 billion in oil. Other exports are almost insignificant. Manufactured goods ($74 million) and crude materials except fuels ($16 million) are the second and third most important categories of exports (Central Bank of Iraq, 2019, Table 46, p. 95). Since Iraqi exports are dominated by oil, its major trading partners are the major energy importers. India and China each received an estimated 24 percent of Iraq's 2018 exports. The United States of America (12 percent) and South Korea (10 percent) accounted for most of the rest. However, because of the usual difficulty of determining the true destination country of any product, the importance of these countries as destinations for Iraqi exports is probably exaggerated. Some of these exports, including oil, are only initially shipped to these countries, before being transshipped elsewhere. Exports to Iraq's Middle East and Central Asian neighbors were less than 2 percent of total exports (IMF 2020a). (See Table 13.1).

Table 13.1   Balance of payments ($ billions)

	2015	2016	2017	2018	2019[e]
Exports (fob)	$56.5	$49.9	$68.0	$92.5	$89.6
(Oil exports)	($56.1)	($49.3)	($67.5)	($91.9)	($86.4)
Imports (cif)	−$56.7	−$47.8	−$59.9	−$63.0	−$66.8
Balance of trade	−$0.2	$2.1	$8.0	$29.5	$22.5
Current account balance	−$11.6	−$12.5	−$9.0	$9.7	$1.1
Errors/omissions	−$8.7	$2.0	$3.7	$3.9	−$4.5
Change in reserves	−$12.6	−$8.6	$3.9	$15.3	$3.3
Reserves (Dec. 31)	$54.1	$45.5	$49.4	$64.7	$68.0
Oil price	$45.9	$35.6	$48.7	$65.2	$61.1

Note: Superscript "e" indicates data is estimated.
Source: World Bank (2020b, Table 1, pp. 16–17), IMF (2019a, Table 1, p. 31 and Table 5, p. 31). and estimates by author.

Almost 90 percent of petroleum exports are loaded in tankers near the al Faw peninsula at the head of the Persian Gulf, while the bulk of the remainder goes through a pipeline to Turkey. Relatively small volumes are exported by tanker trucks. Closure of the Persian Gulf to large oil tankers because of either bad weather or regional political disputes tends to have an immediate impact on Iraq exports.

In addition to smuggling, the low value of non-oil exports has several causes. First, as discussed in Chapter 11, Iraq has an extremely hostile regulatory environment towards private business. In particular, regulations make it very difficult to legally export any item from Iraq. According to the World Bank's (2020d) report on "Doing Business", to legally export a standard container of goods requires 24 days of administrative processing, and costs $2900. For comparison, exporting from the average Middle East and North Africa (MENA) country requires only five days, and costs $680. The complexity, delay, and expense of exporting make Iraq's products uncompetitive on the basis of price.

Second, Iraq's long conflict with Iran and almost a decade of United Nations (UN)-imposed economic sanctions disturbed traditional trade flows and allowed competitors to capture foreign markets that were formerly Iraq's. For example, as discussed in Chapter 8, Iraq's dates were formerly renowned for their quality and Iraq was the world's largest exporter of dates. However, conflict-related destruction of many Iraqi date orchards combined with Iraq's long absence from the market have allowed other nations to greatly increase both the quality and the quantity of their dates. It will be very difficult for Iraq to recover this market.

Third, Iraq suffers from the "Dutch Disease", as large oil export earnings caused an appreciation of the Iraqi dinar (ID) that reduced the competitiveness of non-oil exports. However, the 23 percent devaluation in December 2020 is expected to offset the adverse effects of the Dutch Disease at least for a few years.

Imports

Iraq has much greater diversity with respect to its imports compared to its oil-dominated exports. The country's primary imports are machinery and transportation equipment (accounting for 38 percent of the country's total 2018 imports), miscellaneous manufactured articles (16 percent), manufactured goods (11 percent), and fuels (10 percent). As discussed in Chapter 8, the country imports a large portion of the food products in the Public Distribution System (PDS) baskets, resulting in $2.5 billion of food and live animal imports in 2018.

The country's major import partners are Turkey, which was the source of an estimated 25 percent of Iraq's imports, and China which was 24 percent. Other important sources of Iraq's imports are South Korea and India, which accounted for 6 percent each. The 25 countries included in the Middle East and Central Asia – Iraq's regional neighbors – were collectively the source of 9 percent of Iraq's imports. The absence of Iran from a list of Iraq's major trade partners reflects the fact that most of this trade is smuggled, and therefore unreported or under-reported. Again, because of transshipment, these countries may not be the actual country of origin of Iraq's imports. In particular, there is large-scale transport by trucks across the Turkish–Iraq border into the Kurdish provinces. This is a source of much of Iraq's imported consumer goods from Europe.

With respect to import tariffs, there is a wide gap between perception and reality. Immediately following the US-led invasion in 2003, a 5 percent tariff on most goods was imposed, although little revenue was collected. This was replaced by an extremely complex code that had tariffs from 0 to 50 percent on a detailed list of goods; the Customs Tariff Table was 794 pages long. Again, little tariff revenue was collected, or if collected was not transmitted to Baghdad. More recently, in January 2018, the Iraq General Commission of Customs published a greatly simplified five-page tariff table with only four customs duty rates (0.5 percent, 10 percent, 15 percent, and 30 percent) and 21 main product categories (al Janabi 2018). However, tariff revenue continues to be a fraction of what was expected, for three reasons.

First, Iraq's long and generally lightly populated borders with Iran and Turkey, combined with close relations in families and tribes split by these borders, have led to a long tradition of smuggling. Second, the customs officials and border guards in both the Kurdish Regional Government (KRG) and the rest of Iraq often accept bribes to turn a blind eye to smuggling, or divert tariff fees into their own pockets. This corruption is justified as a recompense for what they consider to be poor official compensation. Finally, officials from various ministries will issue waivers exempting certain imports from paying the published tariffs. This is significant, since government ministries and state-owned enterprises account for over half of Iraq's imports.

In 2020, a new government committed itself to enforcing import regulations, and not only collecting tariff revenues but also ensuring that these revenues end up in the Treasury. This is a commitment that has been made unsuccessfully several times in the recent past. However, the expectation of lower oil prices, combined with the demand for increased spending to increase the quality and quantity of essential services, may strengthen the government's resolve to collect all of the tariff revenue to which it is entitled.

Balance of Trade

Since merchandise exports are mostly determined by the price of oil, Iraq's balance of trade has shown a great deal of year-to-year variance. For example, as shown in Table 13.1, the 2017 oil price rise resulted in a 36 percent increase in exports; an almost $6 billion change in the balance of trade, from a $2 billion deficit in 2016 to a $8.0 billion surplus in 2017. More recently, the oil price fall combined with the Organization of the Petroleum Exporting Countries (OPEC)-guided reduction in the volume of oil exports led to a decline in the balance of trade, from a $22.5 billion surplus in 2019 to a $7.1 billion estimated deficit in 2020. Because the sharply higher trade deficit in 2020 led to a national crisis, it is important to understand its determinates.

Prior to the beginning of 2020, the GoI estimated that 2020 oil exports would fall to 3.3 million barrels per day (mbpd) and that Iraq would be able to earn $56 per barrel (pb). The decline in export volume was based primarily on the expectation that OPEC would call for a reduction in oil supply in order to bring about higher prices. Both the export volume and the oil price predictions were considered conservative since they were substantially less than the oil export volume and average price in 2019. If these predictions had been accurate then Iraq would have earned about $67.5 billion in oil exports or about $18.9 billion – 22 percent less than 2019. Unfortunately, coinciding with the worldwide Covid-19 epidemic, there was a collapse of oil prices from $51.37 pb in February 2020 to $28.18 pb in March, to $13.80 in April. In some world markets, oil prices were negative for a short period of time in April, as the cost of storing and transporting a barrel of oil was greater than the value of the oil.

In response to the decrease in oil prices, OPEC decided that its members would further decrease oil production and exports in an attempt increase oil prices by decreasing oil supply. As a result, Iraq's export volumes and oil prices in 2020 were both substantially less than expected. Instead of the $86.4 billion earned in oil exports in 2019 and the $67.5 billion earnings expected at the beginning of the year, Iraq earned only an estimated $42.3 billion in 2020. As will be discussed in Chapter 14, these earnings were insufficient to pay the salaries and pensions of government workers, much less any other GoI expenditures.

While merchandise trade, especially in oil, dominates the country's international transactions, there were also transactions in tourism and remittances.

INTERNATIONAL TOURISM

Before Covid-19 led to business and border closings in 2020, Iraq received an estimated 5 million international tourists every year, and post-Covid-19 there is great potential for a substantial increase (Mostafa 2018). The tourism contribution to the economy was an estimated 4.3 trillion ID ($3.6 billion) in 2019, and tourism is thought to support over 500 000 jobs (IMF 2020b). Tourists to Iraq can be roughly divided into three groups: religious, historical, and geographic. Of these, religious tourism, primarily Shi'a from Iran, accounted for the greatest volume of visitors.

Several of the most sacred places for Shi'a Muslims are in Iraq. One of the most important is the city of Najaf, about 155 km (95 miles) south of Baghdad on the Euphrates river. There, the Imam Ali Mosque contains the tomb of the First Shi'a Imam, Ali, the cousin and son-in-law of the Prophet Mohammed. While estimates are unreliable, it is safe to say that, since the border with Iraq was reopened after the fall of Saddam, at least 5 million Shi'a tourists and possibly as many as 10 million have visited Najaf. Anecdotal evidence supports the claim that Iranian firms capture much of the financial benefit from such tourism, since they control the travel firms, hotels, and even the restaurants used by visitors to Najaf.

Karbala, about 90 km (55 miles) south of Baghdad, situated just west of the Euphrates river, is also sacred to Shi'a Muslims since both the Imam Husayn Shrine and the Al Abbas Mosque are located there. It is estimated that several million religious tourists have visited Karbala.

The Al-Askari Mosque, also known as the Golden Mosque after its large dome covered in gold, is the major religious site in Samarra, about 115 km (70 miles) north of Baghdad on the Tigris river. This mosque contains the tombs of the tenth and eleventh Shi'a Imams and is considered by many to be the third-holiest site in Shi'a Islam. The Golden Mosque is probably best known to non-Muslims because of the explosion that destroyed the dome on February 22, 2006. Best evidence is that an extreme Sunni-insurgent group associated with the terrorist group al Qaeda detonated the explosives. I was in Baghdad that day, and realized the seriousness of the bombing from the stunned reaction of all Iraqis – Shi'a, Sunni, and Christian – who I spoke to after the blast. To them, the bombing of the Golden Mosque was an unthinkable outrage that had torn apart the fabric of their culture. As al Qaeda had hoped, the bombing of the Golden Mosque lit the fuse to an explosion of savage inter-sectarian violence that continued for almost two years. It was not until April 2009 that security had finally improved to the point where the Golden Mosque with its newly restored dome was reopened to pilgrims.

In Baghdad, the al Kadhimiya Mosque is another site sacred to the Shi'a, since it contains the tombs of the seventh and nineth Shi'a Imams. This mosque attracts large numbers of domestic and international religious persons.

While there is no reliable data on the expenditures, religious tourists in Iraq, even with most of their travel expenses captured by Iranian firms, spent at least $1 billion and possibly as much as $2.5 billion a year. While religious tourism was strong pre-Covid-19, historical and geographic tourism is far from reaching its potential.

As is well known, Iraq – Mesopotamia – has been the home of many great ancient civilizations. Since 6000 BC, these included the Sumerian, Akkadian, Gutian, Amorite, and Elamite civilizations, before one of the greatest of the ancient civilizations – the Babylonian – rose to prominence in about 1800 BC. The land of the two rivers contains the ruins of some of the greatest cities of human literature and history: Ur, Lagash, Uruk, Babylon, Ashur, and Nineveh. Only Egypt can boast of having as much history as Iraq within a relatively small area. The GoI estimated that there are between 50 000 and 200 000 significant archaeological sites, of which only 12 000 have been professionally evaluated. Many of these sites are in isolated areas, which raises security concerns.

During the al Qaeda and ISIS insurgencies, many highly educated and experienced archaeologists fled Iraq, leaving few to train the next generation of archaeologists. In addition to the loss of skilled persons, the insurgents often looted the sites and smuggled artifacts to buyers in other countries. This looting was in addition to the deliberate destruction of religious sites by ISIS. In the six provinces where ISIS was strongest, 37 of 41 historic mosques were damaged, with 22 completely destroyed. Among historic religious shrines in these provinces, 7 out of 25 were completely destroyed (World Bank 2018c, Table 15, p. 41).

The potential for historical tourism with the accompanying expenditures and employment opportunities is great. The major constraints are continuing concerns about security, the necessity of carefully restoring damaged sites, the lack of appropriate infrastructure, especially hotels, and the necessity of preparing the sites so as to protect them from further damage. Some restorations were really desecrations. In the mid-1980s, Saddam embarked on an extensive rebuilding of several ancient cities, mixing – to the horror of archeologists– ancient structures with modern construction. As in Egypt, there is the possibility of fairly large ships carrying foreign tourists up the rivers to stop at various sites.

Geographic tourism is also in its infancy, but the Kurdish region provides a cool green mountainous terrain that offers a welcome contrast to the hot brown flat terrain in much of the region. So far, most tourism in the Kurdish region has been from other parts of Iraq, but the number of foreign tourists is growing rapidly, albeit from a very low starting point.

Formerly, the major barriers to increased tourism were security, especially at the historical sites, the difficulty in obtaining valid tourist visas, and a shortage of hotels. However, a hotel shortage may no longer be a binding constraint. The number of hotels in the 15 non-KRG provinces has increased rapidly since the defeat of al Qaeda in 2007, from less than 500 to 1666 in 2018. Reflecting the importance of religious tourism, over two-thirds of these hotels were in two provinces – Karbala and Najaf – each of which had more hotels than the capital, Baghdad. Only six hotels are considered "Five Star" hotels (CSO 2020, Tables 1, 2, and 3, pp. 5–11). Most hotels are in the private sector or have mixed public–private ownership.

Unfortunately, the increase in the number of hotels has been accompanied by a recent drop in the number of patrons. The number of overnight stays increased over 400 percent from 2007 to 2016 when it reached 16.7 million. However, overnight stays fell to 10.7 million in 2018, about the same as in 2011. In 2020, it is expected that many hotels and other tourist accommodation will face severe difficulties as a result of Covid-19-related restrictions on foreign visitors. In addition, lower oil prices and the accompanying economic slowdown have resulted in fewer foreign businesspersons coming to Iraq.

While the potential for increasing the number of foreign tourists in Iraq is great, currently the outflow of tourism dollars is greater than the inflow. Many well-off and even middle-class Iraqis have traveled extensively in both the Middle East and Europe, seemingly unconstrained by the $10 000 legal limit on the amount of currency that can be taken out of the country. Although there are fewer tourists leaving Iraq each year than enter it, Iraqi's spend a greater amount per person on their foreign trips. In 2019, Iraqis spent $10.9 billion on personal international travel. Combined with other service transactions, the large amount spent by Iraqi tourists contributed to the $15.5 billion services deficit in 2019 (IMF 2020b).

REMITTANCES AND TRANSFER PAYMENTS

Iraq runs a small surplus on transfer payments. Primarily because of official transfers, the surplus reached about $1.1 billion in 2018. It is believed that the actual surplus may be smaller or even a deficit since individuals and non-government entities may avoid processing transfer transactions through the banking system that would report them to the Iraqi authorities. To avoid bureaucratic delays, excessive administrative fees, and possible demands for bribes or other illegal charges, many cross-border transactions are made using *hawala* rather than legal financial intermediaries.

Hawala, which have a long history in Islamic states, is a method of transferring money without actually moving it across borders. For example, assume that you want to send funds to your brother who is in another country. You give funds to a *hawala* broker in your own country and they authorize a *hawala* broker in the other country to pay the same amount, minus a commission, to your brother. There is no contract between the brokers for this specific transfer; it is based on trust. The increasing prevalence of high-speed Internet connections has recently led to high-tech competition for traditional *hawala*. Of course, such transactions would not be reported in the current account statistics of either country. While it is difficult to estimate the value of these unreported transfers, they are believed to be a large and growing component of the capital flight discussed below.

CAPITAL FLOWS

Foreign Direct Investment

Foreign direct investment (FDI) and official borrowing account for the bulk of Iraq's international capital inflows. Foreign purchasing in the Iraq's stock exchange is increasing rapidly but from a very low level and, as a result, is not yet a major source of foreign finance. From less than an estimated $0.1 billion in 2006, FDI increased rapidly to a high of $10.4 billion in 2014, before declining to an estimated $3.3 billion in 2019 (IMF 2020b).

As part of the post-ISIS reconstruction, the GoI sought 104.3 trillion ID ($88.2 billion) in foreign investment to restore housing, industry and commerce, electricity production and transmission, oil and gas production, and to pay for the restoration of the country's cultural heritage (World Bank 2018c, pp. v–vi). Foreign nations pledged only a fraction of this amount, and the actual "new" dollars invested in Iraq have been a fraction of the amount pledged. For example, in 2019, there was only $194 million in new foreign investment in

Iraq; the rest of the $3.3 billion of foreign investment reported that year was an increase in the estimated value of previously existing investment.

It is difficult to obtain accurate data on the sources and uses of FDI, especially since announced investments are rarely executed as described. However, it appears that most of the FDI is related to oil and gas investment. The dominance of energy and "mixed" FDI determines which provinces received the bulk of FDI. Three provinces received more than two-thirds of FDI: Basrah, Anbar, and Baghdad provinces. There was a severe drop in FDI to the KRG, for several reasons. A combination of lower oil prices and mismanagement has adversely affected the KRG economy. The situation worsened following the 2017 independence vote in the KRG which convinced foreign investors that relations between the KRG and Baghdad had deteriorated, especially with respect to the validity of KRG contracts for oil exploration, exploitation, and shipment.

That FDI is expected to account for the bulk of Iraq's foreign capital inflows over the next decade is good for Iraq. Unlike portfolio and debt flows, transfers of managerial and technical skills usually accompany FDI. By working for or with foreign firms, Iraqi managers, engineers, and so on, will be exposed to world standard techniques and procedures. In order to increase the amount of FDI, the 2006 Investment Law (#13) provided for a ten-year tax holiday for new FDI and established a National Investment Commission (NIC) as well as Investment Commissions for individual regions (RICs) or provinces (PICs). In addition to providing information on regulations and possible investments, the NIC provides a "one-stop shop" intended to provide a streamlined process for obtaining an investment license. After a completed application is submitted then an investment license should be obtained within 45 days. The usefulness of the RICs and PICs varies greatly. In mid-2020, about one-third of the PIC websites contain outdated information.

One of the more complicated issues facing FDI in Iraq is obtaining an allocation of land. In some provinces, over 90 percent of the land is controlled by one or more of the ministries under land laws that date from the Saddam regime. Therefore, even when a foreign investor has obtained the necessary investment licenses and other approvals from the appropriate national and provincial authorities, it is necessary to petition the appropriate Baghdad ministry to obtain a land allocation.

Foreign Borrowing and Equity Purchases

Iraqi entities have little debt exposure to foreign banks. At the end of the first quarter of 2020, loans to Iraqi entities from banks in the 48 nations that report bank exposure to the Bank for International Settlements (BIS) was $2.3 billion. This is extraordinarily low exposure to foreign banks for a developing country with the GDP and volume of international trade of Iraq. Iraq's net situation with respect to those banks that report to the BIS was actually positive, since Iraqi entities had $10.2 billion in loans to and deposits at BIS banks (BIS 2020, Iraq, Table A6.1).

Equity investment on the Iraq Stock Exchange (ISX) is open to foreigners, with the exception of limits on foreign ownership of bank stock. However, as discussed in Chapter 9, foreign investors face several difficulties in buying stock on the ISX. First, it is difficult to obtain reliable balance sheet and income statement information on the 104 firms currently listed on the ISX. Second, trading on the ISX is shallow. In June 2020, about one-half of the firms on the ISX did not trade a single share (see Table 9.4 in Chapter 9). Finally, there is little protection for shareholders in the event of corporate mismanagement or bankruptcy. In fact, there is evidence that the net flow of capital, whether equity or debt, is not in but out: capital flight.

CAPITAL FLIGHT

"Capital flight is a large outflow of capital from a low-income less developed country" (Gunter 2008a, p. 434). Some analysts favor a more restrictive definition that capital flight represents a situation where funds are fleeing or propelled illegally across national borders in search of sanctuary. This definition emphasizes that capital flight is often a response to high (or increasing) political or economic risks. Such flight can be seen as a quantifiable measure of how much confidence people have in the political and economic future of their country. On the other hand, some portfolio holders may attempt to secretly move funds out of their country because these funds represent ill-gotten gains of crime or corruption. An alternative view is that capital flight simply represents an attempt to create a diversified international portfolio in the face of unreasonable government restrictions on cross-border financial transactions. If one accepts this view, then one might say that using the term "capital flight" is more of a judgment than a definition: when Americans transfer funds to London, it is international diversification, but when Iraqis transfer funds to London, it is capital flight (Gunter 2004, pp. 63–4).

Regardless of the appropriateness of the definition, is there evidence of large-scale capital flight from Iraq? During the upsurge of violence in 2005–07 related to al Qaeda, and again in 2014–17 related to ISIS, many Iraqis left the country as refugees, and whenever possible took their financial and other assets with them. But even with the more recent sharp decrease in violence and the recent return of refugees, there are signs that Iraqi portfolio holders are still moving large amounts of funds out of the country. Estimating capital flight is difficult since there are many creative options that can be used to avoid government controls. These include manipulating legitimate financial or trade transactions as well as the simple although often dangerous expedient of taking a briefcase full of cash across a border. More sophisticated methods of estimating capital flight involve comparing financial and trade transactions reported by a country to the counterpart data reported by the countries that are its financial and trade partners (see Gunter 2004, pp. 65–9; 2008a, p. 435). Although Almounsor (2005, p. 249) found evidence of over $108 million in mis-invoicing consistent with capital flight during the period 1980–2002, the detailed counterpart data needed for this type of estimate is not yet available for the post-2003 Iraq economy.

In the absence of better data, a crude measure of capital flight is possible using the balance-of-payments numbers reported by the Central Bank of Iraq (CBI). Because of double-entry bookkeeping, the current account balance of a country plus its capital balance must equal any change in its international reserves. For example, if Iraq exports more goods and services than it imports (positive current account balance), then either it must lend some of its current account earnings to entities in other countries (negative capital balance) or Iraq's international reserves must increase. If these numbers do not balance, it might represent capital flight.

The current account balance is the sum of a country's net goods and services trade plus net income and transfers. According to the 2019 data, the most recent available, Iraq's current account balance was in surplus, $1.1 billion; as was its net capital balance – investment, equity, and debt flows. This should have led to a $7.8 billion increase in the country's international reserves. However, Iraq's international reserves increased by only $3.3 billion; $4.5 billion less than expected.

This $4.5 billion gap between expected and actual changes in international reserves might be simply a statistical artifact of accumulated errors and omissions that would disappear with more accurate data gathering. In fact, this is how the CBI classified $4.5 billion of this discrepancy. However, in many developing countries such large negative "errors and omissions" are associated with capital flight; that is, an estimated $4.5 billion in capital was transferred out of country without being reported in 2019 (Cuddington 1987, pp. 85–9). As a matter of scale, $4.5 billion was equal to about 5 percent of Iraq's oil export earnings in 2019. The 2019 estimate appears to be part of a trend of large balance-of-payments discrepancies consistent with capital flight. Iraq may have experienced total capital flight of almost $20 billion from 2015 to 2019. This trend is apparently continuing.

One response to this estimate of capital flight is surprise that it is so low. In view of the combination of large oil revenues, ubiquitous corruption, and great political uncertainty, one might expect much larger volumes of capital secretly leaving the country. There seem to be at least two explanations for the lower than expected capital flight. First, surveys show that the average Iraqi is fairly optimistic about the future of the country. This optimism may translate into a willingness to suffer the current inefficiencies and inequalities since they are thought to be temporary, rather than send their wealth, their children, and themselves to another country. Second, the sheer complexity and inefficiency of Iraq's financial system makes it difficult, risky, and expensive for most Iraqis to move their assets abroad.

In view of Iraq's great need for capital to expand oil production while at the same time developing infrastructure so as to improve the quality of life of the average Iraqi, the possible diversion of more than $20 billion into foreign accounts should be a matter of concern, since this estimated capital flight is equal to about 16 percent of the GoI's total 2015–19 investment expenditures. There are two possible solutions. One option is that Iraq can impose draconian rules in an attempt to stop (or slow) capital flight. This option is probably doomed, since other countries with governments that have lower levels of corruption and a greater willingness to use harsh measures to enforce rules have been unable to make more than a temporary dent in capital flight. (For an example of the successful avoidance of capital controls in the case of the People's Republic of China, see Gunter 2004, pp. 81–2.) The other option is to make Iraq a more attractive place to invest, so as not only to catch the attention of foreign investors, but also to encourage domestic portfolio holders to keep their funds at home. But creating a more favorable investment environment in Iraq will require the careful phasing-in of substantial changes in financial regulations and monetary policies, which were discussed in Chapter 9.

FOREIGN DEBT

By the time of the collapse of Saddam's regime in 2003, Iraq's foreign debt was unsustainable. Not only was Iraq's estimated foreign debt of $133 billion equivalent to almost five times (517 percent) of the war-ravaged country's GDP, but also it was very difficult to determine precisely how much was owed to each creditor. Further complicating the analysis of Iraq's foreign debt burden were the widely varying estimates of the compensation that must be paid for the damage done during Saddam's wars.

One estimate is that servicing the $133 billion of foreign debt would have required almost two-thirds of the GoI total revenues for the foreseeable future. However, as is well known, the major creditor nations not only suspended debt service but also agreed to large-scale debt forgiveness. Since 2003, Iraq's foreign debt to GDP ratio has fallen sharply, to about 107 percent in 2011, and 46 percent in 2012. However, the attention paid to debt forgiveness should not distract from the major cause of the drop in the country's debt to GDP ratio, which was the surprisingly strong growth in GDP. Because of the US-led invasion, the GDP data for 2003 is a crude estimate. But since 2004, Iraq has grown rapidly from an estimated 37.5 trillion ID ($25.8 billion at 2004 exchange rate) to 276.7 trillion ID ($234 billion) in 2019. This is equivalent to a nominal growth rate in the country's GDP of 13.5 percent per year for 15 years. Even if there had been zero debt forgiveness, then the debt to GDP ratio would still have declined from over 500 percent in 2004 to 55 percent at the end of 2019.

Debt forgiveness proceeded in stages. Of Iraq's 2003 external public debt – excluding war reparations – of approximately $133 billion, only a very small amount, $0.9 billion (1 percent of the total), was owed to the World Bank, the International Monetary Fund (IMF), or other multilateral organizations. The bulk of the debt was to foreign governments – about $111.4 billion (84 percent) – with commercial creditors accounting for the remaining $20.7 billion (16 percent) (IMF 2005, Table 1, p. 71; Weiss 2009, p. 1). Of the total owed to foreign governments, $42.6 billion was owed to Paris Club members, and $68.8 billion to foreign countries who are not permanent members of the Club. The Paris Club is a meeting of 19 major creditor nations where disputes concerning government-to-government debt are resolved. In addition to foreign debt, Iraq was also considered liable for war reparations; these reparations were estimated at over $200 billion, mostly from the 1990 invasion and subsequent devastation of Kuwait.

After the collapse of Saddam's regime, it was argued that most of the country's debt should be canceled. Both ethical and practical reasons were given for debt relief (*The Economist* 2003, p. 68). The legal theory of "odious debt" says that a nation should not be held ethically responsible for debts incurred for purposes that are not beneficial to the state. In a sense, these debts were incurred under coercion, since a free public would not have borrowed the funds for this purpose. The theory of odious debt has a long history, extending at least to the eighteenth-century Condorcet–Jefferson theory that, under certain conditions, every generation had the right to repudiate the excessive debts of previous generations (see Gunter 1991). Since Saddam was clearly a tyrant who ruled by terror, and instigated multiple internal and externals conflicts that caused incredible hardships for the Iraqi people, the odious debt argument was a strong one. However, the practical reasons were probably more persuasive.

The scale of Iraq's foreign debt was so large at the end of 2003 that most of the GoI revenue would have to be diverted to debt service. This would prevent using oil export earnings to fund the nation's reconstruction, and possibly add fuel to the smoldering fires of ethnic and religious animosities. On the other hand, Iraq does possess huge oil reserves, and after a decade of the large oil-related infrastructure investments discussed in Chapter 7 it should be able to service its foreign debts. This would seem to support a temporary suspension of debt service rather than debt cancelation. However, in a historically rare case of victors disregarding their own self-interests, the USA and its Paris Club coalition partners canceled most of their Iraqi debt.

The 19 permanent members of the Paris Club approved a total 80 percent reduction in three phases, with the first 30 percent in 2004, the second 30 percent in 2005 when Iraq accepted an agreement with the IMF, and the final 20 percent upon successful completion of the IMF program. To obtain the complete reduction, the GoI and specifically the CBI agreed to make substantial procedural and structural changes intended to improve the efficiency and transparency of governmental operations (IMF 2005, Table 2, p. 21, Table 3, p. 22). In retrospect, the GoI acceptance of the IMF recommendations may have had a stronger favorable impact on Iraq's long-term economic development than the debt decrease itself.

Over half of Iraq's debt is owed to non-Paris Club states, although no authoritative breakdown of this debt exists. However, it is believed that of the approximately $68.8 billion owed to non-Paris Club states, about 85 percent was owed to only two nations: Saudi Arabia and Kuwait. While some non-Paris Club countries reduced their Iraqi debt right after the Paris Club negotiations, negotiations to reduce the rest of the non-Paris Club debt dragged on until 2011, when most of the remaining countries finally accepted – in principle – an 80 percent reduction. However, the signing of legal documents related to this debt reduction continues to be delayed over a variety of technical and political issues. Some commentators believe that the Persian Gulf creditors will continue to delay until they are able to negotiate favorable terms with the GoI on other issues.

Iraq's approximately $20 billion in commercial debt involved many relatively small creditors, with almost two-thirds of the unpaid loans amounting to less than $10 million. Iraq successfully negotiated to buy back with cash the debt of small creditors at 10.25 cents on the dollar, while issuing new bonds to the larger creditors of $200 per $1000 of existing debt (Weiss 2009, p. 9).

The issue of war reparations to Kuwait was also resolved, although both parties fell ill-treated. Under the extremely corrupt UN Oil-for-Food Programme, 25 percent of Iraq's oil export earnings were to go to paying for reparations. With the end of the Oil-for-Food Programme, a UN resolution stated that 5 percent of Iraq's future gross oil export earnings would go to Kuwait until the reparations – plus interest – were completely paid. Kuwait sees this transfer as inadequate to compensate the nation for the terrible damage done by the Iraqi invasion. On the other hand, many Iraqis feel that it was the dictator Saddam who invaded Kuwait and, since he is dead, it is wrong to demand that the Iraqi people pay for his crimes. Until the debt reduction by the remaining non-Paris Club states is finalized, some sources are listing two debt figures. With the reductions, Iraq's 2019 foreign debt was about $54 billion, or 23 percent of GDP (World Bank 2020b, Table 1, p. 16). However, until all of the documentation has been completed, the debt is technically about $35 billion more. Iraq faces challenges over the next decade in amortizing its debt. Such amortization (contracted principal and interest payments) was an estimated 2.8 trillion ID ($2.4 billion) in 2019, but is expected to increase to 9.3 trillion ID ($7.9 billion) in 2022 (IMF 2019b, Table 2, p. 28).

Combined with the decline in oil prices, and widespread civil disorder, the expected increase in amortization payments has reduced investor confidence in Iraq. In mid-2020, Iraq was ranked non-investment grade by the three major rating services. Fitch and S+P consider Iraq to be highly speculative, while Moody's evaluates Iraq's debt as having substantial risks. These low ratings make it unlikely that Iraq will be able to borrow substantial amounts from international private banks and investors. This increases the possibility that the GoI will have to drain its international reserves to finance any future trade deficits.

INTERNATIONAL RESERVES

Iraq's international reserves, mostly held in US Treasury securities, amounted to about $68.0 billion at the end of 2019, equal to about ten months of imports (World Bank 2020b, Table 1, p. 17). One motivation for holding international reserves is to provide an emergency fund that will allow a country to purchase necessary imports even if its ability to export were to be temporarily blocked. This is an important motivation for Iraq, for at least two reasons. First, in view of the fact that petroleum shipments through the Persian Gulf account for most of its exports, and Iran periodically boasts of its capability and intention to close the Gulf. Second, Iraq's export earnings are primarily determined by the extremely variable price of oil. If oil prices fall during a period of low reserves, such as occurred in 2006 and again in 2020, then Iraq finds itself essentially living "hand to mouth", unable to import food each month for the PDS baskets until the payment is received for the previous month's oil exports.

The second motivation for Iraq's large holdings of international reserves is to maintain stability of the ID. Previously, the CBI had managed a gradual appreciation of the ID of almost 20 percent, from 1467 ID per dollar in 2006 to 1170 in January 2009. Since then, there has been a 1 percent depreciation, to 1182 ID per dollar. The exchange rate was almost unchanged for 12 years and was considered one of the great successes of CBI monetary policy. However, the collapse of oil prices in 2020 led to large trade and current account deficits and a loss of Iraq's international reserves. From $68.0 billion at the end of 2019, reserves fell to an estimated $53.8 billion in 2020 (Rabee Research 2020b). To avoid an accelerating depletion of reserves, in December 2020 the GoI devalued the dinar by about 23 percent to 1450 ID per dollar.

As will be discussed in the next two chapters, this dramatic devaluation will have substantial effects on both the government budget and its monetary policy. However, with respect to international trade, the devaluation has reduced the prices of non-oil Iraqi exports, while increasing the prices of non-grain imports. In several years, this should reduce the country's trade deficit.

14. Fiscal, monetary, and exchange rate policies

> The chief limiting factor to the success of development in Iraq may prove to be neither the amount of money for investment, nor even the limits of skilled labour and materials available, but the efficiency of the administrative machine.
>
> (Lord Salter 1955, p. 96)

The collapse of oil prices that began in 2014, combined with the tremendous cost of post-ISIS reconstruction, continues to have adverse impacts on Iraq's fiscal and monetary policies. As this work is being written in late 2020, these problems combined with the Covid-19 shutdown have morphed into a severe political-economic crisis. And in view of the multiple constraints and inefficiencies of Iraq's economy, society, and polity, successfully resolving this crisis will require policy initiatives that would have been considered extreme just a few years ago.

The most serious short-term problem is the need for domestic or international funds to finance the large budget deficit. However, as discussed in Chapter 9, Iraq's financial markets are moribund as a result of state bank dominance and inadequate regulation on financial intermediation. The bond and equity markets are in their infancy and extremely shallow. Also, Iraq is still a two-currency economy: the Iraqi dinar (ID) and the US dollar. Both the inefficient financial markets and dollarization substantially limit Iraqi monetary and exchange rate policies, especially when faced with substantial hard currency loss due, in part, to large-scale capital flight.

Finally, a dysfunctional system of financial intermediation combined with the need for large amounts of foreign infrastructure investment puts the Government of Iraq (GoI) on the horns of a dilemma. By impeding the liberalization of financial markets, the GoI limits destabilizing "hot money" capital flows that could, in the worst-case scenario, force a further devaluation of the dinar. However, the primitive level of financial intermediation in the country reduces financing options for the fiscal deficit as well as acting as a drag on the non-oil economy, hindering economic growth and preventing needed employment growth.

FISCAL POLICY

The demands on the GoI national budget are great. The salaries of about half of Iraq's labor force are dependent on government expenditures. Security expenditures – both police and military – continue to grow, while poverty alleviation in the form of the Public Distribution System (PDS) food baskets and other social safety net expenditures are large. With respect to investment expenditures, not only must the GoI pay for the restoration of a national infrastructure severely degraded by sanctions, war, and the al Qaeda and ISIS insurgencies (see Chapter 12), but also the nation's long-term development strategy calls for investment in oil exploration, production, and shipment (see Chapter 7). Finally, in coordination with monetary and exchange rate policy, fiscal policy must be crafted so as to further the country's macroeconomic goals of real growth, price stability, low unemployment, and external balance.

Unfortunately, Iraq's federal budget process is inadequate and makes achieving the country's fiscal goals difficult. Existing budget law is supposed to ensure that the Iraq budget process is transparent, unified, and comprehensive. In reality, it is none of these. The inadequacies of the budget process are especially obvious with respect to investment expenditures. Iraq ministries "simply lack the capacity to plan, formulate, and execute capital budgets" (Savage 2014, p. 146). The GoI even lacks the capability to determine whether expenditures actually occurred. While the Constitution requires that audited reports of expenditures be published within six months of the end of the fiscal year, such reports were over six years late in 2019. And the amounts in the audited reports are very different than those in budgets approved by the Council of Representatives (CoR). For example, for the period 2003–12, the accumulated budget deficits were projected to total 97 trillion ID ($81.1 billion), but the audited data showed accumulated surpluses of 102 trillion ID ($85.3 billion), a net difference of 199 trillion ID (Al Kafaji and Mahmood 2019, pp. 478, 480). As a result, the data reported in the remainder of this chapter may be only roughly accurate and may undergo substantial revision.

GoI Revenues

Iraq's fiscal policy is hostage to the value of the country's oil exports. Although Iraq's oil export volume tended to gradually increase until at least 2020, there have been dramatic year-to-year changes in oil prices (see Figure 7.2). Oil prices – adjusted for inflation – were over $100 per barrel (pb) in 2014 before falling to about $36 in 2016. As can be seen in Table 14.1, there was a recovery in oil prices in 2018 and 2019, before another collapse in 2020 that saw the price of oil fall to $13.80 pb in April 2020, from $60.14 in January

Table 14.1   *Iraq's fiscal accounts (trillions of ID)*

	2016	2017	2018	2019	2020ᵖ
Total revenues	55.5	77.4	106.5	107.6	52.1
Oil exports	47.2	65.1	95.7	99.2	47.9
Oil price	$35.60	$49.10	$65.50	$61.10	$37.90
Oil export volume	3.8 mbpd	3.3 mbpd	3.5 mbpd	3.5 mbpd	2.9 mbpd
Total expenditures	84.2	74.4	76.9	111.7	68.6
Current expenditures	60.5	58.0	63.0	87.3	67.4
Salaries	32.3	32.8	35.8	41	40.2
Investment expenditures	23 (27%)	20 (25%)	13.8 (18%)	24.4 (23%)	1.2 (2%)
Budget surplus/deficit	-28.7	3.0	29.6	-4 .1	-15.4

Note: Superscript "p" indicates data is projected.
Source: IMF (2019a, Tables 1 and 2, pp. 27 and 28), World Bank (2020b, Table 1,
pp. 16–17), and author's estimate of 2020 based on January–August 2020 data.

of the same year. The revenue impact of the 2020 drop in oil prices was multiplied by an unprecedented decline in the volume of oil exports. Prior to the beginning of the year, it was assumed that Iraq would have been able to export 3.3 million barrels per day (mbpd) at an average price of $56 pb. This would have resulted in 2020 oil export revenues of $67.5 billion. Instead, extrapolating from the first nine months of the year, Iraq is on track to export only 3 mbpd at an average price of $38.60. This will result in only $42.3 billion in oil export earnings; 37 percent less than expected. As a result, Iraq has moved from a large budget surplus in 2018, to about breaking even in 2019, to a large budget deficit in 2020.

The expected 2020 results break a pattern that has held from 2007 through 2019. With a single exception, GoI budget assumptions were subject to offsetting errors. The GoI budget overestimated the volume of oil exports, while it underestimated the world price of oil. In 2020, budget assumptions of both volume and prices were very wrong.

Adding to the challenges of low oil prices is the fact that not only are earnings from crude oil exports the major source of GoI revenues, but also they have grown in importance since 2004. Grants from other nations and international organizations accounted for almost 26 percent of total GoI revenues in the immediate post-war period, but by 2019 were *de minimus* (IMF 2019a, Table 3, p. 29; IMF 2011a, Table 2, p. 17).

In the post-invasion period, revenues from Iraq's oil exports were not immediately deposited in GoI accounts. The GoI faced many lawsuits from foreign governments, entities, and individuals who were injured as a result of the 1990 invasion of Kuwait or other activities of Saddam's regime. When oil sales resumed after the US-led coalition invasion of Iraq in 2003, it was feared that foreign courts would seize earnings from Iraq's oil exports in order to compensate these injured parties. Therefore, in 2003, United Nations Security Council Resolution 1483 sanctioned the establishment of the Development Fund for Iraq (DFI). Various moneys from the UN Oil-for-Food Programme and other programs as well as the earnings of the Iraqi oil sales were deposited into the DFI account managed by the New York Federal Reserve. The DFI received these funds, ensured that Kuwait received 5 percent of these revenues as reparations for the 1990 invasion, and held the remainder until the GoI requested a fund transfer. Although, the UN resolution that created the DFI has expired, the New York Federal Reserve Bank continues to receive Iraq oil export earnings and transfer them to the Ministry of Finance (MoF) when requested (CBI 2020b, p. 51; Newburger 2020).

Current Expenditures

About three-quarters – 104 trillion ID ($88 billion) – of GoI 2019 expenditures were for current activities. The major components of current expenditures were salaries and pensions which accounted for about 50 percent of current expenditures, and transfer payments (social security, PDS food baskets, state-owned enterprise subsidies) which accounted for about 13 percent. The remaining current expenditures were spent on purchases of goods and services, interest, debt payments, war reparations, and so on.

Regardless of the changes in total revenues of the GoI, the amount spent on government salaries and pensions has increased steadily since 2003. As GoI expenditures on salaries and pensions increased faster than the other components of current expenditures, its proportion of current expenditures almost quadrupled from about 13 percent in 2004, to about 50 percent in 2019 when spending on salaries and pensions reached 41 trillion ID ($34 billion).

The increase in salaries and pensions has crowded out spending on goods and services. Annual spending in this category has been cut by one-third from 18 trillion ID ($12.4 billion) in 2005 to 12 trillion ($10.5 billion) in 2019. If the effect of inflation is considered, the decline is even greater. Apparently, freeing up funds for government salaries and pensions is a higher priority than ensuring that those workers have the equipment and consumables necessary to do their jobs.

Debt payments, which have been relatively low since the 2006 Paris Club agreement, are expected to increase substantially over the next several years. The Paris Club agreed to cancel 80 percent of Iraq's debt to its members and extended the maturity of the rest. The time to begin repaying this debt has arrived. Beginning in July 2020, Iraq must pay about $169 million in principal every six months in addition to 5.8 percent interest (Gokoluk and Eder 2020).

The other major components of current expenditures have changed little. Transfer payments accounted for about the same proportion of total expenditures in 2019 – 14 percent – as in 2005. War reparations are a function of the value of oil exports, with Kuwait receiving 5 percent of Iraq's oil export earnings. These reparation payments were suspended during the worst of the fight against ISIS but restarted in 2018. In 2019, Iraq paid an estimated 1.5 trillion ID ($1.3 billion). As a result of Iraq's major creditors canceling substantial amounts of debt and suspending or, at least, reducing interest on the remainder, interest accounts remain a fairly small proportion of GoI expenditures. In 2019, interest was 3 trillion ID ($2.5 billion), or about 3 percent of total expenditures (World Bank 2020b, Table 1, pp. 16–17; IMF 2019a, Tables 2 and 3, pp. 28–9).

In 2019, there were 43 ministries and institutions responsible for actual expenditures, although four ministries accounted for almost 60 percent of expenditures. The Ministry of Finance (24 percent of total expenditures) is responsible for much of the country's social welfare spending, grants, subsidies, and debt service. The ministries of Oil (18 percent), Interior (10 percent), and Electricity (5 percent) are the other large ministries. The Kurdish Regional Government (KRG), the Council of Ministries, and the province of Baghdad rank 6th, 7th, and 8th in total spending. Their spending was dominated by salaries. For scale, the smallest ministry or institution with a separate budget line was the Federal Supreme Court, with expenditures of 4.3 billion ID ($3.6 million) (Ministry of Finance 2020).

Investment Expenditures

Investment expenditures are the shock absorbers of the GoI budget. Adverse oil price shocks translate into large unexpected revenue drops. Reducing salaries and pensions is politically risky, and Iraq's ability to borrow is limited. Therefore, the usual response to fiscal pressure is to cut investment expenditures.

As a result, these investment expenditures vary greatly from year to year. For example, as a result of the fall in oil prices that began in 2014, investment expenditures declined from 32 trillion ID ($27 billion) in 2015 to 14 trillion ID ($12 billion) in 2018. In response to the rise in oil prices in 2018, the GoI budget increased investment expenditures in 2019 to 24 trillion ID ($21 billion) (World Bank 2020b, Table 1, pp. 16–17; IMF 2019a, Tables 2 and 3, pp. 28–9).

As this chapter is being written in late 2020, Iraq has experienced the sharpest reduction in government investment in a decade and a half. During the first eight months of 2020, government investment was an estimated 0.8 trillion ID (less than 2 percent of total expenditures), compared to 6.8 trillion during the same period of 2019 (about 12 percent) (Ministry of Finance 2020).

These rapid unforeseen changes in government investment not only have a substantial impact on capital accumulation but also lead to great inefficiency and encourage corruption. Public investment accounts for almost 90 percent of all fixed capital formation in Iraq. Therefore, when the GoI rapidly cuts investment expenditures, most infrastructure and other investment in roads, electricity, schools, clinics, water supply, and so on, slows or grinds to a stop. Partially completed multi-year building projects are abandoned for months or years until investment spending is restored in a future budget. When projects are restarted, it is often discovered that previous work must be partially or completely redone, due to looting, vandalism, environmental damage, or planned revisions.

Budget Balance

Almost every year, the GoI approves a budget with a substantial deficit, but the budget is rarely a good guide to the actual fiscal outcome. The actual results are usually better than predicted by the budget. In other words, the actual outcome is either a smaller deficit or a larger surplus than budgeted. The Ministry of Finance generally underestimates world oil prices, resulting in actual oil revenues exceeding the budgeted amount. In addition, several accounts have been discovered with hundreds of billions of Iraqi dinar (hundreds of millions of dollars) of what are apparently revenues from oil sales that, for some reason, have not been officially reported to the Ministry of Finance. On a smaller scale, a substantial portion of tariff revenues are collected but not officially recorded. However, the biggest failures to execute the budget are on the expenditure side.

The problem begins with delays by the ministries in developing a consistent budget and submitting it in a timely manner to the CoR. In addition to the usual political disputes that surround any country's budget, the GoI is a relatively new organization, primarily composed of officials and elected members who are new to government. The written and unwritten practices and procedures necessary (but not sufficient) for passing the budget in a timely manner and ensuring that it is executed as planned are only gradually being developed. For example, the fiscal year in Iraq begins on January 1. However, the GoI will often present the budget late in the previous year, leaving little time for discussion and analysis before the fiscal year begins. Several times, this has resulted in budget approval being delayed until after the beginning of the fiscal year. The worst case of delay is 2020. The GoI went at least ten months into the year without a budget, and it appears that the GoI may decide not to officially approve a detailed 2020 budget. The official budgets will jump from 2019 to 2021.

If a budget is not approved before the beginning of a fiscal year, the GoI is permitted each month to spend one-twelfth of the previous year's budget, leading, of course, to two types of errors. Some ministries spend too much early in the year before the budget is passed and have to reduce their spending during the rest of the year. More common is that ministries will spend too little early in the year and have to scramble to increase their expenditures later in the year. Both over- and underexpenditures in the initial months of the fiscal year increase waste. Even when budget funds are released to the various ministries, there is a wide variance in the ability of ministries to actually spend these funds before the end of the fiscal year.

The 2019 Budget called for 128 trillion ID ($108 billion) of total expenditures. However, in the first two-thirds of the fiscal year the ministries were able to spend only 58 percent of budgeted current expenditures and 28 percent of budgeted investment expenditures. As a result, at the end of the year, actual expenditures were only 87 percent of the total budgeted amount, or 112 trillion ID ($95 billion). Of the major categories of spending, all failed to spend their entire budgets except social welfare (Ministry of Finance 2020). This inability to spend is especially serious for investment expenditures. There was an unpublished report that one major ministry was only able to spend less than 5 percent of its 2019 investment budget.

Along with administrative delays and mismanagement, corruption tends to delay expenditures in two somewhat contradictory ways. Corrupt officials will often use bureaucratic methods to delay expenditures until they can divert some of the funds into their own pockets or those of their supporters. At the same time, attempts to reduce corruption may also lead to delays. Some ministries severely constrain the authority of the lower levels of the bureaucracy to make any decisions that could conceivably facilitate corruption. As a result, relatively minor expenditure decisions must be approved at a very senior level, sometimes by the minister himself. Even when the fiscal year has ended, it is difficult to determine actual expenditures, although the Ministry of Finance has adopted new systems that should allow it to more accurately monitor spending in the future.

Financing the Deficit

The GoI has several options to finance a budget deficit. Most directly, the MoF can draw upon its dinar and dollar accounts at the Central Bank of Iraq (CBI). In addition, there are believed to be several trillion dinar (billions of dollars) in various GoI accounts at state-owned banks that are not reported to either the MoF or the CBI. A portion of these funds are "float" and therefore not available to finance a budget deficit. However, a large portion of these funds represent payments made for delayed or discontinued projects that could clawed back by the MoF.

The problem is that Iraq lacks a Treasury Single Account (TSA). A TSA is a single consolidated account in which all revenues and expenditures are recorded. Instead, in addition to their official accounts, every ministry and state-owned enterprise (SOE) has one or more accounts at Rafidain, Rasheed, or the Trade Bank of Iraq. These hidden accounts provide ministries and SOEs with the capability of avoiding GoI mandates. And they make ministerial appointments more valuable. Since 2004, the GoI has made several commitments to the International Monetary Fund (IMF) to establish a TSA, but despite the financial crises of 2009, 2016, and 2020, little progress has been made primarily because of ministerial opposition (Tabaqchali 2020b).

Domestic borrowing to finance the deficit is limited due to lack of a formal government bond secondary market. The few GoI domestic bond sales have been more as a proof of principle rather than as a means of raising funds. For example, only 371 billion ID ($314 million) in Treasury bills with maturities between three and 364 days have been issued (CBI 2020b, Note 44, p. 36).

Although regulatory changes allow the ISX to trade GoI bonds in a secondary market, there is currently a shortage of both bonds and non-bank bond buyers. Another possible barrier to an increase in non-bank public holdings of GoI bonds is concern about the confidentiality of bond purchases. Because of widespread official corruption, many Iraqis think that to openly purchase a government bond will expose them to official harassment and demands for bribes. One proposal is for the GoI to issue bearer bonds at a discount and redeem them – no questions asked – at par when they mature. Since bearer bonds are unregistered and no information is kept concerning the purchaser, they reduce the vulnerability of the purchaser to official interference. Another proposal is that the GoI could issue bonds whose redemption value depends on the world price of oil (Gunter et al. 2020).

In view of the limited demand for GoI debt both domestically and internationally, the government uses the CBI to finance the deficit. Public and private banks purchase government bonds and then resell these bonds to the CBI. This, of course, monetizes the debt and either reduces CBI reserves or increases the dinar money supply. As can be seen in Table 14.2, in 2016, the CBI financed 58 percent of that year's 28 trillion ID ($24 billion) budget deficit (IMF 2019a, Table 2, p. 28). Based on partial year data, the CBI is playing an even greater role in financing the 2020 deficit. Early in 2020, it was estimated that the CBI might have to spend 40 percent of its foreign exchange reserves, approximately 32 trillion ID ($27 billion), to finance the budget deficit (World Bank 2020b, Table 1, p. 17; Rabee Research 2020b, December 24; and author's estimate).

Table 14.2 *GoI budget finance*

	2016	2017	2018ᵉ	2019ᵉ	2020ᵖ
Budget surplus/deficit	−28.7	3.0	29.6	−4.1	−16.5
External financing	2.4	4.0	1.2	3.0	4.7
Budget loans	3.5	4.0	0.3	0	0
Other	−1.1	0	0.9	3.0	4.7
Domestic financing	26.6	−7.0	−30.8	1.1	10.7
CBI	16.7	1.2	−19.1	8.9	12.6
Commercial banks	5.2	−0.8	−2.5	0	1.0
Other	4.7	−7.4	−9.2	−7.8	−2.9
Exchange rate	1180	1182	1182	1182	1182

Note: Superscript "e" indicates data is estimated; superscript "p" indicates data is projected.
Source: 2016: IMF (2019a, Table 2, p. 28); 2017–18: World Bank (2020b, Table 1, p. 16); 2019: MoF (2020); 2020: extrapolated from January–August data from MoF (2020).

However, reductions in both current expenditures and, especially, investment expenditures combined with increasing payment arrears have reduced the deficit. The GoI has delayed making payments to domestic and foreign suppliers, as well as delaying salary and pension payments. The scale of these arrears is unknown but may be as great as 6 trillion ID ($5 billion). These delayed payments have not only created hardship for employees and pensioners but also severely reduced revenues for Iraqi private firms and international oil companies (IOCs) which are suppliers to the GoI. It is expected that the GoI will attempt to settle these arrears when the fiscal situation eases. However, in view of the expected low oil prices over the next decade, settling these arrears may take years. In the meantime, domestic and foreign suppliers to the GoI are expected to reduce their business with the government or cease such transactions entirely. There have already been cases of the IOCs slowing or stopping operations.

As a result of these drastic expenditure reductions, the estimated 2020 deficit is 15.4 trillion ID ($13 billion), less than half of the expected deficit at the beginning of the year. CBI net foreign assets decreased by $13.8 billion, which is consistent with the CBI monetizing the deficit by lending indirectly to the GoI.

The final option for closing the financing gap is devaluation of the dinar. In late December 2020, the GoI devalued the dinar by about 23 percent, from 1182 ID/$ to 1450 ID/$. As a result, in the future, each dollar of oil export earnings will provide the Ministry of Finance with 23 percent more dinars than previously. As a matter of scale, if this devaluation had occurred at the beginning of 2020 then oil export revenues that year would have increased by 11 trillion dinars to an estimated 58.9 trillion. However, this increase in revenues will not result in an equivalent decrease in the country's budget deficit. About one-third of GoI budget expenditures are for imported goods, and it is unlikely that the dollar price of these imports will change as a result of the devaluation. It is only the remaining two-thirds of expenditures, primarily for salaries and pensions, that are affected by the devaluation. As a rough estimate, if the devaluation had occurred at the beginning of 2020, the budget deficit would still have reached about 8 trillion dinars. However, as will be discussed below, this devaluation is expected to have both favorable and adverse impacts on the rest of the economy.

Crisis Price of Oil

Assuming that GoI access to world capital markets continues to be severely constrained, except for some foreign investment in the oil sector, how low will oil prices have to fall before the GoI adjustment process fails? In other words, what is the lowest world oil price that will allow the GoI to pay salaries and pensions, purchase the supplies necessary for the army and police, maintain a minimum social safety net, pay interest and war reparations, and continue the minimum necessary infrastructure maintenance and construction to allow a steady increase in the volume of oil exports? With all the necessary assumptions, this number will be no more than an educated guess.

However, with the devaluation to 1450 ID/\$, assuming average oil exports of 4.0 mbpd, the crisis price of oil is about \$53 pb. If Iraq earns less than this level for an extended period of time, the GoI will be forced to make permanent the 2020 expenditure cuts.

The GoI has predicted that its per barrel earnings will be \$45–\$55 in 2021. And, as discussed in Chapter 7, a strong argument can be made that oil prices will be \$60 pb or less for the next decade. And note that 4.0 mbpd is above the Organization of the Petroleum Exporting Countries (OPEC)-directed production limit for Iraq. For every 10 percent reduction in oil export volumes mandated by OPEC, Iraq's crisis price of oil must increase by 10 percent to compensate. An extended period of low oil prices over the next decade combined with an OPEC-directed production reduction will have dire fiscal implications for Iraq. And the capability of the GoI to offset unfavorable fiscal trends with monetary policy are severely constrained.

MONETARY AND EXCHANGE RATE POLICY

Central Bank Policies

"The primary objectives of the CBI are to achieve and maintain domestic price stability and to foster and maintain a stable and competitive market-based financial system. Subject to these objectives, the CBI shall promote sustainable growth, employment and prosperity in Iraq" (CBI 2020b, p. 5). During the last decade, the CBI has been successful at maintaining price stability. Over the period 2010–20, the average inflation rate was only 1.2 percent a year. However, CBI results in achieving the other objectives has been uneven.

CBI efforts to achieve its objectives are complicated by the country's severe dollarization. Dollarization is the common use of US dollars as both a means of exchange and a store of value in a country where dollars are not the official currency. Iraqis make widespread use of dollars both to make purchases as well as a form of savings. For two currencies to simultaneously circulate in the same country, they must have different characteristics. US dollars are widely accepted in international transactions, the dollar inflation rate is relatively low, and the dollar is available in larger denominations than the dinar. Until 2013, the largest Iraqi dinar note in circulation was only 25 000 ID, which is worth about $21. The largest ID note is currently 50 000 ($42). Therefore, for large transactions, $100 notes are preferred as more convenient and liquid than ID. Among the advantages of dinars, the interest rate paid for banks on dollar deposits is less than that for ID deposits. Another advantage of dinars is that they are legal tender and can be used to pay taxes and other transactions involving the government; although I was told several times that officials prefer to be bribed with dollars.

Dollarization makes it more difficult for the CBI to develop an effective monetary or exchange rate policy. Dollarized economies tend to experience more rapid pass-through of exchange rate changes to domestic inflation. In most countries, there is a lag between depreciation (or appreciation) of the currency and the resulting acceleration (or deceleration) of inflation. However, in dollarized economies the lag is much shorter, reducing the time available for the authorities to adjust policy during a crisis. Also, dollarization breaks the connection between growth in the supply of dinars and the inflation rate. For example, the growth of the broad measure of money has been very uneven in Iraq. Broad money grew 38 percent in 2011, and actually declined by 9 percent in 2015. More recently, as can be seen in Table 14.3, the broad money growth rate averaged 7.7 percent between 2016 and 2019, before shrinking by 2.2 percent in 2020 (World Bank 2020b, Table 1, p. 16; IMF 2019a, Table 1, p. 27).

Table 14.3   Monetary and exchange rate variables

	2016	2017	2018	2019[e]	2020[p]*
Inflation rate	0.5%	0.2%	0.4%	-0.2%	0.1%
Exchange rate (ID/$)	1182	1182	1182	1182	1182
Broad money growth	7.1%	2.6%	2.7%	8.4%	-2.2%
Interest policy rate	4.0%	4.0%	4.0%	4.0%	4.0%
Reserve requirement	15%	15%	15%	15%	15%

Notes: Superscript "e" indicates data is estimated; superscript "p" indicates data is projected.
*Predictions made mid-October 2020.
Source: World Bank (2020b, Table 1, p. 16), IMF (2019a, Table 1, p. 27), CBI Currency
Auction Results.

However, as previously discussed, the large-scale smuggling of dollars from Iraq to both Syria and Iran tends to reduce the dollar component of Iraq's money supply. As a result, a portion of the increase in the dinar money supply simply offsets ongoing dollar loses and therefore should be less inflationary. However, as the CBI continued to expand the dinar supply to replace the dollars lost to Syria and Iran, this increased the pressure for a devaluation of the dinar. But if the CBI reduces money supply growth in order to protect the exchange rate, then this will lead to a money shortage, higher real interest rates, and slowdown of the economic growth.

Since Iraq adopted a fixed exchange rate, it does not have an independent monetary policy. Or, to be more precise, its monetary and exchange rate policy are the same. Any attempt to adopt an expansionary monetary policy during an economic downturn, by reducing interest rates or expanding the monetary base, will increase the rate at which Iraqi portfolio holders exchange dinars for dollars at the CBI or parallel currency auctions. But as the CBI buys these dinars – to maintain the fixed exchange rate – this will take dinars out of circulation, offsetting the attempted monetary expansion. The reverse happens if there is an attempt to adopt a contractionary monetary policy, possibly to fight inflation. (In terms of the standard IS/LM/FE macroeconomic model, a fixed exchange rate results in a horizontal FE curve.)

There are similar difficulties with CBI attempts to encourage growth by lowering interest rates. The policy rate has gradually decreased from 16.75 percent in 2008 to 6 percent, before being reduced to its current rate of 4 percent in 2015. These reductions had little effect on real growth, because of the banking issues discussed at length in Chapter 9. Policy rate reductions did not lead to increased willingness of banks to lend to the private sector. However, lower rates on dinars did tend to decrease the willingness of Iraqi non-bank entities and individuals to hold dinars instead of dollars. This decreased willingness to hold dinars contributed to the pressure on the CBI to devalue the dinar in 2020.

If this pressure continues after the devaluation, the CBI may be forced to raise the policy rate to forestall another devaluation.

The complex relationship between Iraq's monetary and fiscal policy is illustrated using 2018 data in Figure 14.1. Oil export revenues in dollars – $91.9 billion – plus foreign direct investment, and a small amount of foreign aid were received either directly by the MoF or by one of the two MoF accounts at the New York Federal Reserve. Some of these US dollars were used for government spending for imports and some other foreign transactions. However, to obtain dinars needed by the GoI for domestic expenditures, the MoF sold dollars – $62.5 billion in 2018 – to the CBI for dinars at the then official exchange rate of 1182 ID/$. These sales increased the CBI net foreign assets, its dollar holdings.

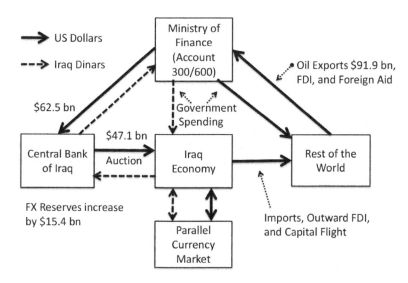

Figure 14.1   Flow of dinars and dollars in Iraqi economy

In order to provide dollars to the private sector, the CBI holds a daily currency auction where it invites a limited group of banks and other financial institutions to buy or sell US dollars at the official exchange rate. As expected, it is a very rare occasion when these institutions seek to sell US dollars to the CBI. In 2018, these institutions purchased $47.1 billion from the CBI. The difference between the US dollar sales to the CBI by the MoF and the dollar sales by the CBI to financial institutions increases or decreases the CBI foreign reserves. In 2018, foreign reserves increased by $15.4 billion: the difference between the $62.5 billion the CBI received from the MoF and the $47.1 sold at the daily auction.

Financial and non-financial entities that are not allowed to buy US dollars at the CBI auction purchase them in a parallel currency market. The difference between the official exchange rate – 1182 ID/$ in 2018 – and the exchange rate in the parallel market provides critical information on the true value of the dinar. A difference of about 1 percent between the two rates is expected. This provides a profit for those financial institutions that have access to the currency auction to buy dollars there and sell them in the parallel currency market.

However, a greater than 1 percent difference between the auction and market price for US dollars indicates that speculators think the dinar is overvalued. The 2020 differential pattern was a cause for concern for the CBI. From mid-2018 through the beginning of 2020, the differential was the expected 1 percent. However, between April and October 2020, the differential averaged about 4 percent, with a spike in the differential reaching 6 percent in mid-2020 (Tabaqchali 2021, pp. 2–3). This widening differential reflected a competition to buy dollars because of the concern that the GoI was considering devaluation: a change in the fixed exchange rate.

With a fixed exchange rate, the CBI promised that it was willing to buy or sell dinars for dollars at the official rate. Of course, there is a fundamental difference between the CBI selling dinars, thereby accumulating dollars, and the CBI buying dinars, experiencing a dollar outflow. There is no practical limit to the amount of dinars that the CBI could create through its purchases of foreign exchange. However, its ability to buy dinars at the fixed exchange rate is limited by its foreign exchange holdings plus those that it is willing and able to borrow.

Like all countries that have adopted fixed exchange rates, Iraq faces the danger of becoming a target for almost riskless speculation. For example, if a speculator expected – accurately, it turned out – that GoI would devalue the dinar in December 2020, they would borrow, say, 1 billion ID. They would use this 1 billion ID to buy dollars from the CBI at a rate near the official exchange rate of 1182 ID/$. This purchase would earn the speculators about $847 000. When the GoI devalued the dinar to 1450 ID/$, this speculator could convert their dollars back into dinars at the new higher rate of 1450 ID/$ and make almost a 23 percent profit. However, if the speculators are wrong and the GoI does not allow a depreciation, then they could convert their dollar holdings back into dinars at the rate of 1182 ID/$ with only a small transaction cost loss.

The possibility of such riskless speculation made the CBI vulnerable to an attack on the dinar. If speculators are able to sell large amounts of dinars for US dollars at the daily auction, the decrease in CBI dollar reserves will be perceived as increasingly the likelihood of a future devaluation of the dinar.

This perception will motivate speculators to buy even more dollars in the next auction, until the devaluation became a self-fulfilling prophecy.

In addition to speculators, Iraq's fixed exchange rate is also under pressure from entities and people in Syria and Iran. Both Iraq's neighbor to the west, Syria, and to the east, Iran, face severe dollar shortages. Since 2011, the government of Syria has used military force in an attempt to crush a widespread popular protest. The resulting violence, combined with international sanctions, has resulted in an economic downturn in Syria and a severe shortage of hard currencies such as the US dollar. Iran is also experiencing economic difficulties and a shortage of hard currency as a result of tightening international sanctions intended to discourage the country from pursuing its nuclear program. As both residents and non-residents lose confidence in the Syrian and Iranian economies, these countries' currencies have collapsed. The Syrian pound has lost almost 90 percent of its value, falling from a pre-crisis rate of 47 Syrian pounds/$ to 513 Syrian pounds/$ at the end of December 2020. The Iranian rial has lost almost 75 percent of its value. It traded at 10 500 rials/$ in December 2011, before falling to 42 060 rials/$ by the end of December 2020. The black market depreciations for both currencies were even greater.

The large depreciations of the Syrian pound and the Iranian rial added to the pressure on the CBI. Entities and individuals in Iraq bought dollars with dinars either at the CBI auction or from the market, and then used these dollars to buy low-cost goods from Syria or Iran.

Devaluation Winners and Losers

The effects on Iraq's economy of the large December 2020 devaluation are mixed. As mentioned above, by increasing the dinars obtained from the sale of every barrel of oil, the devaluation reduced the severe fiscal constraint facing the GoI as a result of the fall in oil prices, OPEC-directed reduction in the volume of oil sales, increased spending for reconstruction, and the expense of fighting Covid-19.

Since Iraq imports most consumer goods, the devaluation is expected to lead to an acceleration of inflation, making the average Iraqi worse off. Provinces such as Dhi Qar, Basrah, and the KRG, which engage in large-scale legal and illegal trade with Iran or Turkey, can expect a trade contraction as higher prices discourage Iraqi demand for imported goods. Gradually, it is expected that Iraqi producers will begin to provide the goods formerly imported from other countries. Iraqi agriculture is expected to increase its domestic market share now that imported food products have become more expensive. But in view of the regulatory barriers discussed in Chapter 11, this adjustment will take time. And while this adjustment is occurring, the proportion of Iraqis living in poverty will increase.

Another adverse impact of the devaluation is on expectations. Speculators are likely to think that another devaluation is coming and therefore they should drain as many dollars as possible out of the CBI. To reduce the likelihood of another speculative attack on the dinar, the GoI has only a few options. The GoI can reduce or cease entirely having the CBI partially finance the budget deficit by purchasing GoI bonds, directly or indirectly. In 2020, it was estimated that the CBI financed almost 90 percent of the deficit by purchasing government debt from state banks. Such purchases leave the total CBI assets unchanged. However, these purchases add illiquid high-risk GoI debt to the CBI portfolio, while reducing liquid low-risk US government bonds. As CBI dollar holdings decrease, the CBI commitment to maintaining a fixed exchange rate becomes less credible.

Another option would be for the CBI to increase the interest policy rate to encourage Iraqis to hold more higher-interest dinar deposits rather than lower-interest dollar deposits. Unfortunately, as was discussed in Chapter 9, the financial system in Iraq is managed by the bureaucracy, not determined by a market. As a result, changes in interest rates are expected to have little effect. Finally, the CBI could once again attempt to reduce the rate at which dollars are leaving by restricting access to the daily currency auction.

CBI Direct Lending

As discussed in Chapter 9, the Iraqi banking system does a poor job at making loans to promising entities, especially those in the private sector. While the CBI is attempting to change the current perverse incentives for bank lending, it is also engaging in directed lending. Such lending substitutes the CBI judgment for that of banks. The CBI intends to lend 5 trillion ID ($3.4 billion at exchange rate of 1450 ID/$) to specialized government banks for financing large projects. The terms of these loans are liberal: 2 percent interest with 5–10 years maturity. The CBI is lending another 1 trillion ID ($6900 million) to commercial banks for financing small and medium-sized projects. The terms are generous for private sector lending in Iraq: a maximum of 5.5 percent interest and three years maturity (CBI 2020b, p. 29). One concern about CBI directed lending is that the choice of loan recipients will be made on political, not economic, criteria.

In summary, Iraq needs a better operating system of financial intermediation to facilitate economic growth and encourage foreign direct and portfolio investment. But creating a more favorable banking and investment environment in Iraq in an increasingly globalized environment will require careful phasing-in of substantial changes in financial regulations and monetary policies.

Table 14.4   *Eichengreen on phasing financial liberalization*

Phase 1:	Liberalize and decontrol domestic financial markets
	Rationalize regulations
	Recapitalize state-owned banks
Phase 2:	Liberalize foreign direct investment
Phase 3:	Liberalize stock and bond markets next
Phase 4:	Liberalize offshore bank borrowing last

Source: Eichengreen (2000, pp. 1107–10).

HORNS OF A POLICY DILEMMA

In the third decade of the twenty-first century, the GoI will be forced to make fundamental policy choices. The moribund financial system slows economic growth, but the complexity and expense of performing even the simplest financial transactions raises the costs of, and therefore discourages, capital flight. If the GoI liberalizes the financial system in order to achieve greater efficiency and economic growth, it will also make it easier for capital to flee. In addition, inadequately regulated and capitalized private financial intermediaries provide incentives for their management to engage in excessive risk-taking using offshore funding (Eichengreen 2000, p. 1108). On the other hand, failure to liberalize will burden the non-oil sector and lead to the continued gross domestic product (GDP) dominance of oil exports and the GoI as the chief employer.

The status quo is not an option. Growing international capital mobility is unavoidable, especially for a country such as Iraq that seeks large amounts of foreign direct and portfolio investment to pay for the modernization of its infrastructure. The challenge is to order the changes in the liberalization of both internal and external financial activities so as to maximize the likelihood of accelerating economic development without increasing the economy's vulnerability to destabilizing capital flows. Eichengreen (2000, pp. 1105–16) argues that there is a phasing of financial liberalization that historically has had the best chance of simultaneously achieving economic development and financial stability. His recommended phasing of financial liberalization is summarized in Table 14.4.

Liberalize and Decontrol Domestic Financial Markets

As discussed in Chapters 9 and 11, there are steps that could be taken to substantially improve access of individuals and entities to the Iraqi debt and equity markets. Probably the most important is the expansion of the credit registry. In addition, the CBI and other regulatory authorities should act to increase the competitiveness of private banks. This would include providing some reasonable amount of deposit and savings account insurance, as well as allowing checks or electronic funds transfer from private banks to be used to make tax and other payments to the GoI. Accounting and auditing standards should require the timely release of accurate information, especially with respect to loan losses and loan loss reserves. At the same time, it is important to ensure that private bank capitalization is adequate, otherwise management will have strong incentives to engage in risky loans and investments (Eichengreen 2000, p. 1108). Only after regulations have been rationalized to support a vibrant banking system should the GoI recapitalize the state-owned banks. If they are recapitalized before the CBI and other regulatory authorities are ready to closely monitor their activities, one could expect to see a return to large-scale loan losses.

Liberalize Foreign Direct Investment

Failure to complete the rationalization of financial regulation and the recapitalization of state-owned banks before opening an economy to foreign direct investment (FDI) led to severe financial difficulties in South Korea and Thailand (Eichengreen 2000, p. 1108). However, once the domestic financial system is prepared, Iraq should seek to further increase FDI, since it provides two advantages over foreign portfolio investment or foreign bank loans. First, FDI tends to be accompanied by managerial and technological expertise that will be needed to achieve the GoI's goals of rapid improvement of both oil and non-oil infrastructure. Second, FDI tends to be more stable than either portfolio or bank finance.

Liberalize Stock and Bond Markets Next

The second most beneficial capital inflow is foreign purchases in the Iraq equity and bond markets. Equity and bond purchases are more stable than bank debt, less likely to reflect volatile "hot money". As discussed in Chapter 9, the Iraq Stock Exchange (ISX) has the potential to be a major financial intermediary between Iraqi companies and foreign investors. However, for this to occur the GoI must adopt a more rational commercial code that provides potential investors with timely audited income statements and balance sheets of non-bank companies listed on the ISX. Until this happens, the ISX will continue to be a shallow market dominated by insiders. With respect to bond markets, if the Ministry of Finance would authorize the sale of short-, medium-, and long-term government bonds then this would not only lead to a more robust secondary market in government debt but also, by providing a base rate, facilitate bond issuance by other entities.

Liberalize Offshore Bank Borrowing Last

Although it can provide substantial benefits, the international transaction with the greatest potential for destabilizing Iraq's current system of financial intermediation is offshore bank borrowing. There are often strong short-term incentives to engage in financial transactions that result in severely mismatched maturities or currency exposures. The 1997 Asian financial crisis and the post-2008 European financial crisis should be warnings on the risks and uncertainties involved in allowing Iraqi financial and non-financial institutions to engage in offshore – cross-border – transactions. In addition, without increased transparency and improved regulation, increased access to offshore borrowing can lead to rapid distortion of the Iraqi economy, including large-scale capital flight (Eichengreen 2000, pp. 1106–7). But regardless of the careful phasing of integration into the global economy, Iraq's current exchange rate system will continue to be vulnerable to external shocks.

SHOULD IRAQ ABANDON ITS FIXED EXCHANGE RATE POLICY?

Iraq's fixed exchange rate facilitates foreign trade and investment. This is especially true since almost all of Iraq's exports, most of its imports, and a large proportion of its FDI are already denominated in US dollars. In addition, about 72 percent of the CBI 2019 asset portfolio was denominated in US dollars, with ID accounting for only 18 percent (CBI 2020b, Note 31.4.2, p. 46). But a fixed exchange rate negates monetarist policy. To protect its fixed exchange rate, the CBI must maintain a domestic Iraqi dinar interest rate above the dollar rate, and this higher real dinar interest rate is a further burden on the non-oil sector. In addition, a fixed rate is vulnerable to the riskless speculation discussed above. For the last several years, the GoI has been free to ignore the adverse effect of higher real interest rates, relying on oil-funded expansionary fiscal policy to grow the country's GDP. But this policy has maintained state sector dominance.

In mid-2020, the gap between the currency auction rate and that in the parallel currency market reached 6 percent (Tabaqchali 2021). This gap represented a loss of confidence in the future willingness of the CBI to sell dollars at the official exchange rate. The large devaluation of December 2020 can be expected to reduce speculative pressure for a period of time.

However, attacks on the dinar will return, and as discussed above, responding to excessive dollar outflows by policies of restricting access to dollars at the daily auction or increasing the policy interest rate are of limited effectiveness and tend to slow economic growth. The longer such restrictions are maintained, the more damage they do to the efficiency of banks and other financial intermediaries. Almost all countries whose fixed exchange rates come under widespread speculative attack are forced to eventually abandon their fixed exchange rate regimes, often after large-scale losses of reserves. Should Iraq adopt a more credible exchange rate regime before the situation again reaches a crisis such as that of 2020? If Iraq decides to abandon its current fixed rate regime, then what are its options?

Floating Exchange Rate

In theory, a country should adopt a floating – market-determined – exchange rate if most of its trade and financial activities are domestic, the international trade and financial transactions that do occur are widely diversified among multiple currencies, and the government can credibly adopt a domestic nominal anchor such as inflation, or nominal income growth. Iraq has none of these characteristics.

With imports equivalent to almost two-thirds of the country's GDP, as well as the large amount of foreign investment, Iraq must be considered a very open economy. In addition, not only are 99 percent of its exports denominated in dollars, but also much of its import trade is in homogeneous products such as grain that also tend to be dollar-denominated. Especially in the oil sector, much of the investment is dollar-denominated. Since there is no world demand for substantial amounts of Iraqi dinar-denominated debt, if public or private Iraqi entities wish to borrow internationally then they will have to borrow in a hard currency, probably dollars. A floating or managed exchange rate would make it more complicated and expensive for Iraq to engage in these international trade or capital transactions.

Also, a country with a floating rate must target another nominal variable – a nominal anchor – such as the inflation rate or nominal GDP in order to maintain price stability. For reasons discussed in Chapter 2, there is a shortage of reliable data on variables such as the inflation rate or nominal GDP that could serve as a nominal anchor for the Iraqi economy. But even if reliable data on a nominal anchor was available to the GoI, neither the average Iraqi nor the average foreign investor trusts the GoI to accurately report these nominal targets. The public can easily observe changes in the exchange rate, whereas changes in the calculation of inflation or nominal GDP are difficult to evaluate even by experts. For all of these reasons, adopting a floating or managed exchange rate would be a mistake.

Return to a Currency Board

While a fixed exchange rate is a better option than a floating one for Iraq, maintaining credibility is the chief challenge facing a fixed exchange rate regime. Regardless of the political or financial surprises that the future may hold, the market must be confident that the CBI will continue to buy or sell dinars at the fixed rate. With the current structure, this commitment is not credible, which is why interest rates on dinar-denominated deposits and loans are higher than those for dollar-denominated transactions. There is a fear in the market that the GoI will again choose, in some future crisis – whether a deep and sustained drop in oil prices or, possibly, another insurgency, or a civil war in Iran – to repeat the 2020 devaluation. There are two options that can increase the credibility of a fixed exchange rate: official dollarization, or establishing an orthodox currency board.

While there are nine non-US countries that have adopted the dollar as their official currency, including Ecuador in 2000, this is probably an unacceptable option for Iraq. In addition to being perceived as an affront to Iraqi sovereignty, official adoption of the dollar would eliminate seigniorage, the revenue that a country obtains from printing its own currency. Printing a 50 000 ID note costs less than 100 ID; the difference between this cost and the face value of the note is revenue to the GoI.

An alternative is for Iraq to return to using a currency board. Iraq had a currency board based on the British pound from 1930 until 1949. This board provided a stable currency for Iraq through a period of both severe internal conflict and World War II. The National Bank of Iraq, that in 1956 became the CBI, succeeded the currency board. The National Bank of Iraq continued some of the conservative actions of its currency board predecessor including maintaining 100 percent reserves behind its currency issuance (CBI 2012a, "History of the CBI", p. 1). But the golden age of conservative monetary policy and a fixed exchange rate came to an end in 1964 with the nationalization of all banks.

How does a currency board differ from a central bank? There are six characteristics of an "orthodox" currency board. These necessary characteristics along with two desirable characteristics are listed in Table 14.5 (Hanke 2002, Table 1, p. 205; see also Hanke and Schuler 1994, Chapters 4 and 5; Walters and Hanke 1992, pp. 558–61). Most central banks with fixed exchange rates stand ready to convert a variety of financial instruments including deposit accounts. But currency boards guarantee convertibility at the fixed rate for physical notes and coins only. A board is able to make this guarantee since it maintains 100 percent reserves for all of the currency and coins in circulation. This is an important distinction because it eliminates riskless speculation.

In a classic speculative raid, the amount of local currency that the speculator could present for conversion into dollars is limited only by their ability to borrow. When the central bank has exhausted its hard currency reserves, it is forced to depreciate its currency, producing a profit for the speculator. But a currency board cannot be overwhelmed in this fashion since its maximum

Table 14.5   Hanke on currency boards

Necessary characteristics increase	Supplies notes and coins only
	Full convertibility
	Foreign reserves of 100–115 percent
	Not a lender of last resort
	Does not regulate commercial banks
	Cannot finance spending by domestic government
Other characteristics to further increase credibility	Director of another nationality
	Conversion office in world capital center, e.g.
	Switzerland

Source: Hanke (2002, Table 1, p. 205).

liability for conversion is both fixed and known, because it is limited to the amount of currency and coins placed in circulation by the board. Also, currency boards tend to be profitable because they earn the equivalent of seigniorage from the earnings on short-term liquid hard currency bonds in their portfolios. These profits – in excess of required reserves – would be paid to the establishing government.

To further reinforce the credibility of a currency board, it does not engage in typical central bank activities of regulating commercial banks, lending to the government, or acting as a lender of last resort in a crisis. Upon the establishment of a currency board, these activities would become fiscal responsibilities; for example, if a lender of last resort was required then the GoI would make the loans as part of its budget.

In addition, if it chooses to establish a currency board, Iraq can further increase the credibility of its currency board by having a director of another nationality and establishing a conversion office in another country. These were both components of Iraq's previous currency board, with an English director and the ability of any dinar holder to exchange dinars for English pounds at the fixed exchange rate in the London office. A foreign director (say, a Swiss banker) and conversion office (say, in New York) improves the credibility of the fixed exchange rate since it prevents or, at least, complicates any future Iraqi government's attempt to confiscate the assets of the currency board, as occurred in 1964. It should be noted that both the IMF and the US government generally oppose the establishment of currency boards. (For a discussion of their hostility to an Indonesian currency board, see Hanke 2002, pp. 215–18).

At the end of 2019, the CBI had sufficient reserves, even with full conversion of bank reserves, to establish a currency board with a fixed exchange rate of 1182 ID/$. As can be seen in Table 14.6, the monetary base was composed of about 51.8 trillion ID in currency and about 27.6 trillion ID in bank reserves, cash equivalent, for a total of 79.4 trillion ID. To provide the 100 percent

Table 14.6   Cost of re-establishing Iraq currency board

	2015	2019
CBI assets – dollars	$52.1 bn	$67.6 bn
CBI liabilities – dinars	62.9 trn ID	79.4 trn ID
Currency in circulation	38.6 trn ID	51.8 trn ID
Bank reserves	24.3 trn ID	27.6 trn ID
CBI liabilities in dollars at current ER	$54.0 bn	$67.2 bn
Surplus at current ER	–$1.9 bn	$0.4 bn
Current ER	1166	1182
"Balancing" currency board rate	1208	1175

Source: Data from CBI, analysis by author.

reserves necessary for a currency board at an exchange rate (ER) of 1182 ID/$ would require about $67.2 billion. At the end of 2019, the CBI reported net foreign international reserves, mostly dollars, of about $67.6 billion. Based on these numbers, the GoI could establish a currency board with almost $0.4 billion to spare. At the end of 2020, Iraq has fewer dollar reserves, but with an exchange rate of 1450 ID/$, each of those dollar reserves will support a greater amount of dinars. A rough estimate is that a current conversion to a currency board would be about neutral: the dollars currently held by the CBI would be sufficient to provide 100 percent reserves of the quantity of dinars in circulation as well as those held by banks.

The status quo is not a viable option for the Iraqi exchange rate. Despite the December 2020 devaluation, the drain on foreign reserves is expected to continue, driven in part by dollar smuggling motivated by the continuing economic troubles in Iran and Syria; sooner or later the Iraqi dinar will again come under attack. When it does, then the CBI will have to decide whether to devalue again or run the risk of a substantial loss of reserves. While a currency board has prevented such adverse outcomes in multiple nations, the GoI appears unwilling to seriously consider this option. When oil export earnings are high and CBI foreign reserves are rising, there is little political incentive to do away with the country's traditional central bank. And when oil export earnings are low, as in 2015 and 2020, the cost of converting to a currency board is substantial without a large dinar devaluation, which is also politically unpopular. A return to a currency board will give Iraq both the fixed exchange rate it needs in view of large foreign trade and investment engagement, as well as policy credibility that could provide the foundation for a rationalization of the country's financial system.

15. Iraq in 2035

Iraq reforming its economy is not a matter for debate. It is essential, and it is vital that reforms are comprehensive and timely.

(Minister of Finance Ali Allawi 2020)

The loss of the strategic compass in the arena of the Iraqi economy has cost us dearly and led us to waste time and money. The country has not prospered and has left the Iraqi people to suffer despite the affluence they could have due to Iraq's wealth and illustrious history.

(Minister of Planning Ali Ghalib Baban 2010, in GoI 2010, p. 3)

What will the political economy of Iraq look like in 2035? That will depend, of course, on both external surprises and the policy decisions made by the Government of Iraq (GoI). At the same time, the Iraqi people are not pawns or simply numbers in a survey. They will not only respond to unexpected exogenous shocks or policy initiatives in unexpected ways but also create new realities that will probably defy prediction. Does this make attempts to predict the future a waste of resources? That depends on its purpose. If it is an attempt to lay out in detail the characteristics of Iraq in 2035, then it is a waste of time. However, it can be a valuable exercise if it forces one to think about trends, priorities, and resources. In this way, it can help private and public Iraqi entities to make choices.

EXOGENOUS DRIVERS OF 2035: OIL PRICES

Oil will remain an important part of the Iraqi economy for the foreseeable future. Oil prices have been on a rollercoaster for almost four decades and there is no reason to think that the future will be any different. In their June 2020 Annual Energy Outlook, the United States Energy Information Administration (EIA) estimated three oil price scenarios for 2035. The 2035 reference oil price – the most likely price – for Brent Blend oil is about $83 per barrel (pb) (in 2019 dollars), while the high oil price is $159 pb. The low oil price scenario posits a 2035 oil price of about $44 pb (EIA 2020).

The reference price – the most likely price according to the EIA – will gradually rise from $38 pb in 2020 to $70 in 2026, and $80 in 2033. Historically, this is a strong price. For example, over the last five decades, the inflation-adjusted price of oil has been less than $70 pb for 75 percent of the time. If this scenario is correct, Iraq should earn more than its 2020 "crisis price of oil" (see Chapter 14) from 2021 through 2035.

Oil prices, export volumes, and population

Table 15.1   Oil earnings scenarios for 2035

	Low $44 pb	Reference $83 pb	High $159 pb
4 mbpd	$58 bn	$109 bn	$209 bn
6 mbpd	$87 bn	$164 bn	$313 bn
8 mbpd	$116 bn	$218 bn	$418 bn

Table 15.1 shows nine possible outcomes for Iraq's oil export earnings in 2035. The columns represent the EIA low, reference, and high prices for Brent Blend in 2035 (in 2019 dollars). Note that Iraq receives about 90 percent of the Brent Blend price and therefore the 90 percent of the Brent Blend price is used in the estimates.

But how much oil will Iraq be able to export in 2035? The National Development Plan: 2018–2022 calls for an export capability of 5.25 mbpd by 2022 (GoI 2018, p. 137). It is unlikely that this goal will be achieved, especially since Iraq has reduced its recent exports to meet its commitment to the Organization of the Petroleum Exporting Countries (OPEC). However, for oil export volumes in 2035, I consider three alternatives. The first row of the table pessimistically assumes that Iraq will be able to export 4 million barrels per day (mbpd), which is equal to the actual 2019 oil exports. The middle row of 6 mbpd reflects a 50 percent increase in oil export volumes over the next 15 years.

Is a 2 mbpd increase in oil export volumes over a decade and a half overly optimistic? It took Iraq about a decade to achieve the most recent 2 mbpd increase. And, as discussed in Chapter 7, there are severe financial, management, and technical challenges to be overcome if Iraq is to significantly increase its oil exports. Among the most serious challenges is the shortage of water. Despite the detailed Ministry of Oil (MoO) plans for a doubling of oil production and exports, I think that 6 mbpd is the most likely level of oil exports in 2035. However, for completeness, I have included estimates for a doubling of current oil exports to 8 mbpd.

Before considering the implications of these scenarios, it should be recognized that the combination of high export volumes of 8 mbpd and high oil price of $159 pb (lower right-hand cell of Table 15.1) is extremely unlikely. It is difficult to conceive of a plausible world situation where there would be a large increase in oil supply at the same time as an even larger increase in oil demand, resulting in high oil prices. In fact, due to Iraq's importance as one of the world's largest oil exporters, the most likely outcome is the diagonal of Table 15.1, extending from high production, low price $116 billion to low production, high price $209 billion.

Low Oil Price Scenario

How likely is $44 pb oil? After adjusting for inflation, average annual oil prices were below this level for almost 20 years, from 1986 through 2005 and again in 2020 (see Figure 7.2). Such a price could result from lower demand combined with increased supply. With respect to demand, it is possible that not only will the economic recovery of the Organisation for Economic Co-operation and Development (OECD) nations after the 2020 Covid-19 crisis require more than a decade, but also there will be a slowdown in the real growth of the BRIC countries (Brazil, Russia, India, and China). Combined with continuing improvements in energy conservation, the reduction in oil demand growth from these economic slowdowns will be substantial. This reduction in the growth rate of energy demand could be accompanied by an increase in energy supply as a result of two trends.

First, technological innovations such as hydraulic fracturing, or fracking, have permitted relatively inexpensive access to natural gas in the United States of America (USA) and other nations. While some countries such as France have banned fracking, many nations are aggressively pursuing this and related technologies. The USA in a fairly short period of time has gone from being a major energy importer to a net exporter of gas. In addition to the direct impact on the natural gas market, in which Iraq is currently a small producer, there is the potential for secondary effects that will seriously impact the oil market. The large gap between the energy equivalent prices of the two petroleum products is leading to an expensive but fairly rapid re-engineering of many electrical generation and plastic production plants from oil to cheaper natural gas.

A second trend that may contribute to lower future oil prices is the current GoI plan to increase the volume of oil exports. As discussed in Chapter 7, demand for oil is inelastic: a relatively small percentage increase in the world supply of oil tends to lead to a larger percentage decrease in oil prices. If the GoI were to succeed in increasing its export volume by even an additional 2 mbpd by 2035, the impact on world prices would be significant even if other oil exporters maintain current levels of production. Whether caused by decreased demand or increased supply, low oil prices will have a severe adverse impact on both economic development and political stability in Iraq.

If Iraq's 2035 economy is still dominated by oil exports, an extended period of low oil prices could lead to a drop in the country's gross domestic product similar to the almost 12 percent contraction in 2020. The accompanying decrease in GoI revenues will lead to a reduction in government investment and transfer payments followed by a rise in unemployment and underemployment. A growing pool of unemployed young men who have despaired of obtaining a good job and being able to start a family will be politically destabilizing. If Iraq still has a fixed exchange rate in 2035, the fall in oil export earnings will further reduce foreigners' confidence that the Iraqi dinar (ID) will keep its value. Especially if the Iraq's international reserves are tapped to compensate for the drop in oil export earnings, the dinar will come under increasing speculative pressure for another large devaluation.

There will also be an adverse effect on government employees. As discussed in Chapter 14, if oil prices fall to $53 pb or less for an extended period of time, the GoI would have to make unprecedented budget cuts that would probably be politically destabilizing. Since the overthrow of the monarchy in 1958, Iraq has joined most modern states in that there is an unwritten understanding between the country's leadership and the bureaucracy that runs the government on a day-to-day basis. The bureaucracy support whoever is in power and, in return, the person in power ensures that the bureaucracy continues to receive high levels of compensation and status. This was exemplified in 2006 and again in 2020, when in the face of a sharp unexpected drop in GoI revenues, almost all components of government expenditures were cut except for salaries and pensions of the already generously compensated current and former government employees.

However, a drop of oil prices below roughly $53 pb for even a few years would force the GoI, in the absence of large-scale foreign borrowing, to cut compensation or employment of government employees. At best, this will lead to creative non-compliance, with the bureaucracy stalling government initiatives until compensation is restored. At worst, government employees will attempt to bring about a change in leadership by unconstitutional means. An angry and disloyal bureaucracy combined with a large number of unemployed young men will turn an economic downturn into a situation of acute political instability.

High Oil Price Scenario

In view of the technological innovations and the GoI plans for rapid expansion of oil exports discussed above, what argument can be made for substantially higher oil prices in the future? A rapid return to moderate growth in the OECD countries and rapid growth in the BRIC countries would increase energy demand, while a failure of the natural gas revolution would reduce supply.

A conflict that leads to the long-term loss of another major oil exporter on the scale of Iran or Venezuela would have a more dramatic effect on world oil supply. Possibly by the time this book is published, Iran's dispute with the rest of the world over its nuclear ambitions will be resolved. However, if the dispute rises to conflict, then the world faces the potential of a continued loss of some or all of Iran's production for an extended period of time.

In the unlikely case – in the opinion of the author – that Iraq can increase its production to 6 mbpd and world prices are high in 2035, Iraqi per capita oil exports would reach $5200 in 2019 dollars. With this massive increase in oil export earnings, everything would be possible: increased government employment and higher wages for government workers, sharp rises in investment, generous provision of free essential services, agricultural restoration, and accelerated construction of homes, factories, and government offices. The Al Rahman Mosque in Baghdad would finally be completed. New soccer stadiums would appear in every town. But there would be an explosion in corruption. Efforts to diversify the Iraqi economy away from its dependency on oil would probably grind to a stop. Without diversification, Iraqi nascent democracy would be threatened.

Concentration of economic power tends to lead to concentration of political power. If the party that controls the government also controls the economy, and therefore access to education, employment, and income, then those in power begin to believe that elections are too important to entrust to the voters. There is also a psychological effect. In most long-lived democracies, the major source of government revenue is tax receipts. People in such countries feel that they are "buying" government services, and if the government fails to provide, at reasonable cost, the services demanded by the taxpayer-citizens then they will attempt to "fire" the government through the ballot box. But in Iraq with $159 pb oil, the GoI will not need to tax citizens. Cynically, the primary role of Iraqi citizens will be to receive – and be grateful for – whatever level of service that the government decides is appropriate. In this sense, Iraqi citizens will not be independent entities who support the GoI. Rather, they will become symbolically, and often in reality, clients of a "beneficent" government. As Huntington (1991, p. 65) stated: "'No taxation without representation' was a political demand; 'no representation without taxation' is a political reality" (see also Ayubi 1995, p. 400).

Reference Oil Price Scenario

If, by 2035, Iraq increases its oil exports to 6 mbpd and the oil hits the EIA reference price of $83 pb (2019 prices), then oil export earnings will reach about $164 billion, which is almost four times greater than in 2020 and 85 percent greater than in 2019. However, if Iraq's population continues to grow at its current rate, the country's population will increase by about 50 percent to 60 million Iraqis in 2035. In other terms, per capita oil export earnings in 2019 were about $2200, but would rise to only $2700 in 2035 as rapid population growth offsets most of the gain from higher export earnings.

While this scenario is not as dire as those that predict low or high oil prices, it serves to emphasize two priorities for Iraq's development. First, higher oil prices alone are insufficient to produce a substantial rise in living standards for Iraq's rapidly growing population. Second, the expansion of the oil industry by itself will not provide sufficient employment opportunities for the large cohort of young people who will be joining the labor force between now and 2035. Note that the young men and women who will be seeking their first jobs in 2035 have already been born. There must be a rapid diversification of the Iraqi economy to provide employment opportunities for the coming cohorts of young Iraqis.

EXOGENOUS DRIVERS OF 2025: REGIONAL CONFLICT

Iran and Syria

By the time this book is published, it is possible that regional circumstances will be very different. However, in late 2020, both of Iraq's neighbors were in crisis. Iran was fighting to remove sanctions imposed to get the country to abandon its nuclear ambitions. These sanctions, combined with misman- agement and corruption, have devastated the Iranian economy and fractured its polity. The adverse impact on Iran's petroleum exports has led to a severe economic recession and loss of confidence in the country, while helping to support a further decline in oil prices by reducing world supply. Syria is racked by a harsh government crackdown on "Arab Spring" groups.

So far, the difficulties in both countries have had a mixed impact on Iraq. As discussed in the previous chapter, both countries' currencies are losing value. This has resulted in lower prices for goods imported by Iraq and, by draining dollars from the Iraq economy, has contributed to the devaluation pressure on the Iraqi dinar. There is also smuggling of weapons and other supplies from western Iraq to Syrian rebels. In addition, some analysts think that the internal difficulties in both countries have forced their governments to focus on these issues, leaving less time to engage in trouble-making in Iraq.

However, if violence increases in either or both countries then the impact on Iraq will be adverse. If the Syrian conflict expands then it will probably spill over the border into the western provinces of Iraq. Iraq can expect both a wave of refugees and fighters seeking temporary sanctuary. On the eastern border, if Iran, the USA, and the other parties involved in the sanctions dispute misjudge each other then there is the possibility of a shooting war. This will not only reduce production from the major oilfields on the Iraq–Iran border but also temporarily close the Persian Gulf, stopping the bulk of Iraq's exports. Even after the US Navy reopens the Gulf to transit, it can be expected that Iraq oil will sell at a deeper discount to world prices to compensate for the increased risk.

In addition, conflict in Syria or Iran will strengthen the perception that Iraq is in a "bad neighborhood", leading to a decline in the willingness of foreign companies to invest in Iraq. As the Latin American debt crisis of the 1980s and the Asian financial crisis of the late 1990s showed, foreign investors tend to think regionally; they are not always careful to distinguish between countries in a region that are in difficulty and their neighbors that might be well managed.

Water Conflict

Both the Euphrates and the Tigris rivers originate in Turkey. While the Tigris then flows directly into Iraq, the Euphrates detours through Syria. As discussed in Chapter 8, while there are informal agreements on water sharing among the three countries, there is no formal treaty with the force of international law. As a result, the extensive dam building programs by both Turkey and Syria have already substantially reduced the cross-border water flow into Iraq, with the potential of even greater reductions over the next decade. If Iraq is unable to negotiate a water agreement with Syria and Turkey, a water shortage will lead to a decrease in agricultural production and increased migration into urban areas. In addition, it will be necessary to increase food imports. The GoI is aware of all of these external or exogenous challenges but, regrettably, its planning for the future is flawed.

GOI STRATEGIC PLANNING

Following its five-decade socialist tradition, the GoI continues to develop national development strategies or plans. In the post-2003 period, plans were published in 2005, 2007, 2010, 2013, and 2018. Over time, these documents have become increasingly comprehensive and detailed with growing participation.

While there is a large overlap among the proposed initiatives listed in these plans, the emphasis may change. For example, the National Development Strategy: 2005–2007 (GoI 2005b) was more private sector-oriented than the National Development Plan: 2010–2014 (GoI 2010). However, the differences among the various plans are generally more of emphasis than of kind. Therefore, it is curious that the later plans generally ignore the progress or lack thereof of the earlier ones. For example, the National Development Strategy:

2007–2010 (GoI 2007) fails to even mention the 2005 National Development Strategy, possibly because of the belief that foreigners dominated the writing of the 2005 document.

The National Development Plan: 2018–2022 (GoI 2018) begins by discussing challenges facing Iraq under four headings: institutional, economic, social, and environmental. This is followed by a list of 11 strategic goals. Each of these following ten chapters provides a useful description of the current situation in each economic sector, followed by specific policy objectives.

However, this most recent plan suffers from four weaknesses. The GoI (2018) plan fails to set priorities, or provide quantifiable goals for all of its initiatives, does not recognize the unique role of the private sector, and was overcome by events.

With respect to the first weakness, although the GoI (2018) plan lists many initiatives in different sectors of the Iraqi economy, there is unwillingness to set priorities. In the listing of 11 strategic goals there is no statement of which is the most important, unless the ordering of the objectives is intended to be meaningful (GoI 2018, p. 20). In fact, in the ten chapters of the plan, dozens of objectives are listed, but one looks in vain for a clear statement that this initiative is more important than that initiative. The proportion of text that is spent on various topics provides little insight into their relative importance to Iraqi development. For example, corruption is the most serious problem facing Iraq and yet is discussed in only a few paragraphs (ibid., p. 69) while there are nine pages discussing transportation initiatives (ibid., pp. 145–53). Improving the country's transportation network and reducing corruption are both worthy goals. However, a future Iraq that significantly reduces corruption but fails to improve its roads and railroads will be in much better shape than a future Iraq where the reverse is true.

Second, the GoI (2018) plan in some sections provides detailed quantifiable goals, and yet, in other sections it is satisfied with non-quantifiable exhortations to virtue. So, while the plan objectives for the energy sector are very specific, for example, Objective 2: "Increase the export capacity of crude oil to 5.25 million bpd" (ibid., p. 137), the plan objectives for improving the status of women in Iraq seem unnecessarily vague, for example, Objective 2: "Empower women economically" (ibid., p. 212).

This is not to say that Iraq's development plans should not be flexible enough to adjust to unexpected changes in domestic or foreign political and economic environments. However, in those sections of the GoI (2018) plan where there is little effort to either announce specific goals or document progress towards achieving these goals, one is forced to conclude that the GoI is not engaged in a serious planning process. Rather, in these sections its purpose seems to be to provide an overview of the current challenges facing Iraq, and present broad objectives to meet those challenges, along with general statements on the best means to achieve each objective by subordinate organizations. But, if advising subordinate organizations was the primary purpose of the National Development Plan: 2018–2022, then the failure to set priorities to guide subordinate organizations is an even greater failing.

The third difficulty with the most recent GoI economic planning exercise is that it simultaneously declares that a main pillar of the plan is: "Developing the private sector as a vital anchor for progress and development and a transformer of economic diversification" (ibid., p. vi), yet seeks to maintain a "government knows best" top-down setting of very specific economic goals. In other words, the private sector is welcome to provide its investment and managerial skills, but only to accomplish the goals delineated in detail by the GoI bureaucracy. This runs the risk of severely limiting private sector entrepreneurship, which is the primary source of economic development and growth (see Chapter 11; also Gunter 2012).

Finally, the 2018 National Development Plan was rapidly overcome by events. The improvement in world oil prices in 2018 was reversed in 2019 and 2020 (see Figure 7.2 in Chapter 7). This oil price fall, combined with continuing widespread corruption, means that most of the policy objectives are unrealistic since they assume reasonably honest government officials and substantial fiscal resources. Iraq currently has neither.

SEVEN KEY POLICY DECISIONS

In 2020, there were two detailed documents that discussed the challenges and possible solutions to what is the most severe economic crisis facing Iraq since the 2003 invasion. The Iraq Britain Business Council (IBBC), registered as an Iraq non-governmental organization, published *Iraq 2020: Country at the Crossroads* (Gunter et al. 2020); while several months later, the Emergency Cell for Financial Reforms of the GoI published its White Paper (GoI 2020b). As a generalization, *Iraq 2020* emphasized how economic reform should be accomplished, while the White Paper focused on what economic reforms should be accomplished.

The IBBC *Iraq 2020* study argued that: "the primary requirements for Iraq to successfully navigate the current chaotic situation are five.

First, the nation's leadership must publicly demonstrate – by actions as well as words – that they are dismantling the barriers between themselves and the Iraqi people" (Gunter et al. 2020, p. 1). As discussed by Renad Mansour, one of the co-authors of *Iraq 2020*, the current fault line in Iraqi politics is not between Sunni and Shi'a but rather between the average Iraqi and the ruling elite. (See also Haddad 2020, p. 6). Unless this fault line can be closed, little progress will be made.

Second, the political community must make difficult decisions in a timely manner. Muddling through is no longer an option.

Third, Iraq must deal with Covid-19 without crippling the existing non-oil economy.

Fourth, priority must be given to the creation of private sector jobs to reduce the economy's dependency on crude oil exports.

Finally, it is critical to deal with the current severe budget crisis without endangering economic growth and job creation. This paper didn't directly address the problems of corruption, physical security, and monetary/exchange rate policy since there are other organizations that are more capable of providing advice on these issues. (Gunter et al. 2020, p. 1)

The statement by the IBBC of the necessity of obtaining broad public support for any initiatives was echoed by the White Paper. It states: "implementation of such set of reforms require political courage, buy in and acceptance of the Iraqi people" (GoI 2020b, p. 2). The White Paper goes on to provide a valuable detailed description, analysis, and policy recommendations for achieving financial stability (ibid., pp. 42–6), employment creation (ibid., pp. 47–64), infrastructure (ibid., pp. 65–71), essential services (ibid., pp. 72–6), and improved governance (ibid., pp. 77–80).

Both of these works reward careful reading. However, among the many policy recommendations in these two works and the 2018 National Development Plan, which are the critical – the most important – policy recommendations? In other words, what are the few policies that, if successfully implemented by 2035, have a high probability of leading to a better quality of life for the average Iraqi even if little or no progress is made dealing with the many other challenges facing the country? I think there are seven: (1) passing the Oil Laws; (2) developing an effective anti-corruption policy; (3) reducing the regulatory hostility towards private business; (4) achieving 100 percent literacy of all Iraqis and 100 percent primary education of the young; (5) restructuring the financial system; (6) adopting a rational water pricing plan; and (7) advancing federalism. Each is discussed below, along with a proposed standard that will allow the measurement of progress towards achieving each goal.

1. Oil Law. Pass and implement the four major components of the Oil Law that have been awaiting approval since 2007. As discussed in Chapter 7, this law is needed not only to efficiently expand oil production but also to clarify the distribution of oil revenues among the different provinces as well as between the current and future generations of Iraqis.
2. Anti-corruption strategy. Develop and execute a multi-pronged anti-corruption strategy until Iraq reaches the top third of Transparency International's Corruption Perception Index. All of the planning doc-

uments note the devastating impact of Iraq's ubiquitous corruption on economic development and political stability. But then, with possibly one exception, they propose anti-corruption strategies that have failed in every country that has tried them. As discussed in Chapter 4, better governance, more arrests for corruption, and expensive publicity campaigns are necessary but not sufficient to achieve a long-term reduction in corruption. The few anti-corruption successes in the world have also included changing incentives to reduce the rents that reward corruption.

3. Regulatory reform. Reduce regulatory hostility towards private business until Iraq reaches the top third of the World Bank's "Ease of Doing Business" survey. When it comes to regulating the private sector; the GoI is simultaneously ubiquitous, weak, opaque, contradictory, arbitrary, and corrupt. As discussed in Chapter 11, Iraq not only provides the worst overall regulatory environment for private business in the Middle East and North Africa (MENA) countries, but in two subcategories it actually provides the worst regulatory environment in the world (Table 11.2). Unless there is a serious effort to reduce regulatory hostility towards private business, then the GoI's attempts to reduce dependence on oil by diversifying the economy will fail. Among the reasons for maintaining a hostile environment towards private businesses in Iraq is to prevent competition to state-owned enterprises (SOEs), which tend to be low-quality, high-cost producers. Closing or privatizing the SOEs might have been possible a decade ago, but that ship has sailed. The best option might be the Chinese solution of neither closing nor expanding SOEs, but letting growth of the private sector gradually reduce SOE dominance of production and employment.

4. Primary education and literacy. Educational resources should be directed towards literacy training for all ages and primary education for those younger than 12 years of age, until Iraq achieves 100 percent literacy among all Iraqis and 100 percent primary education among the young. The high level of illiteracy, especially among older Iraqis, those in rural areas, and women, not only slows economic development and facilitates corruption but also weakens democracy. As discussed in Chapter 3, achieving 100 percent literacy and 100 percent primary education will require not only more local schools but also more female teachers.

5. Financial liberalization. The December 2020 devaluation of the dinar should reduce the gap between the CBI auction rate and the parallel market rate to less than 1 percent and reduce the pressure for further devaluation for at least a year or two. However, in the absence of an unlikely large increase in oil prices or a dramatic reduction in fiscal expenditures that will restore fiscal balance, the pressure for devaluation will return. The GoI should seriously consider a return to the kind of currency board

that provided Iraq with decades of monetary stability. The regulation of financial intermediation should be rationalized until Iraq is in the top-third worldwide with respect to the "getting credit" subcategory of the World Bank's "Ease of Doing Business" survey. Finally, recapitalize state-owned banks until they meet Basel III standards. As discussed in Chapter 13, reform of the financial sector requires deliberate phasing in of initiatives to prevent an acceleration of capital flight or large loan losses among state-owned banks.

6.   Water pricing. Above a certain minimum water right – say 125 m³/yr per person – the government should price water at its average cost. For most of its history, Iraq was one of the few countries in the region with enough water to support large-scale agribusiness. However, the country's access to water is becoming increasingly precarious. Turkey, Syria, and Iran are diverting water from the Euphrates and Tigris rivers in order to generate electricity and provide irrigation water to their own farmers. At the same time, increased oil production and industrialization in Iraq will require increasing quantities of water, while pollution in the Euphrates and Tigris rivers continues to worsen, especially south of Baghdad. Ensuring adequate quantity and quality of water is a complex issue involving not only many entities in Iraq but also difficult negotiations with its neighbors. But without pricing water at its average cost, not only is water planning almost impossible, but also such planning leads to corruption and huge amounts of waste, because if something is free then people act as if it has zero cost.

7.   Iraqi federalism. Thirty percent of gross oil export earnings should be transferred to the provinces on the basis of population, to be spent solely according to the priorities set by provincial governments. Also, administration of business regulations, with the exception of the regulation of financial intermediaries, should be delegated to the provinces. Attempting to manage a 40 million person economy from Baghdad has been an exercise in inefficiency and corruption. With the geographic, economic, historical, and cultural differences across Iraq, a "one size fits all" strategy is not the best method for achieving rapid economic development. Because they are physically and emotionally closer to the average Iraqi, provincial and local governments are probably more responsive to their needs. Also, it can be expected that competition among provinces will lead to more creative solutions to the many challenges facing Iraq. Finally, provincial and local governments will provide training and experience for persons who might later enter the national government.

In most cases, the seven proposed policies reinforce each other. A reduction of the regulatory hostility towards the private sector tends to reduce corruption. Achieving 100 percent primary education will strengthen private business and should also reduce corruption. Passage of the Oil Laws, reduced corruption, reduced regulatory hostility, and improved banking regulation will encourage inward capital flows, thereby reducing pressure on the exchange rate.

There will be opposition to some of these initiatives on the grounds that they weaken GoI control of the economy. There is a tendency in Iraq to think that if the GoI does not do something then it will not be done. Among the GoI leadership, this attitude reflects a socialist mindset reinforced by a lack of experience with market operations. Anti-market attitudes among the Iraqi leadership are reinforced during discussions with advisors from foreign governments and international organizations who themselves generally favor detailed state control over the economy.

While this view of the primacy of the state in economic development was widely held among development specialists several decades ago, the collapse of the socialist states throughout the world, combined with the rapid acceleration of real growth and resulting higher living standards in countries that pursued economic liberalization, has led to its intellectual abandonment. Iraq should not try to build a twenty-first-century political economy guided by mid-twentieth-century knowledge.

My final thought is this. Current conditions are less auspicious than at any other time in the last two decades. More than in previous periods, the future of Iraq is in the hands of the Iraqis. Over the next decade, Iraq will make – or fail to make – critical irrevocable decisions. Rich countries with long histories of stable government can afford to make stupid decisions. Iraq cannot.

Bibliography

Abrol, I.P., J.S.P. Yadav, and F.I. Massoud (1988). *Salt-Affected Soils and Their Management*. Rome: FAO Land and Water Development Division, Bulletin 39.

Acemoglu, Daron and James A. Robinson (2006). Paths of Economic and Political Development. In Barry R. Weingast and Donald A. Wittman (eds), *The Oxford Handbook of Political Economy*. Oxford: Oxford University Press, pp. 673–92.

Acemoglu, Daron, James A. Robinson, and Simon Johnson (2001). The Colonial Origins of Comparative Development: An Empirical Investigation. *American Economic Review*, 91: 1369–401.

Acs, Zoltan J. (2006). How is Entrepreneurship Good for Economic Growth? *Innovations*, Winter: 97–107.

Ades, A. and R. Di Tella (1999). Rents, Competition and Corruption. *American Economic Review*, 89: 982–94.

AFP (2010). Iraq Signs Electricity Deal With Alstom. August 1.

Agnew, Clive and Ewan Anderson (1992). *Water Resources in the Arid Realm*. New York: Routledge.

Ahmad, Mahmood (2002). Agricultural Policy Issues and Challenges in Iraq: Short- and Medium-term Options. In Kamil Mahdi (ed.), *Iraq's Economic Predicament*. Reading: Ithaca, pp. 169–99.

Ajami, Fouad (2006). *The Foreigner's Gift: The Americans, The Arabs, and The Iraqis in Iraq*. New York: Free Press.

Ajrash, Kadhim and Dahlia Kholaif (2012). Iraq to Raise Crude Oil Production to 4 Million Barrels in 2013. *Bloomberg News*, February 23.

Al Janabi, Ahmed (2018). Client Alert: New Customs Rates to be Applied in Iraq as of 1 January 2018. *Amereller*. At: http://amereller.com/wp-content/uploads/2018/01/20180114-Client-Alert_Iraq_New_customs_rates.pdf#:~:text=The%20Iraq%20General%20Commission%20of%20Customs%20"IGCC"%29%20has,customs%20duty%20rates%20into%20smaller%20groups%20of%20categories.

Al Mawlawi, Ali (2018). Iraq's State-Owned Enterprises: A Case Study for Public Spending Reform. At: https://blogs.lse.ac.uk/crp/2018/10/18/iraqs-state-owned-enterprises/.

Ala'Aldeen, Dlawer (2020). Decentalization in Iraq: Process, Progress and a New Tailor-Made Model. *Middle East Research Institute*.

Ali, Abdullah Yusuf (1985). *The Holy Qur-an*. Medina: King Fahd Holy Qur-an Printing Complex.

Al Kafaji, Yass and Hameed Shukur Mahmood (2019). Iraq's Budgetary Practices Post US Invasion: A Critical Evaluation. *Public Money and Management*, 39(7): 478–85.

Allawi, Ali A. (2007). *The Occupation of Iraq: Winning the War, Losing the Peace*. New Haven, CT: Yale University Press.

Allawi, Ali A. (2009). *The Crisis of Islamic Civilization*. New Haven, CT: Yale University Press.

Allawi, Ali A. (2020). The Political Economy of Institutional Decay and Official Corruption – The Case of Iraq. *Iraqi Economists Network*, May 19.

Almounsor, Abdullah (2005). A Development Comparative Approach to Capital Flight: The Case of the Middle East and North Africa. In Gerald A. Epstein (ed.), *Capital Flight and Capital Controls in Developing Countries*. Cheltenham, UK and Northampton, MA, USA: Edward Elgar Publishing, pp. 234–61.

Alnasrawi, Abbas (1994). *The Economy of Iraq: Oil, Wars, Destruction of Development and Prospects, 1950–2010*. Westport, CT: Greenwood Press.

Alnasrawi, Abbas (2002). *Iraq's Burdens: Oil, Sanctions, and Underdevelopment. The Economy of Iraq: Oil, Wars, Destruction of Development and Prospects, 1950–2010*. Westport, CT: Greenwood Press.

Ansary, Khalid al- (2011). Iraqi Journalists Face Sacks of Gold, Fists of Fire. *Reuters*, January 26.

Ansary, Khalid al- and Kadhim Ajrash (2020). Iraq to Honor OPEC Oil Cut Decisions as Monthly Exports Rise. Bloomberg. At: https://www.msn.com/en-us/money/markets/iraq-to-honor-opec-oil-cut-decisions-as-monthly-exports-rise/ar-BB1bG234.

Ansary, Tamim (2009). *Destiny Disrupted*. New York: Public Affairs Books.

Arab News (2011). Iraq's Cabinet Approves $692 Million Cement Factory. June 16.

Arab Weekly (2019). TV Drama "Al Funduq" Stirs Controversy in Iraq. June 9. At: https://thearabweekly.com/tv-drama-al-funduq-stirs-controversy-iraq.

Arab Youth Survey (2020). *A Voice for Change*. Arabyouthsurvey.com.

Arango, Tim and Clifford Krauss (2012). Oil Output Soars as Iraq Retools, Easing Shaky Markets. *New York Times*, June 2.

Arraf, Jane (2009). Iraqi Voters Show Preference for Can-Do Over Creed. *Christian Science Monitor*, January 23.

Ashwarya, Sujata (2020). Iraq's Power Sector: Problems and Prospects. At: https://gjia.georgetown.edu/2020/01/13/iraqs-power-sector-problems-and-prospects/.

Askari, Hossein (2010). *Corruption and its Manifestation in the Persian Gulf*. Cheltenham, UK and Northampton, MA, USA: Edward Elgar Publishing.

Aswat al-Iraq (2010). China and France to Set Up Car and Truck Assembling Plants in Iraq. August 23. At: http://en.aswataliraq.info.

Ayubi, Nazih N. (1995). *Over-stating the Arab State*. New York: I.B. Tauris.

Banerjee, Abhijit V. (2000). Prospects and Strategies for Land Reform. In Boris Pleskovic and Joseph E. Stiglitz (eds), *Annual World Bank Conference on Development Economics*. Washington, DC: World Bank, pp. 253–72.

Bank for International Settlement (BIS) (2012). BIS Quarterly Review, June.

Bank for International Settlement (BIS) (2020). BIS Quarterly Review, June.

Bardhan, Pranab (1997). Corruption and Development: A Review of Issues. *Journal of Economic Literature*, 35(3): 1320–46.

Barkey, Henri J., Scott B. Lasensky, and Phebe Marr (eds) (2011). *Iraq: Its Neighbors and the United States*. Washington, DC: United States Institute of Peace.

Bauer, Peter T. and Basil S. Yamey (1957). *The Economics of Under-Developed Countries*. Chicago, IL: University of Chicago Press.

Baumol, William J. (1990). Entrepreneurship: Productive, Unproductive and Destructive. *Journal of Political Economy*, 98(5): 893–921.

Baumol, William J., Robert E. Litan, and Carl J. Schramm (2007). *Good Capitalism, Bad Capitalism, and the Economics of Growth and Prosperity*. New Haven, CT: Yale University Press.

Bayati, Hamid al- (2011). *From Dictatorship to Democracy*. Philadelphia, PA: University of Pennsylvania Press.

Benhaida, Sarah (2018). Conflict and Drought Ravage Iraq's Prized Data Palms. *Phys. org*, September 28.

Bennathan, Esra and Alan A. Walters (1979). *Port Pricing and Investment Policy*. New York: Oxford University Press.

Berdal, Mats and David Malone (eds) (2000). *Greed and Grievance: Economic Agendas in Civil Wars*. Boulder, CO: Lynn Rienner Publishers.

Black, Edwin (2004). *Banking in Baghdad*. Hoboken, NJ: John Wiley.

Black, Jeff (2009). Background: Key Provinces in Iraq Provincial Elections 2009. *Middle East News*, January 27.

Blanchard, Christopher M. (2018). Iraq: In Brief. Washington DC: Congressional Research Center. March 5.

Blas, Javier and Laura Hurst (2020). Cash-Strapped Iraq Seeks $2 Billion Upfront Payment for Oil. *Bloomberg*. At: https://www.msn.com/en-us/money/markets/cash-strapped-iraq-seeks-2-billion-upfront-payment-for-oil/ar-BB1bi4c2.

Blattman, Christopher and Edward Miguel (2010). Civil War. *Journal of Economic Literature*, 48(1): 3–57.

Bliss, Christopher and Rafael Di Tella (1997). Does Competition Kill Corruption? *Journal of Political Economy*, 105(5): 1001–23.

Bloomberg (2020). Iraq Set to Reduce Food Imports by Boosting Wheat Production. July 27.

Bobroff-Hajal, Anne (2006). Why Cousin Marriage Matters in Iraq. *Christian Science Monitor*. December 26.

Brookings Institution (2007, 2009, 2010, 2011, 2012). *Iraq Index*. At: http://www.brookings.edu/about/centers/saban/iraq-index.

Brunetti, Aymo and Beatrice Weder (2003). A Free Press is Bad News for Corruption. *Journal of Public Economics*, 87(7–8): 1801–24.

Casson, M.C. (1987). Entrepreneur. In John Eatwell, Murray Milgate, and Peter Newman (eds), *The New Palgrave Dictionary of Economics*. New York: Palgrave Macmillan, pp. 151–3.

Casson, M., B. Yeung, A. Basu, and N. Wadeson (eds) (2006). *The Oxford Handbook of Entrepreneurship*. Oxford, UK and New York, USA: Oxford University Press.

Catherwood, Christopher (2004). *Churchill's Folly: How Winston Churchill Created Modern Iraq*. New York: Carroll & Graf Publishers.

Center for International Private Enterprise (CIPE) (2008a). Business Attitudes Towards Political and Economic Reconstruction in Iraq. February 29, 1–38.

Center for International Private Enterprise (CIPE) (2008b). Iraq Business Owner Survey. Project 08081, February 29, 1–13.

Center for International Private Enterprise (CIPE) (2011). Attitudes on the Economy, Government, and Business Organizations: 2011 Iraqi Business Survey, Final Report. 1–49.

Central Bank of Iraq (CBI) (2004). Press Communique. At: www.cbi.iq/index.php?pid=Thecbi.

Central Bank of Iraq (CBI) (2012a). Iraqi and Foreign Financial Institutions. At: www.cbi.iq/index.php?pid=IraqFinancialinst.

Central Bank of Iraq (CBI) (2012b). Balance of Payments. At: www.cbi.iq/documents/bop.pdf.

Central Bank of Iraq (CBI) (2019). Annual Statistical Bulletin. At: https://cbi.iq/static/uploads/up/file-156740307438821.pdf.

Central Bank of Iraq (CBI) (2020a). Economic and Statistical Data. At: https://cbiraq.org.

Central Bank of Iraq (CBI) (2020b). *Financial Statements: 31 December 2019.* Baghdad: Central Bank of Iraq.

Central Organization for Statistics and Information Technology (COSIT) (2008). *Iraq Household Socio-Economic Survey: 2007.* Amman: National Press.

Central Organization for Statistics and Information Technology (COSIT) (2012). *Annual Abstract of Statistics: 2010–2011.* At: http://cosit.gov.iq/english/index.php.

Central Statistical Organization (CSO) (2014). *Iraqi Woman and Man in Statistics.* Baghdad: Central Statistical Organization.

Central Statistical Organization (CSO) (2019). *Industrial Large Establishments.* Baghdad: Central Statistical Organization.

Central Statistical Organization (CSO) (2020). *Hotels and Tourism Resorts Complexes for 2018.* Baghdad: Central Statistical Organization.

Chalabi, Fadhil (2002). The Oil Capacity of Post-War Iraq: Present Situations and Future Prospects. In Kamil A. Mahdi (ed.), *Iraq's Economic Predicament.* Reading: Ithaca Press, pp. 141–68.

Chattopadhyay, Raghabendra and Esther Duflo (2005). Women as Policy Makers: Evidence from a Randomized Experiment in India. In Gerald M. Meier and James E. Rauch (eds), *Leading Issues in Economic Development.* New York: Oxford University Press, pp. 284–91.

Chehade, Nadine (2016). 20 Years of Financial inclusion in the Arab World. At: https://www.cgap.org/blog/20-years-financial-inclusion-arab-world.

Chehade, Nadine, Peter McConaghy, and Teymour Abdel Aziz (2016). Towards a More Enabling Legal Framework for Microfinance in Iraq. CGAP Blog, January 21.

Chêne, Marie (2015). Successful Anti-Corruption Reforms. *Transparency International,* April 30, https://www.transparency.org/whatwedo/answer/successful_anti_corruption_reforms.

Cheung, S.N. (1996). Simplistic General Equilibrium Theory of Corruption. *Contemporary Economic Policy,* 14(3): 1–5.

China News Asia (2020). As Iraqis Rally Against Corruption, Ministries Up "For Sale". February 21.

Christen, Evan W. and Kasim Saliem (2013). Managing Salinity in Iraq's Agriculture: Report 1 Situation Analysis. International Center for Agriculture Research in Dry Areas.

Coles, Isabel and Ali Nabhan (2018). Worried About a Global Markets Meltdown: Try Iraqi Stocks. *Wall Street Journal,* April 1.

Collier, Paul (2000). *Economic Causes of Civil Conflict and their Implications for Policy.* Washington, DC: Development Research Group, World Bank.

Collier, Paul (2009). *Wars, Guns, and Votes: Democracy in Dangerous Places.* New York: Harper.

Commission of Integrity (2018a). Summary of Annual Report: 2017. Baghdad: Republic of Iraq. At: http://nazaha.iq/en_body.asp?id=1778.

Commission of Integrity (2018b). Anti-Corruption Roadmap. Baghdad. At: http://nazaha.iq/pdf_up/1844/Anti-corruption%20Roadmap-2018.pdf.

Commission of Integrity (2020). Summary of Annual Report: 2019. Baghdad. At: https://nazaha.iq/en_default.asp.

Cooper, John C.B. (2003). Price Elasticity of Demand for Crude Oil: Estimates for 23 Countries. OPEC. March.

Cooprider, Zoë, Merriam Mashatt, and James Wasserstrom (2007). State-Owned Enterprises: Post-Conflict Political Economy Considerations. USI Peace Briefing, March.

Cordesman, Anthony H. and Khalid R. al-Rodhan (2006). *The Changing Dynamics of Energy in the Middle East*, Vol. 2. Westport, CT: Praeger Security International.

Cordesman, Anthony H. and Grace Hwang (2020). *Strategic Dialogue: Shaping a US Strategy for the "Ghosts" of Iraq*. Draft. Center for Strategic and International Studies. June 8.

Cordesman, Anthony and Sam Khazai (2012). *Iraq After US Withdrawal: US Policy and the Iraqi Search for Security and Stability*. Washington, DC: Center for Strategic and International Studies.

Cordesman, Anthony H., Charles Loi, and Adam Mausner (2011). Iraq's Coming National Challenges: Economy, Demographics, Budget, and Trade. Center for Strategic and International Studies. Slides. January 5.

Crisis Group (2009). Iraq's Provincial Elections: The Stakes. Crisis Group Middle East Report No. 82. January 27.

Cuddington, J.T. (1987). Macroeconomic Determinants of Capital Flight: An Econometric Investigation. In Donald R. Lessard and John Williamson (eds), *Capital Flight and Third World Debt*. Washington, DC: Institute for International Economics, pp. 85–96.

Curran, Timothy W. (2006). *Success and Failure Factors for Iraq State Owned Enterprises*. Mimeo.

De Soto, Hernando (1989). *The Other Path: The Invisible Revolution in the Third World*. New York: Harper & Row.

De Soto, Hernando (2000). *The Mystery of Capital: Why Capitalism Triumphs in the West and Fails Everywhere Else*. New York: Basic Books.

Del Castillo, Graciana (2008). *Rebuilding War-Torn States: The Challenges of Post-Conflict Economic Reconstruction*. Oxford: Oxford University Press.

Desai, Sameeksha (2009). Measuring Entrepreneurship in Developing Countries. Research Paper No. 2009/10. United Nations University – World Institute for Development Economics Research. March, 1–12.

Desai, Sameeksha (2011a). Remittances from Iraqis in the US (conversation with the author).

Desai, Sameeksha (2011b). A Tale of Entrepreneurship in Two Iraqi Cities. *Journal of Small Business and Entrepreneurship*, 24(2): 283-92.

Desai, Sameeksha, Soltan J. Acs, and Utz Weitzel (2010). A Model of Destructive Entrepreneurship. Research Paper No. 2010/34. United Nations University – World Institute for Development Economics Research. April, 1–16.

Djankov, Simeon, Rafael La Porta, Florencio Lopez-di-Silanes, and Andrei Shleifer (2002). The Regulation of Entry. *Quarterly Journal of Economics*, 117(1): 1–34.

Dobbins, James, Seth G. Jones, Benjamin Runkle, and Siddharth Mohandas (2009). *Occupying Iraq: A History of the Coalition Provisional Authority*. Santa Monica, CA: Rand.

Dolatyar, Mostafa and Tim S. Gray (2000). *Water Politics in the Middle East: A Context for Conflict or Cooperation?* New York: St Martin's Press.

Downey, Morgan (2009) *Oil 101*. Albany, NY: Wooden Table Press.

Duflo, Esther (2001). Schooling and Labor Market Consequences of School Construction in Indonesia: Evidence from an Unusual Policy Experiment. *American Economic Review*, 91(4): 795–813.

Dunia Frontier Consultants (2009). Private Foreign Investment in Iraq. November.

Duvanova, Dinissa (2014). Economic Regulation, Red Tape, and Bureaucratic Corruption in Post-Communist Economies. *World Development*, 59(July): 298–312.

Easterly, William (2001). The Middle Class Consensus and Economic Growth. *Journal of Economic Growth*, 6(4): 318–26.

Easterly, William (2006). *The White Man's Burden*. New York: Penguin Books.

Economist Intelligence Unit (2012). Iraq. At: http://country.eiu.com/Iraq.

Efird, Neil (2010). The State Owned Enterprise as a Vehicle for Stability. US Army, Strategic Studies Institute.

Eichengreen, Barry (2000). Taming Capital Flows. *World Development*, 28(6): 1105–16.

El-Gamal, Mahmoud A., Nihal El-Megharbel, and Hulusi Inanoglu (2000). Beyond Credit: A Taxonomy of SMEs and Financing Methods for Arab Countries. Paper presented at ECES Workshop in Cairo, March 6–8.

Energy Information Administration (EIA) (2010). *Country Analysis Briefs: Iraq*. September.

Energy Information Administration (EIA) (2012). *Annual Energy Outlook: 2012*. June.

Energy Information Administration (EIA) (2019). *Annual Energy Outlook: 2019*. June.

Energy Information Administration (EIA) (2020). *Annual Energy Outlook: 2020*. June.

Estrin, Saul, Klaus E. Meyer, and Maria Bytchkova (2006). Entrepreneurship in Transition Economies. In Mark Casson, Bernard Young, Anuradha Basu, and Nigel Wadeson (eds), *The Oxford Handbook of Entrepreneurship*. Oxford: Oxford University Press, pp. 693–725.

Evans, Ray, Richard Soppe, Ed Barrett-Lennard, and Kasim Saliem (2013). Managing Salinity in Iraq's Agriculture: Report 2, Potential Solutions. International Center for Agriculture Research in Dry Areas.

Fairbanks, Waylon (2019). "We Have Very Little Time": MEES Sits Down with Iraq's Electricity Minister. MEES. At: https://www.mees.com/2019/2/15/power-water/we-have-very-little-time-mees-sits-down-with-iraqs-electricity-minister/63226520-3147-11e9-88ab-b16d8ba91504

Farole, Thomas and Gokhan Akinci (eds) (2011). *Special Economic Zones: Progress, Emerging Challenges and future Directions*. Washington, DC: World Bank.

Faruqi, Anwar (2011). Fuel Theft Hits Iraq Power Grid: Inspector. *Agence France-Presse*, June 23.

Fathallah, Hadi (2020). Iraq's Governance Crisis and Food Insecurity. Carnegie Endowment for International Peace, June 4.

Food and Agricultural Organization of the United Nations (FAO) (2012a). *Faostat: Iraq*. At: http://faostat3.fao.org/home/index. html#VISUALIZE_BY_AREA_IRAQ.

Food and Agricultural Organization of the United Nations (FAO) (2012b). *Aquastat: Iraq*. http://www.fao.org/nr/water/aquastat/ counties_regions/IRQ/Index.stm.

Food and Agricultural Organization of the United Nations (FAO) (2020a). Iraq: Statistical Yearbook.

Food and Agricultural Organization of the United Nations (FAO) (2020b). Iraq: Revised Humanitarian Response: Coronavirus Disease 2019 (Covid-19).

Food and Agricultural Organization of the United Nations (FAO) (2020c). Country Brief: Iraq. May 15.

Food and Agricultural Organization of the United Nations (FAO) (2020d). FAOStat: Iraq. At: http://www.fao.org/faostat/en/#data/QC.

Freedom House (2021). Freedom in the World 2021/Iraq. At: https://freedomhouse.org/country/iraq/freedom-world/2021.

Frye, Timothy and Andrei Shleifer (1997). The Invisible Hand and the Grabbing Hand. *American Economic Review*, 87(2): 354–8.

Fukuyama, Francis (2011). *The Origins of Political Order: From Prehuman Times to the French Revolution*. New York: Farrar, Straus & Giroux.

Fund for Peace (2020). Fragile States Index. At: https://fragilestatesindex.org/country-data/.

Galal, Ahmed, Leroy Jones, Pankaj Tandon, and Ingo Vogelsang (1994). *Welfare Consequences of Selling Public Enterprises: An Empirical Analysis*. New York: Oxford University Press.

Galiani, Sebastian, Paul Gertler, and Ernesto Schargrodsky (2006). Water for Life: The Impact of the Privatization of Water Services on Child Mortality. At: www.iadb.org/res/publications/pubfiles/pubS-233.pdf.

Gallup (2010). Religiosity Highest in World's Poorest Nations. August 31. At: www.gallup.com/poll/142727/Religiosity-Highest-World-Poorest-Nations.aspx.

Gimbel, Barney (2007). In Iraq, One Man's Mission Impossible. *Fortune*. September 4.

Glaeser, Edward, Rafael La Porta, Florencio Lopez-de-Silanes, and Adnrei Shleifer (2004). Do Institutions Cause Growth? *Journal of Economic Growth*, 9(3): 271–303.

Global Findex (2017). The Global Findex Database. At: https://openknowledge.worldbank.org/bitstream/handle/10986/29510/211259ov.pdf.

Godolphin, Francis R.B. (1942). *The Greek Historians*, Vol. 1, Book VI. New York: Random House.

Gokoluk, Selcuk and Marton Eder (2020). Iraq Budget Battle Looms as Saddam-Era Bond Plan Falters. Bloomberg. May 7. At: https://www.bloombergquint.com/global-economics/iraq-premier-faces-budget-battle-as-saddam-era-bond-plan-falters.

Goldman, David P. (2011). *How Civilizations Die*. Washington, DC: Regency Publishing.

Government of Iraq (GoI) (2005a). Constitution. At: http://www.uniraq.org/documents/iraqi.constitution.pdf.

Government of Iraq (GoI) (2005b). *National Development Strategy: 2005–2007*. Baghdad: Government of Iraq.

Government of Iraq (GoI) (2007). *National Development Strategy: 2007–2010*. Baghdad: Government of Iraq.

Government of Iraq (GoI) (2010). *National Development Plan: 2010–2014*. Baghdad: Government of Iraq.

Government of Iraq (GoI) (2012). *Industrial Strategy 2030*. At: https://www.academia.edu/42242147/Industrial_strategy_for_Iraq_until_2030_and_implementation_mechanisms الاستراتيجية_الصناعية_للعراق_حتى_عام_2030_وآليات التنفيذ.

Government of Iraq (GoI) (2013). *National Development Plan: 2013–2017*. Baghdad: Government of Iraq.

Government of Iraq (GoI) (2016). Performance and Fiscal Risks from Non-Financial State-Owned Enterprises in the Republic of Iraq. Baghdad: Government of Iraq. December.

Government of Iraq (GoI) (2018). National Development Plan: 2018–2022. Baghdad: Government of Iraq.

Government of Iraq (GoI) (2020a). Iraq's Parliament Approves Government Program. Baghdad: Government of Iraq.

Government of Iraq (GoI) (2020b). White Paper: Final Report. Emergency Cell for Financial Reforms, October.

Grigorian, David A. and Udo Kock (2010). Inflation and Conflict in Iraq: The Economics of Shortages Revisited. IMF Working Paper WP/10/159.

Gunter, Frank R. (1991). Thomas Jefferson on the Repudiation of Public Debt. *Constitutional Political Economy*, 2(3): 283–301.

Gunter, Frank R. (2004). Capital Flight from China: 1984–2001. *China Economic Review*, 15(1): 63–85.

Gunter, Frank R. (2007). Economic Development during Conflict: The Petraeus–Crocker Congressional Testimonies. *Strategic Insights*, 6(6): 1–15.

Gunter, Frank R. (2008a). Capital Flight. In William A. Darity, Jr (ed.), *International Encyclopedia of the Social Sciences*. Detroit, MI: Macmillan Reference, pp. 434–6.

Gunter, Frank R. (2008b). Corruption. In Vincent N. Parrillo (ed.), *Encyclopedia of Social Problems*. New York City: SAGE Publications, pp. 173–6.

Gunter, Frank R. (2009a). Liberate Iraq's Economy. *New York Times*, November 16, A25.

Gunter, Frank R. (2009b). Microfinance During Conflict: Iraq, 2003–2007. In Todd A. Watkins and Karen Hicks (eds), *Moving Beyond Storytelling: Emerging Research in Microfinance*. Bingley: Emerald Group Publishing, pp. 183–214.

Gunter, Frank R. (2012). A Simple Model of Entrepreneurship for Principles of Economics Courses. *Journal of Economic Education*, 43(4): 386–96.

Gunter, Frank R. (2013). *The Political Economy of Iraq: Restoring Balance in a Post-Conflict Society*. Cheltenham, UK and Northampton, MA, USA: Edward Elgar Publishing.

Gunter, Frank R. (2015). ISIS and Oil: Iraq's Perfect Storm. In Tally Helfont (ed.), *The Best of FPRI's Essays on The Middle East: 2005–2015*. Philadelphia, PA: Foreign Policy Research Institute, pp. 157–64.

Gunter, Frank R. (2018a). Immunizing Iraq Against al Qaeda 3.0. *Orbis*, Summer: 389–408.

Gunter, Frank R. (2018b). *Rebuilding Iraq's Public Works Infrastructure Following the Defeat of ISIS*. Foreign Policy Research Institute.

Gunter, Frank R. (2020). Iraq: Asking the Right Questions About Civil Disorder. Foreign Policy Research Institute.

Gunter, Frank R., Mohammed al Uzri, Renad Mansour, Hani Akkawi, Hussein al Uzri, Shwan Aziz Ahmed, and Christophe Michels (2020). *Iraq 2020: Country at the Crossroads*. London: Iraq Britain Business Council.

Gutman, Roy (2011). Dysfunctional Banking Sector Helps Keep Iraq in Economic Shambles. *Kansas City Star*, December 25.

Gvenetadze, Koba and Amgad Hegazy (2015). Iraq: Selected Issues. International Monetary Fund, August, 15–29.

Haddad, Fanar (2011). *Sectarianism in Iraq: Antagonistic Visions of Unity*. New York: University of Chicago.

Haddad, Fanar (2019). The Waning Relevance of the Sunni–Shia Divide: Receding Violence Reveals the True Colours of "Sectarianism" in Iraqi Politics. The Century Foundation, April 10. At: https://tcf.org/content/report/waning-relevance-sunni-shia -divide/?agreed=1.

Haddad, Fanar (2020). *Understanding "Sectarianism": Sunni–Shi'a Relations in the Modern Arab World*. London, UK and New York, USA: Hurst and Oxford University Press.

Hafedh, Mehdi, Ibrahim Akoum, Imad Zbib, and Zafar Ahmed (2007). Iraq: Emergence of a New Nation from the Ashes. *International Journal of Emerging Markets*, 2(1): 7–21.

Hafidh, Hassan (2011). Iraq Tackles Its Next Oil Bottleneck. *Wall Street Journal*, April 25.

Haiss, Peter, Katharina Steiner, and Markus Eller (2005). How do Foreign Banks Contribute to Economic Development in Transition Economies – How Much do we Know about Challenges and Opportunities? IMDA. At: http://fgr.wu-wien.ac.at/institut/ef/nexus.html.

Hall, Peter A. and David Soskice (eds) (2001). *Varieties of Capitalism: The Institutional Foundations of Comparative Advantage*. Oxford: Oxford University Press.

Hamilton, Alexander (2020a). *The Political Economy of Economic Policy in Iraq*. London School of Economics: Middle East Center. Series 32. March.

Hamilton, Alexander (2020b). *Is Demography Destiny? The Economic Implications of Iraq's Demography*. London School of Economics: Middle East Center. Series 41. November.

Hanke, Steve H. (2002). On Dollarization and Currency Boards: Error and Deception. *Policy Reform*, 5(4): 203–22.

Hanke, Steve H. and Kurt Schuler (1994). *Currency Boards for Developing Countries: A Handbook*. San Francisco, CA: ICS Press.

Hanushek, Eric A. (2005). Interpreting Recent Research on Schooling in Developing Countries. In Gerald M. Meier and James E. Rauch (eds), *Leading Issues in Economic Development* (8th edn). New York: Oxford University Press, pp. 201–5.

Harris, John R. (1970). Some Problems in Identifying the Role of Entrepreneurship in Economic Development: The Nigerian Case. *Explorations in Economic History*, 7(2): 347–69.

Hayes, Brian (2005). *Infrastructure: A Field Guide to the Industrial Landscape*. New York: W.W. Norton.

Helfont, Tally (ed.) (2015). *The Best of FPRI's Essays on The Middle East: 2005–2015*. Philadelphia, PA: Foreign Policy Research Institute.

Hertog, Steffen (2010). Defying the Resource Curse: Explaining Successful State-Owned Enterprises in Rentier States. *World Politics*, 62(2): 261–301.

Hettige, Hemamala, Mainul Huq, Sheoli Pargal, and David Wheeler (1996). Determinates of Pollution Abatement in Developing Countries: Evidence from South and Southeast Asia. *World Development*, 24(12): 1891–1904.

Hirmis, Amer K. (2018). *The Economics of Iraq: Ancient Past to Distant Future*. Surbiton: Grosvenor Publishing.

Hofstede Insights (2020) Country Comparison: Iraq. At: https://www.hofstede-insights.com/country-comparison/iraq/.

Huntington, Samuel P. (1968). *Political Order in Changing Societies*. New Haven, CT: Yale University Press.

Huntington, Samuel P. (1991). *The Third Wave: Democratization in the Late Twentieth Century*. Norman, OK: University of Oklahoma Press.

Hussein, Aqeel and Colin Freeman (2007). US to Reopen Iraq's Factories in $10 million U-Turn. *Sunday Telegraph*, January 29.

Hyne, Norman J. (2001). *Nontechnical Guide to Petroleum Geology, Exploration, Drilling, and Production*. Tulsa, OK: Penn Well Corporation.

Ibn Khaldun ([1384] 1967). *The Muqaddimah: An Introduction to History*. Princeton, NJ: Princeton University Press.

IEA (2019). Iraq's Energy Sector: A Roadmap to a Brighter Future. At: https://www.iea.org/reports/iraqs-energy-sector-a-roadmap-to-a-brighter-future.

Integrated Regional Information Network (IRIN) (2010a). Iraq: Iraqis Welcome WFP Role in State Food Aid System. January 6. At: http:// www.wfp.org/content/IRAQ -IRAQIS-WELCOME-WFP-ROLE-STATE-FOOD-AID-SYSTEM.

Integrated Regional Information Network (IRIN) (2010b). Iraq: Streamlining the State Food Aid System. February 9. At: http://www.irinnews.org/Report/88041/IRAQ -STREAMLING-THE-STATE-FOOD-AID-SYSTEM.

Integrated Regional Information Network (IRIN) (2010c). Iraq: Killing for Water. June 23. At: http://www.irinnews.org/Report/89586/IR AQ-KILLING-FOR-WATER.

International Labour Organization (ILO) (2019) ILOStat Iraq. At: https://www.ilo.org/ global/research/global-reports/weso/2019/WCMS_670542/lang--en/index.htm.

International Monetary Fund (IMF) (2005). *Iraq – Request for Stand-By Arrangement.* Washington, DC: International Monetary Fund. December 8.

International Monetary Fund (IMF) (2008a). *Iraq: First Review Under the Stand-by Arrangement and Financing Assurances Review.* IMF Country Report No. 08/303. Washington, DC: International Monetary Fund.

International Monetary Fund (IMF) (2008b). *Iraq: Second Review Under the Stand-by Arrangement and Financing Assurances Review.* IMF Country Report No. 08/383. Washington, DC: International Monetary Fund.

International Monetary Fund (IMF) (2010a). *Regional Economic Outlook: Middle East and Central Asia.* Washington, DC: International Monetary Fund.

International Monetary Fund (IMF) (2010b). *Iraq: Staff Report for the 2009 Article IV Consultation and Request for Stand-By Arrangement.* Washington, DC: International Monetary Fund.

International Monetary Fund (IMF) (2010c). *Annual Report on Exchange Arrangements and Exchange Restrictions: 2010.* Washington, DC: International Monetary Fund.

International Monetary Fund (IMF) (2011a). *Iraq: Second Review Under the Stand-By Arrangement, Requests for Waiver of Applicability, Extension of the Arrangement, and Rephasing of Access.* IMF Country Report No. 11/75. Washington, DC: International Monetary Fund. March.

International Monetary Fund (IMF) (2011b). *Iraq IMF Country Report.* No. 11/7. Washington, DC: International Monetary Fund.

International Monetary Fund (IMF) (2015). *Staff Report for the 2015 Article IV Consultation.* Washington, DC: International Monetary Fund. July.

International Monetary Fund (IMF) (2017). *Iraq: Selected Issues.* Washington, DC: International Monetary Fund. August.

International Monetary Fund (IMF) (2019a). *Iraq: 2019 Article IV Consultation.* Washington, DC: International Monetary Fund.

International Monetary Fund (IMF) (2019b). *Iraq: Selected Issues.* IMF Country Report, No. 19/249. Washington, DC: International Monetary Fund. July.

International Monetary Fund (IMF) (2020a). *Direction of Trade Statistics: Iraq.* At: https://data.imf.org/regular.aspx?key=61013712.

International Monetary Fund (IMF) (2020b). *Balance of Payments Statistics.* At: https://data.imf.org/?sk=7A51304B-6426-40C0-83DD-CA473CA1FD52&sId= 1542635306163.

International Organization for Migration (IOM) (2020). Impact of Covid-19 on SMEs in Iraq. At: https://iraq.iom.int/publications/impact-covid-19-smes-iraq.

Iraq Business News (2019). WFP Supports Iraq in Modernizing its Public Distribution System. January 13.

Iraq Partners Forum (2010). *The Iraq Briefing Book.* Washington, DC: Iraq Partners Forum. December.

Iraq Stock Exchange (ISX) (2012). Subscribed Shares and Closing Price, Market Capitalization, and Turnover Ratio for the Listed Companies in ISX. At: http://www .isx-iq.net/isxportal/portal/sectors Details.html.

Iraq Stock Exchange (ISX) (2020). Subscribed Shares and Closing Price, Market Capitalization, and Turnover Ratio for the Listed Companies in ISX. At: http://www .isx-iq.net/isxportal/portal/sectorsDetails.html.

Integrated Regional Information Network (IRIN) (2009). Iraq: HIV-positive Persons Fear Reprisals. News Release of UN Office for the Coordination of Humanitarian Affairs. January 14. At: http://www.irinnews.org/Report/82357/IRAQ-HIV -POSITIVE-PERSONS-FEAR-REPRISALS.

Issawi, Charles (1966). *The Economic History of the Middle East: 1800–1914.* Chicago, IL: University of Chicago Press.

Issawi, Charles (1982). *An Economic History of the Middle East and North Africa.* New York: Columbia University Press.

Issawi, Charles (1988). *The Fertile Crescent: 1800–1914.* New York: Oxford University Press.

Jadara (2020). *Iraq – Tragedy to Catharsis.* Sept 28. https://jadarablog.com/tag/ catharsis/.

Jawari, Hadeel al- (2009). Iraq Government Hit by Graft: 4,000 Forged University Degrees Uncovered. *Azzaman.* At: http://www.assaman.com.

Johnson, Kirk A. (2007). Understanding the Unemployment and Underemployment Situation in Iraq (unpublished).

Joint Anti-Corruption Council (2010). The National Anti-Corruption Strategy: 2010–2014. Baghdad. http://www.track.unodc.org/LegalLibrary/LegalResources/ IraqAuthorities/National%20Anti-Corruption%20Strategy%202010-2014.pdf.

Jurewitz (1987). The Deregulation of Electricity. Working Paper.

Kami, Aseel and Yara Bayoumy (2012). Banks that Can't Cash a Cheque Slow Iraqi Economy. *Reuters,* March 7.

Kami, Aseel and Serena Chaudhry (2011). Iraqi Banks Struggle with Limited Services, Capital. *Reuters,* July 20.

Kane, Sean (2010). Iraq's Oil Politics: Where Agreement Might Be Found. *Peaceworks,* 64, January.

Kaufmann, Daniel, Aart Kraay, and Massimo Mastruzzi (2003). Governance Matters III: Governance Indicators for 1996–2002. World Bank Policy Research Working Paper No. 3106. Washington, DC: World Bank.

Khaleej Times Online (2011). Iraq Cancels Direct Cash Purchases of Food Items. May 30. At: http://iraqdailytimes.com/iraq-cancels-direct-cash-buying-of-essential-food.

Khaleej Times Online (2012). Iraq Seeks DED Expertise. June 27. At: http:// www.khaleejtimes.com/biz/inside.asp?xfile=/data/bankingfinance/2012/June/ bankingfinance_june24.xm/&section=bankingfinance.

Kirzner, I.M. (1979). *Perception, Opportunity, and Profit.* Chicago, IL: University of Chicago Press.

Klein, Matthias S. (2006). Development of the Iraqi Public Pension System. *Global Action on Aging.* September 8. At: http://www.globalaging.org/pensions/world/ 2006/iraqenglish.htm.

Klitgaard, Robert (1988). *Controlling Corruption.* Berkley, CA: University of California Press.

Knack, Stephen and Philip Keefer (1995). Institutions and Economic Performance: Cross-Country Tests Using Alternative Institutional Measures. *Economics and Politics,* 7(3): 207–27.

Knight, Frank H. ([1921] 1971). *Risk, Uncertainty and Profit*. Chicago, IL: University of Chicago Press.

Krishnan, Nandini, Sergio Olivieri, and Racha Ramadan (2019). Estimating the Welfare Costs of Reforming the Iraq Public Distribution System. *Journal of Development Studies*, 55(S1): 91–106.

Krueger, Anne (1974). The Political Economy of the Rent Seeking Society. *American Economic Review*, 64(3): 291–302.

Kuran, Timur (2004). *Islam and Mammon: The Economic Predicaments of Islamism*. Princeton, NJ: Princeton University Press.

Kuran, Timur (2010). The Scale of Entrepreneurship in Middle Eastern History: Inhibitive Roles of Islamic Institutions. In David S. Landis, Joel Mokyr, and William J. Baumol (eds), *The Invention of Enterprise*. Princeton, NJ: Princeton University Press, pp. 62–87.

Kuran, Timur (2011). *The Long Divergence: How Islamic Law Held Back the Middle East*. Princeton, NJ: Princeton University Press.

La Porta, Rafael, Florencio Lopez-de-Silanes, Andrei Shleifer, and Robert Vishny (1999). The Quality of Government. *Journal of Law, Economics and Organization*, 15(1): 222–79.

Lambsdorff, Johann G. (2006). Causes and Consequences of Corruption: What do we Know from a Cross-Section of Countries? In Susan Rose-Ackerman (ed.), *International Handbook on the Economics of Corruption*. Cheltenham, UK and Northampton, MA, USA: Edward Elgar Publishing, pp. 3–51.

Landes, David (1998). *The Wealth and Poverty of Nations*. New York: W.W. Norton.

Latif, Hammam (2020). Kadhimi Going After Big Fish in Anti-Corruption Crackdown. *Arab Weekly*, September 19. At: https://thearabweekly.com/kadhimi-going-after-big-fish-anti-corruption-crackdown.

Lawler, Alex (2020). Iraq Increases Oil Exports in July, Pumps Above OPEC+ Target. *Reuters*, July 30.

Lawless, R.I. (1972). Iraq: Changing Population Patterns. In J.I. Clark and W.B. Fisher (eds), *Populations of the Middle East and North Africa*. New York: Africana Publishing Corporation, pp. 97–129.

Leite, C. and J. Weidemann (1999). Does Mother Nature Corrupt? Natural Resources, Corruption and Economic Growth. International Monetary Fund Working Paper, 99/85. July.

Levinson, Micah (2020). Iraq. *Middle East Insider*, July 31.

Lewis, Bernard (2002). *What Went Wrong?* New York: Oxford University Press.

Lingelbach, David, Lynda de la Viña, and Paul Asel (2005). What's Distinctive About Growth-Oriented Entrepreneurship in Developing Countries? At: http://papers.ssrn.com/sol3/papers.cfm?abstract_id=742605.

Looney, Robert E. (2006). Economic Consequences of Conflict: The Rise of Iraq's Informal Economy. *Journal of Economic Issues*, 60(4): 991–1007.

Looney, Robert E. (2008). Reconstruction and Peacebuilding Under Extreme Adversity: The Problem of Pervasive Corruption in Iraq. *International Peacekeeping*, 15(3): 424–40.

Machiavelli, Niccolo (1988). *Florentine Histories*. Trans. Laura F. Banfield and Harvey C. Mansfield, Jr. Princeton, NJ: Princeton University Press.

Maddison, Angus (2003). *The World Economy: Historical Statistics*. Paris: OECD.

Maddison, Angus (2007). *Contours of the World Economy, 1–2030 AD*. New York: Oxford University Press.

Mahdi, Kamil A. (2000). *State and Agriculture in Iraq: Modern Development, Stagnation and the Impact of Oil*. Reading: Ithaca Press.

Mahdi, Kamil A. (ed.) (2002). *Iraq's Economic Predicament*. Reading: Ithaca.

Mahdi, Kamil A. (2007). Iraq's Oil Law: Parsing the Fine Print. *World Policy Institute*, June: 11–23.

Mardini, Ramzy (2020). Preventing the Next Insurgency: A Pathway for Reintegrating Iraq's Sunni Population. In Aaron Stein (ed.), *Iraq in Transition: Competing Actors and Complicated Politics*. Philadelphia, PA: Foreign Policy Research Institute, pp. 77–97.

Marr, Phebe and Ibrahim al Marashi (2017). *The Modern History of Iraq*, 4th edn. Boulder, CO: Westview Press.

Mauro, Paolo (1995). Corruption and Growth. *Quarterly Journal of Economics*, 110(3): 681–712.

McMillan, John and Christopher Woodruff (2002). The Central Role of Entrepreneurs in Transition Economies. *Journal of Economic Perspectives*, 16(3): 153–70.

Meagher, Patrick (2005). Anti-corruption Agencies: Rhetoric Versus Reality. *Journal of Policy Reform*, 8: 69–103.

Meese, Michael J. (2007). Electrical Frequency and Underfrequency Load-Shedding. Working Paper. Baghdad.

Megginson, William L. and Jeffrey M. Netter (2001). From State to Market: A Survey of Empirical Studies on Privatization. *Journal of Economic Literature*, June, 39: 321–89.

Merza, Ali (2008). Policies and Economic and Social Trends in Iraq: 2003–2007. Paper presented at International Association of Contemporary Iraqi Studies, July 16–17.

Merza, Ali (2012). Budget 2012: Financial, Economic and Institutional Issues in Iraq. Memo, January 26.

Microfinance Information Exchange (2018). *Global Outreach and Financial Performance Benchmark Report – 2017–2018*. At: www.themix.org.

Ministry of Finance (MoF) (2005). Joint letter with Director of Central Bank of Iraq to International Monetary Fund. December 6.

Ministry of Finance (MoF) (2020). *Open Budget Survey*. At: http://www.mof.gov.iq/obs/en/pages/default.aspx.

Ministry of Finance and Ministry of Planning (2003). Iraq Budget 2003: State-Owned Enterprises, July–December.

Ministry of Industry and Minerals (2009). Master Strategic Plan. Ministry of Industry and Minerals Study.

Ministry of Oil (2012). Crude Oil Exports and Domestic Consumption. Baghdad: Ministry of Oil. At: http://www.oil.gov.iq/moo/.

Ministry of Oil (2020). Ministry of Oil Website. At: https://oil.gov.iq/upload/2611240027.jpg.

Monitor, al- (2019). Iraqi President Gives Priority to Job Opportunities for Youth, Calls Corruption "Political Economy of Conflict". September 28. At: https://www.al-monitor.com/originals/2019/09/iraq-president-barham-salih-unga-jobs-youth-corruption.html#:~:text=Salih%3A%20Corruption%20is%20the%20political%20economy%20of%20conflict.,terrorism%20cannot%20be%20sustained%20without%20funding%20and%20financing.

Mookherjee, Dilip and I.P.L. Png (1995). Corruptible Law Enforcers: How Should They Be Compensated? *Economic Journal*, 105(428): 145–59.

Mostafa, Mohamed (2018). Five Million Tourists Visit Iraq Annually. *Iraqi News*. At: https://www.iraqinews.com/features/five-million-tourists-visit-iraq-annually-govt/.

Murphy, Kevin, Andrei Shleifer, and Robert V. Vishny (1993). Why is Rent Seeking So Costly to Growth? *American Economic Review*, 83(2): 409–14.

Müller, Quentin and Sebastian Castelier (2018). Drought, Dams, and Dry Rivers: Iraqi Farmers are Giving Up Hope. *Middle East Eye*, September 18.

Nagl, John and Daniel Rice (2009). A Jump Start for Iraq's Private Sector. *Wall Street Journal*, July 7.

National Iraqi News Agency (2011). Telecommunications. March 15. At: http://www .ninanews.com/english.

National US–Arab Chamber of Commerce (NUSACC) (2011). US Exports to the Arab World to Reach $117 Billion by 2013. February. At: www.nusacc.org/assets/library/ 12_trdln0211outlook13a.pdf.

Newburger, Emma (2020). Trump Administration Warns Iraq Could Lose New York Fed Account if US Troops Forced to Leave. *CNBC*, January 11. At: https://www .cnbc.com/2020/01/11/trump-administration-warns-iraq-could-lose-new-york-fed -account-wsj.html.

Niqash (2011a). Iraq Gets Tough on Fake Qualifications, Up to 50,000 Jobs at Risk. April 11. At: www.niqash.org/articles/?id=2821.

Niqash (2011b). Basra's Big Business: Fake Building Contractors Defaulting on Jobs. July 20. At: www.niqash.org/articles/?id=2866.

Noland, Marcus and Howard Pack (2007). *The Arab Economies in a Changing World*. Washington, DC: Peterson Institute for International Economics.

North, Douglass C. (1990). *Institutions, Institutional Change and Economic Performance*. New York: Cambridge University Press.

Oldfield, Jackson (2017). Overview of National Approaches to Anti-Corruption Packages. *Transparency International*, May 23.

Olivier, G.A. (1988). Trade of Iraq, 1790s. In Charles Issawi (ed.), *The Fertile Crescent: 1800–1914*. New York: Oxford University Press, pp. 178–82.

Oman Tribune (2011). Iraq Begins Pumping Oil Through Al Ahdab. July 24. At: http://omantribune.com/index.php?page=new&id=97092&heading=news%20in %20detail.

Omer, Tara Mohamed Anwar (2011). Country Pasture/Forage Resource Profile. Rome: Food and Agricultural Organization of the UN.

Omer, Tara Mohamed Anwar (2020). OPEC Monthly Oil Market Report. July 14.

Organization of Petroleum Exporting Countries (OPEC) (2009). *OPEC Annual Statistical Bulletin*. At: www.opec.org/opec_web/static_files_ project/media/down-loads/publications/ASB2009.pdf.

Organization of Petroleum Exporting Countries (OPEC) (2020). *OPEC Monthly Oil Market Report*. July 14. At: https://www.opec.org/opec_web/en/publications/338 .htm.

Owen, Roger and Şevket Pamuk (1999). *A History of Middle East Economies in the Twentieth Century*. Cambridge, MA: Harvard University Press.

Oxfam International (2007). Rising to the Humanitarian Challenge in Iraq. At: http:// news.bbc.co.uk/2/shared/bsp/hi/pdfs/18_07_07_oxfam_iraq.pdf.

Ozlu, Onur (2006). *Iraqi Economic Reconstruction and Development*. Washington, DC: Center for Strategic and International Studies.

Parker, Ned and Usama Redha (2009). Last Minute Campaigning in Najaf. *Los Angeles Times*, January 31.

Parker, Simon C. (2009). *The Economics of Entrepreneurship*. New York: Cambridge University Press.

Parkinson, C. Northcote (1957). *Parkinson's Law and Other Studies in Administration.* London: Houghton Mifflin.

Patai, Raphael (2002). *The Arab Mind.* New York: Hatherleigh Press.

Pearson, Bryan (2009). Theater Groups Reappear in Iraq. *Variety*, January 7.

Persson, Torsten and Guido Tabellini (2004). Constitutions and Economic Policy. *Journal of Economic Perspectives*, 18(1): 75–98.

Pickthall, Marmaduke ([1930] 1992). *The Glorious Koran.* New York: Alfred A. Knopf.

Planning, Ministry of (2006). SOE's Economic Reform Commission. December 26.

Planning, Ministry of and Ministry of Finance (2004). Iraq Budget 2003, July–December, State-Owned Enterprises.

Pollack, Kenneth M. (2012). Reading Machiavelli in Iraq. *National Interest*, November/December. At: http://nationalinterest.org/article/ reading-machiavelli-iraq-7611.

Previté-Orton, C.W. (1952). *The Shorter Cambridge Medieval History*, Vol. 2. London: Cambridge University Press.

Pring, Coralie (2015). Iraq: Overview of Corruption and Anti-Corruption. Transparency International, 7–10. At: https://www.transparency.org/whatwedo/answer/iraq _overview_of_corruption_and_anti_c orruption1.

Przeworski, Adam, Michael E. Alvarez, Jose Antonio Cheibub, and Fernando Limongi (2000). *Democracy and Development: Political Institutions and Well-Being in the World.* New York: Cambridge University Press.

Psacharopoulos, George (1991). *The Economic Impact of Education: Lessons for Policymakers.* San Francisco, CA: ICS Press.

Psacharopoulos, George (2005). Economic Impact of Education. In Gerald M. Meier and James E. Rauch (eds), *Leading Issues in Economic Development.* New York: Oxford University Press, pp. 189–92.

Psacharopoulos, George and Harry Anthony Patrinos (2002). Returns to Education: A Further Update. World Bank Policy Research Working Paper 2881. September.

Qian, Yingi (2003). How Reform Worked in China. In Dani Rodrick (ed.), *In Search of Prosperity.* Princeton, NJ: Princeton University Press, pp. 297–333.

Quah, Jon S.T. (2013). Different Paths to Curbing Corruption: A Comparative Analysis. In Jon S.T. Quah (ed.), *Different Paths to Curbing Corruption: Lessons from Denmark, Finland, Hong Kong, New Zealand, and Singapore.* Bingley: Emerald Publishing, pp. 219–55.

Rabee Research (2020a). ISX Monthly Report, June. At: http://www.rs.iq/research _reports/preview/2020_12_202011_RS_-_ISX_Monthly_Report_(English).pdf.

Rabee Research (2020b). ISX Monthly Report, December. At: http://www.rs.iq/ research_reports/preview/2020_12_202011_RS_-_ISX_Monthly_Report_(English) .pdf.

Rahdi, Rahdi Hamza al- (2007). Testimony to the House Committee on Oversight and Government Reform on the Status of Corruption in the Iraqi Government. October 4.

Raphaeli, Nimrod; S. Ali, and Ze'eve B. Begin (2020). Between Baghdad and Erbil: Some Outstanding Challenges for the Government of Iraqi Prime Minister al-Kadhimi in Its Relations with the Kurdistan Regional Government. The Middle East Media Research Institute, July 1.

Rauch, James E. and Peter B. Evans (2000). Bureaucratic Structure and Bureaucratic Performance in Less Developed Countries. *Journal of Public Economics*, 75(1): 49–62.

Reporters Without Borders (2021). World Press Freedom Index. At: https://rsf.org/en/ ranking.

Reuters (2011a). Iraqi Banks Struggle with Limited Services, Capital. July 20.

Reuters (2011b). Iraq Names Trade Bank President. July 17.

Reuters (2011c). Baghdad Subsidizes Electricity. February 13.

Reuters (2012). Iraq Approves $100.5 Bln Budget for 2012. February 23.

Reuters (2018). Iraq Transfers Ownership of Nine State Oil Companies to New National Oil Company. At: https://www.reuters.com/article/us-iraq-oil-idUSKCN1MS27E.

Reuters (2019). Iraq Suspends US-Funded Broadcaster al Hurra Over Graft Investigation. September 2.

Reuters (2020). Iraqi Kurdistan Agrees with Baghdad on 2020 Budget Share. February 5.

Richards, Alan and John Waterbury (2008). *A Political Economy of the Middle East.* Boulder, CO: Westview Press.

Ridha, Fadhel (2020). Iraq's Wheat Production Economics: To What Extent Is It Possible to Close the Current Yield Gap? *Iraqi Economists Network,* May 28.

Rivlin, Paul (2009). *Arab Economies in the Twenty-First Century.* Cambridge: Cambridge University Press.

Robin-D'Cruz, Benedict and Renad Mansour (2020). Making Sense of the Sadrists: Fragmentation and Unstable Politics. In Aaron Stein (ed.), *Iraq in Transition: Competing Actors and Complicated Politics.* Philadelphia, PA: Foreign Policy Resaerch Institute, pp. 4–38.

Robinson, Linda (2008). *Tell Me How This Ends: General David Petraeus and the Search for a Way Out of Iraq.* New York: Public Affairs.

Rose-Ackerman, Susan (1999). *Corruption and Government: Causes, Consequences and Reform.* New York: Cambridge University Press.

Rousseau, J.B. (1966). Description du Pachalik de Baghdad. In Charles Issawi (ed.), *The Economic History of the Middle East: 1800–1914.* Chicago, IL: University of Chicago Press, pp. 135–6.

Rubin, Alissa J. (2008). Iraqi Trade Officials Ousted in Corruption Sweep. *New York Times,* September 24.

Rudaw (2017). Kurdistan referendum commission reveals four-language sample ballot. September 6.

Rudolf, Inna (2020). The Future of the Popular Mobilization Forces after the Assassination of Abu Mahdi al Muhandis. In Aaron Stein (ed.), *Iraq in Transition: Competing Actors and Complicated Politics.* Philadelphia, PA: Foreign Policy Research Institute, pp. 61–76.

Saadoun, Mustafa (2020). Closing Borders Due to Covid-19 Revives Iraqi Agriculture. *Al Monitor,* July 21.

Saeed, Addulbasit Turki (2016). Rebuilding the Board of Audit in a Shattered Country. In Roel Janssen (ed.), *The Art of the Audit.* Amsterdam: Amsterdam University Press, pp. 91–100.

Salam, Dara (2020). It is Time for a System Overhaul in Iraq's Kurdish Region. *Aljazeera.* At: https://www.aljazeera.com/opinions/2020/12/30/it-is-time-for-an -overhaul-of-the-system-in-iraqs-kurdish-region.

Salem, Paul (2019). Middle East Civil Wars: Definitions, Drivers and the Record of the Recent Past. In Paul Salem and Ross Harrison (eds), *Escaping the Conflict Trap: Toward Ending Civil Wars in the Middle East.* Washington, DC: Middle East Institute, pp. 1–31.

Salem, Raad (2016). Head-Ons and Handicaps: Bad Inter-City Roads in Southern Iraq Causing More Accidents. *Nigash,* January 7.

Salter, Lord (1955). *The Development of Iraq: A Plan of Action*. London: Iraq Development Board.

Saqr, Amal (2019). Corruption Sidetracks Projects Intended to Make Baghdad a "Capital of Arab Culture". Al-Fanar Media, August 24. At: https://www.al -fanarmedia.org/2019/08/corruption-sidetracks-projects-intended-to-make-baghdad -a-capital-of-arab-culture/.

Sattar, Omar (2012). Integrity Committee to Investigate Corruption in Iraqi Ministries. *Al-Monitor*, July 7.

Sattar, Omar (2019). Will Iraq Close All Inspectors General Offices? *Al Monitor*, November 2.

Savage, James D. (2014). *Reconstructing Iraq's Budgetary Institutions: Coalition State Building after Saddam*. Cambridge: Cambridge University Press.

Savello, Paul A. (2009a). Water Issues in Iraq: Water Availability and Its Usage. Paper #1. Baghdad: Iraq Transition Assistance Office.

Savello, Paul A. (2009b). Water Issues in Iraq: Future Water Projections. Paper #2. Baghdad: Iraq Transition Assistance Office.

Schoon, Natalie (2008). *Islamic Banking and Finance*. London: Spiramus Press.

Schumpeter, Joseph A. (1911). *Theory of Economic Development*. Cambridge, MA: Harvard University Press.

Shamsi, Pishko (2020). The Future of the Kurdistan Region After the Defeat of ISIS and the Failure of the 2017 Independence Referendum. In Aaron Stein (ed.), *Iraq in Transition: Competing Actors and Complicated Politics*. Philadelphia, PA: Foreign Policy Research Institute, pp. 39–59.

Shilani, Hiwa (2020). Iraqi Sentences Government Officials for $1.68 million Check Cashing Scheme. *Kurdistan24*, February 28.

Slim, Randa (2019). Iraq: A Conflict Over State Identity and Ownership. In Paul Salem and Ross Harrison (eds), *Escaping the Conflict Trap: Toward Ending Civil Wars in the Middle East*. Washington, DC: Middle East Institute, pp. 161–86.

Smith, Adam (1776). *An Inquiry into the Nature and Causes of the Wealth of Nations*. New York: Modern Library.

Smith, Grant, Salma El Wardany, and Javier Blas (2020). OPEC+ Finalizes Deal on Compensation for Iraq Cuts Cheating. *Bloomberg*, June 18.

Sowell, Kirk H. (2020). Continuity and Change in Iraq's Sunni Politics: Sunni Arab Political Trends, Factions, and Personalities. In Aaron Stein (ed.), *Iraq in Transition: Competing Actors and Complicated Politics*. Philadelphia, PA: Foreign Policy Research Institute, pp. 98–124.

Special Inspector General for Iraq Reconstruction (SIGIR) (2009a). Full Impact of Department of Defense Program to Restart State-Owned Enterprises Difficult to Estimate. SIGIR-09-09. January 30.

Special Inspector General for Iraq Reconstruction (SIGIR) (2009b). *Hard Lessons: The Iraq Reconstruction Experience*. Washington, DC: US Government Printing Office.

Special Inspector General for Iraq Reconstruction (SIGIR) (2009c). Cost, Outcome, and Oversight of Iraq Oil Reconstruction Contract with Kellogg Brown & Root Services, Inc. SIGIR-09-008. January 13.

Special Inspector General for Iraq Reconstruction (SIGIR) (2010). *Quarterly Report to the United States Congress*. July 30. Washington, DC: SIGIR.

Special Inspector General for Iraq Reconstruction (SIGIR) (2012). *Quarterly Report to the United States Congress*. April 30. Washington, DC: SIGIR.

Speville, Bertrand de (1997). *Hong Kong: Policy Initiatives Against Corruption*. Paris: Development Center of OECD.

Springborg, Robert (2007). *Oil and Democracy in Iraq*. London: SAQI.

Stallings, Barbara and Wilson Peres (2000). *Growth, Employment and Equity: The Impact of Economic Reforms in Latin America and the Caribbean*. Washington, DC: Brookings Institution.

Stansfield, Gareth and Hashem Ahmadzadeh (2008). Kurdish or Kurdistanis? Conceptualizing Regionalism in the North of Iraq. In Reidar Visser and Gareth Stansfield (eds), *An Iraq of Its Regions: Cornerstones of a Federal Democracy?* New York: Columbia University Press, pp. 123–49.

Stein, Aaron (2020). *Iraq in Transition: Competing Actors and Complicated Politics*. Philadelphia, PA: Foreign Policy Research Institute.

Stevenson, Lois (2010). *Private Sector and Enterprise Development*. Cheltenham, UK and Northampton, MA, USA: Edward Elgar Publishing.

Stiglitz, Joseph E. (1987). Some Theoretical Aspects of Agricultural Policies. *World Bank Research Observer*, 2(1): 43–60.

Sundahl, Mark J. (2007). Iraq Secured Transactions and the Promise of Islamic Law. *Vanderbilt Journal of Transnational Law*, 40: 1302–43.

Svensson, Jakob (2005). Eight Questions about Corruption. *Journal of Economic Perspectives*, 19(3): 19–42.

Szirmai, Adam (2005). *The Dynamics of Socio-Economic Development: An Introduction*. New York: Cambridge University Press.

Tabaqchali, Ahmed (2020a). Iraq's Power Conundrum: How to Secure Reliable Electricity While Achieving Energy Independence. *LSE Blog*, March 24. At: https://blogs.lse.ac.uk/mec/2020/03/24/iraqs-power-conundrum-how-to-secure-reliable-electricity-while-achieving-energy-independence/.

Tabaqchali, Ahmed (2020b). The Accounts that Didn't Bark: Iraq's Hidden State Balances. *LSE Blog*, July 16. At: https://blogs.lse.ac.uk/mec/2020/07/16/the-accounts-that-didnt-bark-iraqs-hidden-state-balances/.

Tabaqchali, Ahmed (2020c). Will Covid-19 Mark the Endgame for Iraq's Muhasasa Ta'ifia? *Arab Reform Initiative*, April 24. At: https://www.arab-reform.net/wp-content/uploads/pdf/Arab_Reform_Initiative_en_will-covid-19-mark-the-endgame-for-iraqs-muhasasa-taifia_10247.pdf?ver=1be8032ad3f587a9c448a3c1cd9e8066.

Tabaqchali, Ahmed (2021). AFC Iraq Fund December 2020 Update. *Asia Frontier Capital*. At: https://asiafrontiercapital.com.

Tanzi, V. (1998). Corruption around the world: Causes, Consequences, Scope and Cures. *IMF Staff Papers*, 45: 559–94.

Task Force for Business and Stability Operations (TFBSO) (2011). Iraq Final Impact Summary. Washington, DC: Task Force for Business and Stability Operations.

Teti, Andrea and Pamela Abbott (2016). Relative Importance of Religion and Region in Explaining Differences in Political, Economic and Social Attitudes in 2014. Arab Transformations, Working Paper No. 1. At: https://www.abdn.ac.uk/cgd/documents/Arabtrans/WP11_Working_Paper_1_-_Teti_and_Abbott_2016_-_ArabTransitions_-_Region_and_Religion_in_Iraq_20140525_AT.pdf.

The Economist (2003). Paying for Saddam's Sins. May 17, p. 68.

The Economist (2007). Islam vs. Science. September 10.

The Economist (2009a). Global Heroes: A Special Report on Entrepreneurship. March 14.

The Economist (2009b). Iraq's Elections: A Real Choice for the People. January 22.

The Economist (2010). Hard to Get Out. August 21, p. 38.

The Economist (2015). Not So Special. April 4.

Thede, Susanna and Nils-Åke Gustafson (2010). *The Multifaceted Impact of Corruption on International Trade*. Mimeo.

Timmer, C. Peter (1988). The Agricultural Transformation. In H. Chenery and T.N. Srinivasan (eds), *Handbook of Development Economics*. Amsterdam: Elsevier Science Publishers, pp. 321–8.

Timmer, C. Peter (2002). Agriculture and Economic Development. In Bruce L. Gardner and Gordon C. Rausser (eds), *Handbook of Agricultural Economics*. Amsterdam: Elsevier Science Publishers, pp. 1520–24.

Timmer, C. Peter (2005). Agricultural Development Strategies. In Gerald M. Meier and James E. Rauch (eds), *Leading Issues in Economic Development*. New York: Oxford University Press, pp. 394–6.

Toynbee, Arnold (1972). *A Study of History*. New York: Barnes & Noble Books.

Transparency International (2013a). Global Corruption Barometer: 2013, Iraq. https://www.transparency.org/geb2013/country?country=iraq.

Transparency International (2013b). Iraq: Overview of Corruption and Anti-corruption. Anti-Corruption Resource Centre. No. 374. April.

Transparency International (2015). Iraq: Overview of Corruption and Anti-corruption. Anti-Corruption Resource Centre. No. 2015:3. March.

Transparency International (2020). Corruption Perceptions Index. At:http://www.transparency.org/identity#IRQ.

Treisman, Daniel (2000). The Causes of Corruption: A Cross-National Study. *Journal of Public Economics*, 76(3): 399–457.

Tripp, Charles (2000). *A History of Iraq*. Cambridge: Cambridge University Press.

Tullock, Gordon (1967). The Welfare Costs of Tariffs, Monopolies, and Theft. *Western Economic Journal*, 5: 224–32.

Tullock, Gordon (1986). Industrial Organization and Rent Seeking in Dictatorships. *Journal of Institutional and Theoretical Economics*, 142: 4–15.

Tullock, Gordon (2005a). Rent Seeking: The Problem of Definition. In Charles K. Rowley (ed.), *The Rent Seeking Society*. Indianapolis, IN: Liberty Fund, pp. 3–10.

Tullock, Gordon (2005b). Rent Seeking. In Charles K. Rowley (ed.), *The Rent Seeking Society*. Indianapolis, IN: Liberty Fund, pp. 11–81.

UNCT Iraq (2010). *United Nations Development Assistance Framework for Iraq 2011–2014*. New York: United Nations.

UNDP (2003). *Arab Human Development Report*. New York: United Nations.

UNDP (2020). Human Development Index. At: http://hdr.undp.org/en/countries/profiles/IRQ.

United Nations (2004). *Convention Against Corruption*. New York: United Nations.

United Nations (2012). Population Division. At: http://esa.un.org/unpd/wpp/documentation/publications.htm.

United Nations and World Bank (2003). Joint Needs Assessment: State-Owned Enterprises. Working Paper. October.

United Nations Industrial Development Organization (UNIDO) (2010). UNIDO Supports Iraq's Efforts to Rehabilitate State-Owned Enterprises. March 22. At: http://www.unido.org/index.php?id=7881&tx_ttnews%5Btt_news%5D=457&cHash=5f0eb07d6c.

United Nations International Children's Emergency Fund (UNICEF) (2012). Iraq Statistics. At: http://www.unicef.org/infobycountry/iraq_statistics.html.

United Nations Office for Coordination of Humanitarian Affairs (UN OCHA) (2020). Iraq: Covid -19 Situation Report No. 12.

United Press International (UPI) (2012). Iraq Oil Exports Rise But Problems Remain. May 9. At: http://www.upi.com/Business_News/ Energy-Resources/2012/05/09/Iraq-oil-exports-rise-but-problems-rema in/UPI-20161336580985/#axzz2CwsqPBDr.

United States Agency for International Development (USAID) (2005). Land Registration and Property Rights in Iraq. January.

United States Agency for International Development (USAID) (2006a). A Presentation on Privatization. Baghdad: USAID, October 29.

United States Agency for International Development (USAID) (2006b). *Agricultural Reconstruction and Development Program for Iraq: Assessment of Rangelands in Iraq*. Preliminary Report. February. Baghdad: USAID.

United States Agency for International Development (USAID) (2006c). *Excess Employment in State Owned Enterprises*. Iraq Private Sector Growth and Employment Generation Series. June. Washington, DC: USAID.

United States Agency for International Development (USAID) (2006d). *Iraq Competitiveness Analysis*. Iraq Private Sector Growth and Employment Generation Series. Washington, DC: USAID.

United States Agency for International Development (USAID) (2007a). *An Overview of the Iraqi Banking System*. Washington, DC: USAID.

United States Agency for International Development (USAID) (2007b). *An Overview of the Iraq Cement Industry*. Iraq Private Sector Growth and Employment Generation Series. November 25. Washington, DC: USAID.

United States Agency for International Development (USAID) (2009). *Iraq: Economic Recovery Assessment*. February. Washington, DC: USAID.

United States Agency for International Development (USAID) (2011). *State of Iraq's Microfinance Industry*. Washington, DC: USAID.

United States Agency for International Development (USAID) (2019). *Rural Areas in Ninewa: Legacies of Conflict*. Washington, DC: USAID.

United States Agency for International Development (USAID) (2020). Food Assistance Fact Sheet: Iraq. May 14.

Unruh, Jon D. (2018). The Geography of Water and Oil Resource Governance in Post-Conflict Iraq. *Arab World Geographer*, 21(4): 261–78.

US Army (2006). *Counterinsurgency Field Manual*. US Army Field Manual 3-24, Marine Corps Warfighting Publication No. 3-33.5.

US Bureau of Census (2020), Iraq: Population. At: https://www.census.gov/data-tools/ demo/idb/informationGateway.php.

US Central Intelligence Agency (US CIA) (2012). *Fact Book: Iraq*. At: https://www.cia .gov/library/publications/the-world-factbook/geos/ iz.html.

US Central Intelligence Agency (US CIA) (2019). *Factbook Iraq*. At: https://www.cia .gov/the-world-factbook/countries/iraq/.

US Chamber of Commerce (2020). Recommendations to the Government of Iraq and the Government of the United States in Support of the US–Iraq Strategic Dialogue. US-Iraq Business Council. June.

US–Iraq Business Council (2020). *Recommendations to the Government of Iraq and the Government of the United States in Support of the US–Iraq Strategic Dialogue*. Washington DC: US Chamber of Commerce.

US Congress (2004). Review Iraqi Agriculture: From Oil for Food to the Future of Iraqi Production, Agriculture and Trade. Testimony before Committee on Agriculture, House of Representatives. Serial No. 108-33. June 16.

US Department of State (2006). Iraq Railroads. Mimeo.

US Department of State (2008). Iraqi Government Charges More than 300 Officials with Corruption in 2008. News Release. November 26.

US Department of State (2019). *2019 Investment Climate Statements: Iraq*. At: https://www.state.gov/reports/2019-investment-climate-statements/iraq/

US Department of Treasury (2009). Iraqi Banking System: An Overview. Washington, DC: Department of Treasury.

US Embassy (2015). Iraqi Republic Railway: Strategic Analysis. Draft. US Embassy, Baghdad, Iraq.

US Government Accountability Office (US GAO) (2007). Rebuilding Iraq: Integrated Strategic Plan Needed to Help Restore Iraq's Oil and Electricity Sectors. Washington, DC: GAO.

US News and World Report (2007). Islam and Science. September 10.

Urdal, Henrik (2012). Youth Bulges and Violence. In Jack A. Goldstone, Eric P. Kaufmann, and Monica Duffy Toft (eds), *Political Demography: How Population Changes are Reshaping International Security and National Politics*. Boulder, CO: Paradigm Publishers, pp. 117–32.

Visser, Hans (2009). *Islamic Finance: Principles and Practice*. Cheltenham, UK and Northampton, MA, USA: Edward Elgar Publishing.

Visser, Reidar (2008). Introduction. In Reidar Visser and Gareth Stansfield (eds), *An Iraq of Its Regions: Cornerstones of a Federal Democracy?* New York: Columbia University Press, pp. 1–26.

Visser, Reidar (2011). Anti-Corruption Measure Sparks Constitutional Confusion in Iraq. May 10. At: http://gulfanalysis.wordpress.com/2011/05/10/anti-corruption-measure-sparks-constitutional-confusion-in-iraq/.

Visser, Reidar and Gareth Stansfield (eds) (2008). *An Iraq of Its Regions: Cornerstones of a Federal Democracy?* New York: Columbia University Press.

Vogler, Gary (2017). *Iraq and the Politics of Oil*. Lawrence, KS: University Press of Kansas.

Voltaire (2010). *La Begueule, Conte Moral (1772)*. Whitefish, MT: Kessinger Publishing.

Wall Street Journal (2011). Crude Oil: All Barrels are Equal, but some are More Equal Than Others. April 9–10, A6.

Walters, Alan A. (1968). *The Economics of Road User Charges*. Baltimore, MD: Johns Hopkins University Press.

Walters, Alan A. and Steve H. Hanke (1992). Currency Boards. In Peter Newman, Murray Milgate, and John Eatwell (eds), *The New Palgrave Dictionary of Money and Finance*. New York: Stockton Press, pp. 558–61.

Warren, Count Edward de (1966). European Interests in Railways in the Valley of the Euphates. In Charles Issawi (ed.), *The Economic History of the Middle East, 1800–1914*. Chicago, IL: University of Chicago Press, pp. 137–45.

Watkins, Simon (2019). The Real Reason Why ExxonMobil Won't Go Ahead with $53 billion Iraqi Megaproject. Oilprice, July 3. At: https://oilprice.com/Energy/Energy-General/The-Real-Reason-Why-ExxonMobil-Wont-Go-Ahead-With-53-Billion-Iraqi-Megaproject.html.

Weber, Max ([1918] 2015). Politics as a Vocation. In Tony Waters and Dagmar Waters (transl. and eds), *Weber's Rationalism and Modern Society*. New York: Palgrave Macmillan, p. 136.

Weber, Max ([1920] 2002). *The Protestant Ethic and the Spirit of Capitalism*. Los Angeles, CA: Roxbury Publishing Company.

Weiss, Martin A. (2009). Iraq's Debt Relief: Procedure and Potential Implications for International Debt Relief. Congressional Research Service, 7-5700. January 26.

White, Thomas E., Robert C. Kelly, John M. Cape, and Denise Youngblood Coleman (2003). *Reconstructing Eden*. Houston, TX: Country Watch.

Wille, Belkis (2020). Iraq's New Government Should Life Barriers to Free Speech. *Hill*, May 6.

Woertz, Eckart (2017). Food Security in Iraq: Results from Quantitative and Qualitative Studies. *Food Security*, 9: 511–22.

World Bank (1995). *Bureaucrats in Business*. New York: Oxford University Press.

World Bank (2004). State Owned Enterprises Reform in Iraq. Reconstructing Iraq Working Paper No. 2. Washington, DC: World Bank.

World Bank (2005). *Pensions in Iraq: Issues, General Guidelines for Reform, and Potential Fiscal Implications*. Washington, DC: World Bank.

World Bank (2006a). Rebuilding Iraq: Economic Reform and Transition. Report No. 35141-IQ. Washington, DC: World Bank.

World Bank (2006b). Iraq Country Water Resource Assistance Strategy. Report No. 36297-IQ. Washington, DC: World Bank.

World Bank (2009). *Doing Business in the Arab World*. Washington, DC: World Bank.

World Bank (2010). *World Development Indicators*. Washington, DC: World Bank.

World Bank (2011a). *Financial Sector Review: Republic of Iraq*. Washington, DC: World Bank.

World Bank (2011b). *Accelerating Reform Within Iraq's National Board of Pensions*. MENA Quick Note. Washington, DC: World Bank.

World Bank (2011c). *Ease of Doing Business 2011*. Washington, DC: World Bank.

World Bank (2011d). *Confronting Poverty in Iraq*. Washington, DC: World Bank

World Bank (2011e). *Doing Business in the Arab World: 2011*. Washington, DC: World Bank.

World Bank (2012a). *Economy Profile: Iraq*. Doing Business Series 2012. Washington, DC: World Bank

World Bank (2012b). World *Development Indicators 2012*. Washington, DC: World Bank.

World Bank (2017). *Iraq: Systematic Country Diagnostic*, Washington, DC: World Bank. February 3.

World Bank (2018a). *Statistical Capacity Indicator*. http://datatopics.worldbank.org/statisticalcapacity/SCIdashboard.aspx

World Bank (2018b). *Iraq Economic Monitor: Toward Reconstruction, Economic Recovery and Fostering Social Cohesion*. Washington, DC: World Bank. Fall.

World Bank (2018c). Iraq Reconstruction and Investment, Part 2. Washington, DC: World Bank.

World Bank (2018d). *Reforming to Create Jobs: Economic Profile, Iraq*. Washington, DC: World Bank.

World Bank (2019a). World Development Indicators. http://wdi.worldbank.org.

World Bank (2019b). *Project Appraisal Document on a Proposed Loan in the Amount of US$200 Million to the Republic of Iraq for an Electricity Reconstruction and Enhancement Project*. Washington, DC: World Bank. April 19.

World Bank (2020a). *Iraq Economic Monitor: Navigating the Perfect Storm (Redux)*. Washington, DC: World Bank. Spring.

World Bank (2020b). *Iraq Economic Monitor: Protecting Vulnerable Iraqis*. Washington, DC: World Bank. Fall.

World Bank (2020c). World Development Indicators. At: http://wdi.worldbank.org.

World Bank (2020d). Doing Business 2020: Economy Profile Iraq. At: https://www
.doingbusiness.org/content/dam/doingBusiness/country/i/iraq/IRQ.pdf.
World Bank (2020e). *Doing Business 2020*. Washington, DC: World Bank.
World Food Programme (WFP) (2010). Capacity Development to Reform the Public
Distribution System (PDS) and Strengthen Social Safety Nets for Vulnerable Groups
in Iraq. World Food Programme Development Project Iraq 200104.
World Food Programme (WFP) (2015). Iraq: Food Consumption Deteriorates in
Anbar. Bulletin 7, June 15. At: https://documents.wfp.org/stellent/groups/public/
documents/ena/wfp275868.pdf?_ga=2.125929933.1925707260.1611238405
-630364261.1610759990.
World Health Organization (WHO) and United Nations International Children's
Emergency Fund (UNICEF) (2006). Meeting the MDG Drinking-Water and
Sanitation Target: The Urban and Rural Challenge of the Decade. At: http://www
.who.int/water_sanitation_health/monitoring/jmp2006/en/index.html.
Wunsch, Cornelia (2010). Neo-Babylonian Entrepreneurs. In David S. Landis, Joel
Mokyr, and William J. Baumol (eds), *The Invention of Enterprise*. Princeton, NJ:
Princeton University Press, pp. 40–61.
Yilmaz, Mehsmet (2003). The War that Never Happened: The Sharing of Euphrates
– Tigris Rivers' Water between Turkey, Syria and Iraq. Naval Postgraduate Thesis,
June.
Zedalis, Rex J. (2009). *The Legal Dimensions of Oil and Gas in Iraq*. New York:
Cambridge University Press.
Zulal, Shwan (2012). The Battle for Iraqi Oil: Can There Ever Be a Winner? *Niqash*.
At: http://www.niqash.org/articles/?id=3043.
Zulfikar, Yavuz Fahir (2012). Do Muslims Believe More in Protestant Work Ethic than
Christians? Comparison of People with Different Religious Background Living in
the US. *Journal of Business Ethics*, 105(4): 489–502.

Index

Printed and bound by CPI Group (UK) Ltd, Croydon, CR0 4YY

27/10/2024

14580411-0004